Visual Culture and Mathematics in the Early Modern Period

During the early modern period there was a natural correspondence between how artists might benefit from the knowledge of mathematics and how mathematicians might explore, through advances in the study of visual culture, new areas of enquiry that would uncover the mysteries of the visible world. This volume makes its contribution by offering new interdisciplinary approaches that not only investigate perspective but also examine how mathematics enriched aesthetic theory and the human mind. The contributors explore the portrayal of mathematical activity and mathematicians as well as their ideas and instruments, how artists displayed their mathematical skills and the choices visual artists made between geometry and arithmetic, as well as Euclid's impact on drawing, artistic practice and theory. These chapters cover a broad geographical area that includes Italy, Switzerland, Germany, the Netherlands, France and England. The artists, philosophers and mathematicians whose work is discussed include Leon Battista Alberti, Nicholas Cusanus, Marsilio Ficino, Francesco di Giorgio, Leonardo da Vinci and Andrea del Verrocchio, as well as Michelangelo, Galileo, Piero della Francesca, Girard Desargues, William Hogarth, Albrecht Dürer, Luca Pacioli and Raphael.

Ingrid Alexander-Skipnes is Lecturer in Art History at the Kunstgeschictliches Institut at Albert-Ludwigs-Universität Freiburg, Germany. She is an Associate Professor Emerita, University of Stavanger, Norway.

Visual Culture in Early Modernity
Series Editor: Kelley Di Dio

A forum for the critical inquiry of the visual arts in the early modern world, Visual Culture in Early Modernity promotes new models of inquiry and new narratives of early modern art and its history. The range of topics covered in this series includes, but is not limited to, painting, sculpture and architecture as well as material objects, such as domestic furnishings, religious and/or ritual accessories, costume, scientific/medical apparata, erotica, ephemera and printed matter.

51 **Genre Imagery in Early Modern Northern Europe**
New Perspectives
Edited by Arthur J. DiFuria

52 **Material Bernini**
Edited by Evonne Levy and Carolina Mangone

53 **The Enduring Legacy of Venetian Renaissance Art**
Edited by Andaleeb Badiee Banta

54 **The Bible and the Printed Image in Early Modern England**
Little Gidding and the pursuit of scriptural harmony
Michael Gaudio

55 **Prints in Translation, 1450–1750**
Image, Materiality, Space
Edited by Suzanne Karr Schmidt and Edward H. Wouk

56 **Imaging Stuart Family Politics**
Dynastic Crisis and Continuity
Catriona Murray

57 **Sebastiano del Piombo and the World of Spanish Rome**
Piers Baker-Bates

58 **Early Modern Merchants as Collectors**
Edited by Christina M. Anderson

59 **Visual Culture and Mathematics in the Early Modern Period**
Edited by Ingrid Alexander-Skipnes

Visual Culture and Mathematics in the Early Modern Period

Edited by Ingrid Alexander-Skipnes

Routledge
Taylor & Francis Group
LONDON AND NEW YORK

First published 2017 by Routledge

2 Park Square, Milton Park, Abingdon, Oxfordshire OX14 4RN
52 Vanderbilt Avenue, New York, NY 10017

Routledge is an imprint of the Taylor & Francis Group, an informa business

First issued in paperback 2019

Library of Congress Cataloging in Publication Data
Names: Alexander-Skipnes, Ingrid, editor.
Title: Visual culture and mathematics in the early modern period / edited by Ingrid Alexander-Skipnes.
Description: New York : Routledge, 2017. | Series: Visual culture in early modernity | Includes bibliographical references and index.
Identifiers: LCCN 2016034996 | ISBN 9781138679382 (alk. paper)
Subjects: LCSH: Art—Mathematics. | Mathematics in art.
Classification: LCC N72.M3 V575 2017 | DDC 700/.46—dc23
LC record available at https://lccn.loc.gov/2016034996

ISBN: 978-1-138-67938-2 (hbk)
ISBN: 978-0-367-33416-1 (pbk)

Typeset in Sabon
by Apex CoVantage, LLC

Contents

Figures

Cover: Raphael. *School of Athens*. Detail of Euclid and his pupils. Stanza della Segnatura, Vatican Palace. Photo copyright: Vatican Museums.

Acknowledgments

I would like to thank the College Art Association for its support of the two panels upon which these chapters are based, Erika Gaffney for her enthusiastic support of this project in its early stages, and the anonymous reader for the helpful comments on the volume's chapters. I would also like to acknowledge the kind help of the staff of the Kunstgeschichtliches Institut Bibliothek, the Mathematisches Institut Bibliothek, and the Universitätsbibliothek of the Albert-Ludwigs-Universität Freiburg. I am grateful to Felisa Salvago-Keyes and Isabella Vitti, my editor at Routledge, as well as the series editor, Kelley Di Dio, and others from the editorial staff, particularly Christina Kowalski and Nicole Eno, who were so generous with their help at every stage of this project. I thank the production editor, Adam Guppy and the project manager, Kerry Boettcher for skillfully shepherding the book through to production. Heartfelt thanks go also to the contributors for their insightful studies and pleasant cooperation.

1 Introduction

Ingrid Alexander-Skipnes

Mathematics has had a significant role in the cultural production of early modern Europe. It can also be argued that ideas from the world of visual representation led to advances in mathematics. Visual culture and mathematics were more closely connected in the fifteenth, sixteenth and seventeenth centuries than they are today. In our specialist-focused universities, these areas of study are commonly thought to be worlds apart and are rarely brought together.[1] In the early modern period, however, the study of visual culture together with mathematics was by no means considered impractical, as it might be thought of today, as these fields shared common concerns and approaches. While specialization has worked against a more holistic approach, recently there has been a growing interest in the interdisciplinary relationships between the two areas.

The importance of geometry was emphasized already in antiquity. Plato had insisted that the study of geometry was necessary in order to comprehend higher things, as the inscription "Let no one ignorant of geometry enter here" above the portal of his academy reaffirmed. For Aristotle, geometry held pride of place within the mathematical sciences because of its irrefutable proofs. In Book 35 of his *Natural History*, Pliny the Elder praised the skill of the artist Pamphilus, who was "highly educated in all branches of learning, especially arithmetic and geometry, without the aid of which he maintained art could not attain perfection."[2] Euclid's *Elements* and *Optics* and manuscripts of the work of Archimedes were fundamentals texts for artists. Book XIII of Euclid's *Elements* was of particular interest for artists as it dealt with the geometry of rectilinear and circular figures. Archimedes' work on circular and spherical geometry showed his familiarity with π; his balancing the lever and even his work with parabolic mirrors were of interest to early modern artists and engineers.

The hunt for Greek mathematical texts was an important part of a revival of interest in the culture of antiquity. There was a renewed confidence that what was excellent in Greek literature, architecture, art and mathematics could be recovered and strengthen the skills and knowledge of the ancient Greek world. A chair in Greek studies was established at the University of Florence in 1397, when Manuel Chrysoloras was invited to teach Greek. Tracking down manuscripts became a lively pursuit. Emissaries were sent out to explore Byzantium to find manuscripts, sometimes without patronage. Florence quickly became a vibrant center for the acquisition, translation and the dissemination of classical texts. Some mathematical texts were brought to Italy by humanists such as the Sicilian Giovanni Aurispa, who brought back a cache of some 238 Greek manuscripts from his second voyage in the East (1421–23) that included a manuscript of the *Mathematical Collection* of Pappus.[3]

One story that demonstrates the eagerness with which mathematical manuscripts were sought is that of Rinuccio da Castiglione and his purported acquisition of a manuscript by Archimedes.[4] The story is told by Ambrogio Traversari, a Florentine monk and Greek scholar, who in 1424 was anxious to obtain a copy of an Archimedes manuscript which Rinuccio claimed to be in possession of. Intrigued by the rumor that had spread of the existence of the Archimedean text, Traversari tried in vain to see the manuscript. He invited Rinuccio to his monk's cell, but the manuscript hunter babbled on incoherently on topics as varied as the perfidy of the Greeks to denouncing Tuscany's hostility to learning.[5] In the end, Traversari never saw the manuscript of Archimedes, and it is doubtful whether Rinuccio had it after all.

Much of the knowledge of mathematics was transmitted through and beyond the confines of a university education and into cultural circles. Humanist courts enjoyed the presence of mathematicians where other scientists, along with poets, painters, musicians and philosophers mingled comfortably, mostly due to the heterogeneity of their interests. Furthermore, an understanding of mathematics, feigned or learned, held a certain prestige in the humanist milieu.

In the fifteenth century, mathematically minded artists like Leon Battista Alberti and Piero della Francesca had frequented humanist courts where mathematics was central in a revival of interest in Greek culture. Interestingly, both artists were at the papal court of Pope Nicholas V (r. 1447–55) who not only commissioned one of the first translations of Archimedes but also was one of the few popes who lent his Greek mathematical texts. Thus the pope assisted in the spread of interest in Greek mathematics throughout the Italian peninsula. Leonardo da Vinci investigated questions of proportionality and optics extensively, and his drawings as well as his notes offer insight into his knowledge and use of mathematics, particularly geometry.

According to Giorgio Vasari, Piero della Francesca wrote "many" treatises on mathematics. The three known treatises reveal his knowledge of both Euclid and Archimedes. His *Trattato d'abaco* (Abacus treatise) covers arithmetic, algebra and geometry, while the *Libellus de quinque corporibus regularibus* (Short book on the five regular solids) goes further into a study of the Archimedean solids.[6] Piero paraphrased parts of Euclid's *Optics* in his treatise, *De prospectiva pingendi* (On perspective for painting), where his investigations into visual angles attempt to characterize the proportional relationships that Euclid had left undefined.[7] Furthermore, it has been shown that he made a copy of an Archimedean text.[8] Luca Pacioli, mathematician and compatriot of Piero della Francesca, published, among several mathematical texts, one of the first Latin editions of Euclid's *Elements* (1509). Leonardo da Vinci drew the illustrations for Pacioli's important *De divina proportione* (On Divine Proportion).

On the other side of the Alps, the sixteenth-century German painter and engraver Albrecht Dürer contributed to the role that Greek mathematics played for visual artists in northern Europe through his study of proportion and perspective. He was instrumental in the dissemination of Italian theories of perspective in northern Europe, which had wide-reaching effects on other fields such as cartography and mathematics.[9] Dürer purchased a copy of Euclid's *Elements* in Italy and wrote a treatise on mathematics, *Underweysung der Messung mit dem Zirckel und Richtscheyt* (A Course in the Art of Measurement with Compass and Ruler) and *Vier Bücher von Menschlicher Proportion* (Four Books on Human Proportion). Dürer demonstrated his profound interest in geometric figures and instruments in a memorable way in his engraving *Melencolia I* (1514). In their writings, Albrecht Dürer, Piero della Francesca and Leonardo da Vinci had revitalized early modern interest in the so-called Archimedean solids, polyhedra with surfaces made up of plane polygonal surfaces whose sides make up their edges and the corners their vertices.

Linear perspective, rediscovered in fifteenth-century Florence, developed out of a need to depict a three-dimensional space on a two-dimensional surface.[10] Linear perspective was arguably the most important technique for representation at the disposal of artists and architects of the early modern period. The Florentine architect and engineer Filippo Brunelleschi is credited with the invention, or rather rediscovery, of linear perspective. He had studied with the mathematician Paolo dal Pozzo Toscanelli. Brunelleschi invented a technique, essentially mathematical, whereby objects projected on a surface acquire a three-dimensional appearance. According to his biographer, Antonio di Tuccio Manetti, Brunelleschi painted two demonstrations of perspective (now lost), one of a view of the Baptistery seen from the cathedral door and the other from the Palazzo della Signoria, viewed a short distance away. His famous demonstration, which involved a perspectival construction and showed how perspective would work, took place in front of the Florence cathedral. Linear perspective is remarkably illustrated in Masaccio's Trinity fresco (c. 1426) in Santa Maria Novella. Filippo Brunelleschi's ground-breaking discovery and his important work in proportion for building as well as his adoption of mathematics would play an important role in the development of the period's architecture. In his *De pictura* (1435), Leon Battista Alberti elaborated on the importance of linear perspective and optics. His call for a more naturalistic treatment of painting may have been inspired by a trip he made to northern Europe.

Studies in the geometry of vision, or optics, had their origins in ancient Greece. Euclid wrote on optics. His ideas were advanced by Ptolemy (c. 100–170) and further expanded on by Galen. Further experiments on the nature of light and its reception by the eye by Ibn al-Haytham (Alhazen, 965–c. 1040) remained central to the understanding of vision in the Middle Ages. Artistic practice and theory, which contributed to a better understanding of the sensory perception of light, color and form, had much in common with the scientific interest in empirical discovery. As a practical and theoretical tool, mathematics not only informed painters, draftsmen, architects, musician and philosophers, but it was also a way to connect with the classical past and unlock the mysteries of the natural world. Irrational ratios like the golden section and irrational numbers like π had their own cultural currency. Artists turned to mathematics to resolve questions of proportion and vision and to arrive at a clearer understanding of nature, while mathematicians sought to analyze natural phenomena; the behavior of numbers could be seen as an aesthetic question as much as a utilitarian one.

In the seventeenth century, it was believed that nature was mathematical in structure. A quest that had dogged natural philosophers in the seventeenth century was how to bring a quantity into a hitherto qualitative study of nature.[11] Galileo Galilei, Johannes Kepler and Isaac Newton advanced the study of optics through their study of astronomy. Galileo famously wrote:

> Philosophy is written in this grand book, the universe, which stands continually open to our gaze. But the book cannot be understood unless one first learns to comprehend the language and read the letters in which it is composed. It is written in the language of mathematics, and its characters are triangles, circles and other geometric figures without which it is humanly impossible to understand a single word of it; without these, one wanders about in a dark labyrinth.[12]

The study of light and theories of vision advanced in the seventeenth century, particularly through the work of Kepler, who had uncovered laws of planetary motion, were grounded in mathematical patterns.[13] Newton's work on light and color, and reflection and refraction, went even further. Uncovering nature's mysteries through mathematics had a parallel with the studies early modern artists engaged in. The discovery of the advances made in optics and the invention of the telescope in the Netherlands played a role in the naturalistic way northern European artists described reality.[14] Mirrors and lenses could provide unusual optical effects and create spatial tensions within a painting. Architects also could achieve unusual light effects by manipulating forms and geometric shapes. Perspective, the advent of devices like the microscope and the telescope, optics and mathematics brought together representation and a new understanding of nature in unprecedented ways.

This volume has its genesis in two sessions I organized at the 100th College Art Association annual conference held in Los Angeles in February 2012. The interest in the conference papers and the lively discussions that ensued suggested to me that, in spite of the fact that visual culture and mathematics are rarely studied together in our specialist-focused universities, there exist common areas of interest that bring them together.

This volume consists of eight chapters that explore ways in which visual culture and mathematics interacted in the period from the fifteenth to the seventeenth centuries in Europe. The present volume makes no claim to cover the topic comprehensively or provide a parallel history of visual culture or the history of mathematics in the period but rather complements earlier studies such as Martin Kemp's *The Science of Art: Optical Themes in Western Art from Brunelleschi to Seurat*, J.V. Field's *The Invention of Infinity: Mathematics and Art in the Renaissance* and, more recently, Mark Peterson's *Galileo's Muse: Renaissance Mathematics and the Arts*, Alexander Marr's *Between Raphael and Galileo: Mutio Oddi and the Mathematical Culture of Late Renaissance Italy*, and Robert Felfe, *Naturform und bildnerische Prozesse: Elemente einer Wissensgeschichte in der Kunst des 16. und 17. Jahrhunderts.*[15]

The contributors represent a broad range of disciplines that include art history, architectural history, mathematics, history of science, philosophy and economics. Together the chapters explore three main areas of focus—the role mathematics played in the period's art theory; painters and the language of mathematics (painters express their mathematical knowledge); how Euclid offered not only practical solutions for structure, the representation of space and line for architects, painters and draftsmen, but how his geometry and optics engaged the viewer. The chapters cover a broad geographical area that includes Italy, Germany, Switzerland, the Netherlands, France and England.

The first section of the book, "The Mathematical Mind and the Search for Beauty," addresses the role that mathematics played in defining Renaissance aesthetics, theories of vision and proportion in the search for beauty and harmony. It explores mathematical principles that enhanced both the liberal and the mechanical arts. This section opens with an chapter by John Hendrix, "Renaissance Aesthetics and Mathematics," that examines the writings of Leon Battista Alberti, Nicolas Cusanus, Marsilio Ficino, Piero della Francesca and Luca Pacioli. Hendrix explores the underlying mathematical theories of these writers, derived from ancient authors such as Plato and Vitruvius, which contained concepts that could link nature, the human mind and the divine mind and which resulted in a kind of aesthetics and artistic creation. He returns to a concept mentioned earlier, that is, that mathematics played a different role than it does today. For Hendrix, mathematics played a fundamental role in defining the human mind as a microcosm of the cosmos and in cultural definitions of beauty and harmony.

In the next chapter, "Design Method and Mathematics in Francesco di Giorgio's *Trattati*," Angeliki Pollali addresses the interplay between Euclidean geometry and arithmetical solutions in two richly illustrated treatises by the fifteenth-century Sienese architect, sculptor and painter. In the following chapter, Matthew Landrus informs us that Leonardo helped his friend and improved Francesco's method of technical illustrations. Earlier scholarship has emphasized Francesco's use of arithmetic proportions derived from the human body, although *Trattati I* is informed by geometry. Pollali traces the scholarly preference for arithmetic back to the historiography of Renaissance architectural proportion and points out the limitations of Vitruvius' *De architectura* as a model for Francesco di Giorgio's method of architectural design. She demonstrates that Francesco di Giorgio's use of a modular system, through examples such as a double-aisled basilica and oblong and circular temples, reveals the architect's preference for solutions derived from practical geometry.

In the next chapter, "Mathematical and Proportion Theories in the Work of Leonardo da Vinci and Contemporary Artist/Engineers at the Turn of the Sixteenth Century," Matthew Landrus turns our attention to the artist/mathematicians; here the emphasis is on their theories of proportion. For the early modern *uomini practici* "practical man," proportional geometry was essential for their approaches to natural philosophy and the practical arts at the turn of the sixteenth century. Through a study of the proportion exercises of artists such as Verrocchio, Michelangelo, Leonardo da Vinci, Giovanni Antonio Amadeo and Raphael, to name a few, Landrus explores how proportion theory enriched their pictorial, mechanical and architectural projects. Landrus focuses primarily on the work of Leonardo and explores his preference for geometry and proportion over arithmetic. For the most part, Leonardo favored visual solutions to numerical ones. For technical projects like the Sforza Horse proposal, Leonardo used a combination of geometry and arithmetic. In addition, Landrus examines Leonardo's large machines in the context of his *De re militari*. By tracing the mathematical and artistic approaches of several turn of the sixteenth-century artist/engineers, Landrus demonstrates their particular interest in mathematical spaces and universal laws.

The next group of chapters, "Artists as Mathematicians," looks at artists who successfully expressed themselves both as visual artists and mathematicians and how their mathematical knowledge enriched their artistic theory and practice. Both Albrecht Dürer and Piero della Francesca wrote treatises on mathematics. In her chapter, "Dürer's *Underweysung der Messung* and the Geometric Construction of Alphabets," Rangsook Yoon looks at applied geometry in the Third Book of the *Treatise on Measurement* (1525) and Dürer's detailed instructions on how to construct Roman and Gothic letters. Yoon uncovers parallels between the artist's geometric construction of alphabets and his ideas on ideal human proportion. Yoon examines Dürer's use of an Albertian perspectival system in order to establish the dimensions of the lettering on columns, towers or high walls. Furthermore, she investigates the power of the gaze and the privileged status that mathematical knowledge held in Dürer's cultural environment.

In the next chapter, "Circling the Square: The Meaningful Use of φ and π in the Paintings of Piero della Francesca," Perry Brooks explores the work of an artist who was as skilled in writing on mathematics as he was in painting extraordinary works. Brooks builds on the earlier work of Rudolph Wittkower and B.A.R. Carter in their examination of perspective and proportion in Piero's *Flagellation*, which reveals a symbolic use of π. Brooks examines the fascination the painter Piero della Francesca had for irrational numbers. In fact, the irrational number π and another topic related to the circle—squaring the circle—have baffled mathematicians for millennia. Whether cultural or computational, π has had a significant role in artistic design and mathematical enquiry. In paintings such as the *Resurrection* and the *Nativity*, Brooks also looks at Piero's use of φ, essentially the golden section, a proportion whose incommensurability fascinated the Franciscan friar Luca Pacioli (a compatriot of Piero), of whom we will hear more in Renzo Baldasso and John Logan's chapter.

Brooks considers the writing of several contemporary authors who dismiss a connection between metaphysical associations and artistic relevance, but he prefers to return to early modern writers such as Luca Pacioli and the philosopher-theologian Nicolaus Cusanus, who expressed an ontological significance in the study of φ and π. For Brooks, it is important to look at this in a historical context and less with a contemporary view.

The book's last section, "Euclid and Artistic Accomplishment," takes a closer look at the Greek mathematician. Euclid's *Elements* became a particularly important text for mathematicians, artists, architects and engineers as Latin and vernacular editions became available throughout Europe.[16] In her chapter, "The Point and Its Line: An Early Modern History of Movement," Caroline Fowler examines how geometry engages with drawing through a comparative study of various interpretations of Euclid's *Elements* in relationship to printed drawing manuals, which reveals the engagement of both geometry and drawing not only with each other but also with seventeenth-century philosophical discourses that explored bodies moving through space. Fowler traces how the shifting definitions of the Euclidean point and line impacted the pedagogy of drawing in printed drawing manuals and vice versa, which resulted in a transformation of the teaching of drawing from a study of proportion to a study of movement. While the author begins with Leon Battista Alberti and how he addresses the discrepancy between the definition of the mathematical figure and its visual representation, Fowler's study focuses on the seventeenth-century divisions between practical geometry and theoretical geometry through an examination of treatises written in France, Germany, England and the Netherlands.

The next chapter in this section, "Between the Golden Ratio and a Semiperfect Solid: Fra Luca Pacioli and the Portrayal of Mathematical Humanism," examines the fascination in early modern Italy with the representation of polyhedra. The authors, Renzo Baldasso and John Logan, focus on the *Portrait of Luca Pacioli and Gentleman* (Capodimonte, Naples) and the interpretation of the Euclidean figures in the painting. Baldasso and Logan explain the significance of the mathematical items in the painting, which hitherto have been largely ignored. They posit that the geometrical figures depicted in the painting also challenge the mathematical knowledge of the viewer. Furthermore, they see the painting as a display of mathematics and uncover the meaning of the diagram that Pacioli is drawing in terms of the golden ratio. Their study is not limited to this painting, however, but also examines a range of depictions of polyhedra such as those in Dürer's *Melencolia I* and the *mazzocchio*, a fancy hat, represented in paintings by Paolo Uccello and a geometric solid in the floor mosaic of the Basilica of St. Mark in Venice.

In the last chapter in the section, "Mathematical Imagination in Raphael's *School of Athens*," Ingrid Alexander-Skipnes examines the increase in mathematical manuscripts in the Vatican Library and the mathematical themes in the fresco. Although Raphael is not known to have written a treatise on mathematics, as Albrecht Dürer or Piero della Francesca had, nevertheless, and as successor to Bramante, he was undoubtedly well versed in Euclidean geometry. Furthermore, Raphael would have had the possibility to study mathematical texts within the courtly circles he frequented. Alexander-Skipnes argues that Raphael has reinterpreted the traditional representations of the *uomini illustri* and the Seven Liberal Arts, and she examines the prominence of the quadrivial disciplines, with a focus on the fresco's right foreground. Alexander-Skipnes offers an interpretation of the geometric drawing on Euclid's slate and the significance of Raphael's presence among a group of mathematicians.

These chapters collectively examine the interaction between visual culture and mathematics in several ways. This includes perspective but goes beyond the defining of space that has dominated discussions in this area. In this volume, the relationship between visual culture and mathematics extends also to the depiction of mathematicians along with their scientific knowledge and the engagement of the viewer with mathematical ideas and symbols. Each of the chapters, with their interdisciplinary focus, expands our knowledge of how both visual culture and mathematics enriched the human mind in the early modern period and in that way also reveals how these areas share a common ground of intellectual activity with impulses for creativity and perception.

Notes

1 C. P. Snow, *The Two Cultures: And a Second Look* (Cambridge: Cambridge University Press, 1964).
2 Pliny the Elder, *Natural History*, trans. H. Rackham (Cambridge, MA and London: Harvard University Press, 1995), 35, 76.
3 Luis Radford, "On the Epistemological Limits of Language: Mathematical Knowledge and Social Practice During the Renaissance," *Educational Studies in Mathematics* 52(2) (2003): 135, n. 12; Paul Lawrence Rose, *The Italian Renaissance of Mathematics: Studies on Humanists and Mathematicians from Petrarch to Galileo* (Geneva: Librairie Droz, 1975), 28.
4 For the fundamental study on Rinuccio da Castiglione see D. P. Lockwood, "De Rinucio Aretino Graecarum Litterarum Interprete," *Harvard Studies in Classical Philology* 24 (1913): 51–119.
5 James Hankins, *Plato in the Italian Renaissance* (Leiden, New York and Cologne: E.J. Brill, 1994), 86.
6 J. V. Field, "Mathematics and the Craft of Painting: Piero della Francesca and Perspective," in *Renaissance and Revolution: Humanists, Scholars, Craftsmen and Natural Philosophers in Early Modern Europe*, ed. J. V. Field and Frank A.J.L. James (Cambridge: Cambridge University Press, 1993), 81.
7 Martin Kemp, *The Science of Art: Optical Themes in Western Art from Brunelleschi to Seurat* (New Haven and London: Yale University Press, 1990), 27–28; Menso Folkerts, "Piero della Francesca and Euclid," in *Piero della Francesca: tra arte e scienza. Atti del convegno internazionale di studi*, Arezzo, 8–11 ottobre 1992, ed. Marisa Dalai Emiliani and Valter Curzi (Venice: Marsilio, 1996), 293–312; Ingrid Alexander-Skipnes, "Greek Mathematics in Rome and the Aesthetics of Geometry in Piero della Francesca," in *Early Modern Rome, 1341–1667*, ed. Portia Prebys (Ferrara: Edisai, 2011), 178.
8 James R. Banker, "A Manuscript of the Works of Archimedes in the Hand of Piero della Francesca," *Burlington Magazine* 147 (March, 2005): 165–169.

9 The first printed book on perspective in northern Europe was published in 1505 by Jean Pélerin (Viator), richly illustrated, it was of particular interest for architects. See Kirsti Andersen, *The Geometry of an Art: The History of the Mathematical Theory of Perspective from Alberti to Monge* (New York: Springer, 2007), 161–163.

10 For a study on the importance of mirrors for linear perspective theories in the antiquity, see Rocco Sinisgalli, *Perspective in the Visual Culture of Classical Antiquity* (Cambridge: Cambridge University Press, 2012); Samuel Y. Edgerton, *The Mirror, the Window, and the Telescope: How Renaissance Linear Perspective Changed Our Vision of the Universe* (Ithaca and London: Cornell University Press, 2009).

11 R. W. Serjeantson, "Proof and Persuasion," in *The Cambridge History of Science*, vol. 3, *Early Modern Science*, ed. Katherine Park and Lorraine Daston (Cambridge: Cambridge University Press, 2006), 155–156.

12 Stillman Drake, trans., *Discoveries and Opinions of Galileo* (Garden City, NY: Doubleday, 1957), 237–238.

13 Ewa Chojecka, "Johann Kepler und die Kunst: Zum Verhältnis von Kunst und Naturwissenschaften in der Spätrenaissance," *Zeitschrift für Kunstgeschichte* 30 (1967): 55–72.

14 Svetlana Alpers, *The Art of Describing: Dutch Art in the Seventeenth Century* (Chicago: University of Chicago Press, 1983).

15 Kemp, *The Science of Art*, see note 7; J. V. Field, *The Invention of Infinity: Mathematics and Art in the Renaissance* (Oxford: Oxford University Press, 1997, reprinted 2005); Mark A. Peterson, *Galileo's Muse: Renaissance Mathematics and the Arts* (Cambridge, MA and London: Harvard University Press, 2011); Alexander Marr, *Between Raphael and Galileo: Mutio Oddi and the Mathematical Culture of Late Renaissance Italy* (Chicago and London: University of Chicago Press, 2011); Robert Felfe, *Naturform und bildnerische Prozesse: Elemente einer Wissensgeschichte in der Kunst des 16. und 17. Jahrhunderts* (Berlin and Boston: Walter de Gruyter, 2015). See also, William M. Ivins Jr., *Art & Geometry: A Study in Space Intuitions* (Cambridge, MA: Harvard University Press, 1946, reprinted 1964); Sabine Rommevaux, Philippe Vendrix and Vasco Zara, eds., *Proportions: Science-Musique-Peinture & Architecture. Actes du LIe Colloque International d'Études Humanistes, 30 juin–4 juillet 2008* (Turnhout: Brepols, 2011).

16 Sabine Rommevaux, "La réception des *Éléments* d'Euclide au Moyen Âge et à la Renaissance. Introduction," Revue d'histoire des sciences, 56.2 (2003): 267–273.

Part I

The Mathematical Mind and the Search for Beauty

2 Renaissance Aesthetics and Mathematics

John Hendrix

Mathematics played a key role in Renaissance aesthetics, in concepts such as *analogia*, *lineamenti*, *concinnitas*, *commensuratio*, polygonal and polyhedral geometries, Pythagorean harmonies, the tetractys, and the Platonic Lambda, as developed in the writings of Leon Battista Alberti (*De pictura*, *De re aedificatoria*), Nicolas Cusanus (*De docta ignorantia*, *De coniecturis*, *De circuli quadratura*), Marsilio Ficino (*De amore*, *Opera Omnia*), Piero della Francesca (*Trattato d'abaco*, *De prospectiva pingendi*), and Luca Pacioli (*De divina proportione*). For these writers, mathematics was seen as the most fundamental concept that could link nature, the human mind, and the divine mind in the humanist project, which resulted in a particular kind of aesthetics and artistic creation. The mathematical principles were taken from ancient writers such as Plato and Vitruvius, so their validity was not questioned; nor was their role in defining relations between human and divine, and humanity and nature. This chapter hopes to show that mathematics played a role in the humanist world view and Renaissance aesthetics.

The word "aesthetics" is used here to mean "philosophy of art." The argument of this chapter is that mathematics was essential to a humanist philosophy of art, based on classical philosophy, as distinguished from a theory of art applied to practice. Mathematics played a fundamental role, as it still does, in the understanding of the cosmos and nature and in cultural definitions of beauty and harmony. A philosophy of art with mathematics at its basis is of importance in art and architecture to the present day, as art and architecture connect the human mind with nature and the cosmos.

Leon Battista Alberti designed the façade of the Palazzo Rucellai in Florence for Giovanni Rucellai around 1455 (Figure 2.1). The façade consists of seven vertical bays divided into three tiers, with two doors. The proportion of the door bays is 3:2; the proportion of the bays above the doors is 7:4; the proportion of the other bays is 5:3. The bays of the façade are seen by Alberti as areas, each being a square that is proportionally enlarged according to a consistent ratio. Seen as extended squares, the bays on the façade of the Palazzo Rucellai are one plus a half, one plus two-thirds, and one plus three-fourths. These three ratios are the octave or *diapason* (1:2), fifth or *diapente* (2:3), and fourth or *diatessaron* (3:4) of the Pythagorean harmonies. Alberti explained in his treatise on architecture, *De re aedificatoria* (1452), that in architectural design "an area may be either short, long or intermediate. The shortest of all is the quadrangle. . . . After this come the *sesquialtera* [*diapente*], and another short area is the *sesquitertia* [*diatesseron*]" (IX.6).[1] Alberti explained that "the musical numbers are 1, 2, 3, and 4. . . . Architects employ all these numbers in the most convenient

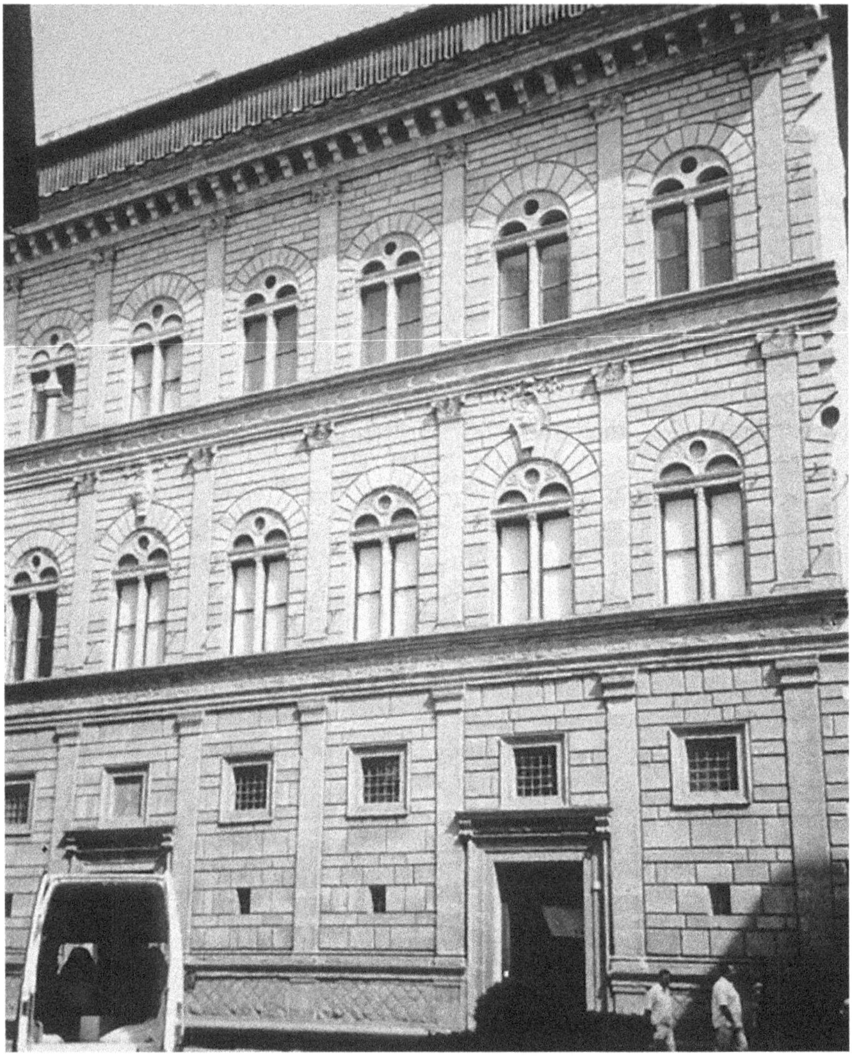

Figure 2.1 Leon Battista Alberti. Palazzo Rucellai, Florence.

Photo credit: John Hendrix.

manner possible" (IX.5), because "the numbers by means of which the agreement of sounds affects our ears with delight, are the very same which please our eyes and our minds." Marsilio Ficino, an acquaintance of Alberti's at the Platonic Academy in Florence in the 1460s, called Alberti a "Platonic mathematician."[2]

Alberti was 29 years older than Ficino.[3] Ficino wrote that during his adolescence, he and Alberti became correspondents, as mentor and pupil. They became partners in a "ritual correspondence" and exchanged "noble wisdom and knowledge."[4] Between the years 1443 and 1465, Alberti spent little time in Florence, being occupied by the papal curia in Rome. When Alberti returned to Florence in the 1460s, he stayed at

Ficino's house in Figline Valdarno. By 1468 he was recorded by Cristoforo Landino in the *Disputations at Camaldoli* as being active in discussions at the Academy. Landino described conversations between Ficino and Alberti on the subject of Platonic philosophy.

The placement of the pilasters in the façade of the Palazzo Rucellai, dividing the bays, is determined by the proportions of the Pythagorean harmonies; the proportions of the pilasters are determined by the harmonies as well, as they are related to the human body. Alberti described the proportioning of the classical column in *De re aedificatoria*. The proportions of the Doric column correspond to the proportions of the male body, according to Vitruvius. Vitruvius described in *De architectura* (IV.1.6):[5]

> When they wished to place columns in that temple, not having their proportions, and seeking by what method they could make them fit to bear weight, and in their appearance to have an approved grace, they measured a man's footstep and applied it to his height. Finding that the foot was a sixth part of the height in a man, they applied this proportion to the column. Of whatever thickness they made the base of the shaft they raised it along with the capital to six times as much in height. So the Doric column began to furnish the proportions of a man's body, its strength and grace.

But the actual proportions of the Doric column are seven to one rather than six to one, according to Vitruvius, as he noticed of the classical Greek architects: "Advancing in the subtlety of their judgments and preferring slighter modules, they fixed seven measures of the diameter for the height of the Doric column, nine for the Ionic" (IV.1.8). The proportions of Ionic and Corinthian columns are adjusted to their more feminine nature and are made more slender and graceful, thus nine modules high in relation to their thickness rather than six.

Alberti had a mathematical explanation for this in *De re aedificatoria*. In Book IX, he began by acknowledging the proportioning of columns based on the human body. He explained (IX.7):

> The shapes and sizes for the setting out of columns, of which the ancients distinguished three kinds according to the variations of the human body, are well worth understanding. When they considered man's body, they decided to make columns after its image. Having taken the measurements of a man, they discovered that the width, from one side to the other, was a sixth of the height, while the depth, from navel to kidneys, was a tenth.

But Alberti then explained that for some reason, which he described as an innate sense of *concinnitas*, the proportions taken from the body were not completely graceful and pleasing for the column, so further adjustments had to be made. Alberti defined beauty as *concinnitas*, which is "a harmony of all the parts . . . fitted together with such proportion and connection that nothing could be added, diminished, or altered for the worse" (VI.2). In Book IV Alberti explained, "In this we should follow Socrates' advice, that something that can only be altered for the worse can be held to be perfect" (IV.2). The adjustment to be made entailed finding a pleasing mean between the extremes of six and ten, so the two are added together and divided in half, and the result, eight, becomes the height of the Ionic column, in modules in relation to its width. Alberti did not include the capital, the height of which Vitruvius described as "one third of the thickness of the column" (*De architectura*, IV.1.1), so for Vitruvius the height of the column is nine modules, including the capital. For the Doric column, the mean is between six and eight, thus seven, as Vitruvius described. For the Corinthian column, the mean is between eight and ten, thus nine. So Alberti explained the subtlety of the refinements of the proportions of columns by the ancients, as reported by Vitruvius.

Vitruvius described how the modules of the column, including six, eight, and ten, were derived from the human body (III.1.2):

> For nature has so planned the human body that the face from the chin to the top of the forehead and the roots of the hair is a tenth part; also the palm of the hand from the wrist to the top of the middle finger is as much; the head from the chin to the crown, an eighth part; from the top of the breast with the bottom of the neck to the roots of the hair, a sixth part.

Further, "the foot is a sixth of the height of the body," the proportion in the body applied to the column, and thus, "by using these, ancient painters and famous sculptors have attained great and unbounded distinction."

The numbers six and ten had philosophical significance for Vitruvius as well as practical significance. As the number ten is taken from various places in the human body, thus "the ancients determined as perfect the number which is called ten" (III.1.5) and "Plato considered that number perfect, for the reason that from the individual things which are called *monads* among the Greeks, the decad is perfected." For Pythagoreans, ten was the number of the tetractys, as the sum of the four digits; one, two, three, and four, which comprise the Pythagorean harmonies. As musical harmony contained the principles underlying the order of the universe, the tetractys thus revealed those principles. The tetractys "embraced the whole nature of number," as Aristotle explained in *Metaphysics*, and "contained the nature of the universe."[6] Plato appropriated a version of the tetractys in the *Timaeus* to symbolize the harmonic constitution of the world soul, as it contains the "musical, geometrical, and arithmetical ratios of which the harmony of the whole universe is composed."[7] Ten was the perfect number for the Pythagoreans because, as in Egyptian cosmology, it symbolizes completion and signals the reversion to unity. The numerical process is an allegory of the process of creation, as all things originate from a state of unity into multiplicity, as in the Egyptian Ennead, the group of nine creator gods who, along with Horus, the so-called "tenth Ennead," symbolized the completion of the cycle of creation.

But, according to Vitruvius, the number six is perfect as well, because the foot is the sixth of a part of a man's height and because "this number has divisions which agree by their proportions" (*De architectura* III.1.6). This was understood by the ancient Egyptians, who divided time into multiples of six. The sum of ten and six divided by ten is the Golden Ratio, 1.6 to 1, a proportion found throughout the human body and many works of classical and Renaissance art and architecture. The Golden Ratio is present in illustrations of the so-called "Vitruvian Man" made during the Renaissance by the likes of Leonardo da Vinci, Francesco di Giorgio, and Cesare Cesariano. Vitruvius described the male body as a model for proportioning in architecture (III.1.3):

> In like fashion the members of temples ought to have dimensions of their several parts answering suitably to the general sum of their whole magnitude. Now the navel is naturally the exact center of the body. For if a man lies on his back with hands and feet outspread, and the center of a circle is placed on his navel, his figure and toes will be touched by the circumference.

The ratio between the distance from the navel to the bottom of the feet and the distance between the navel to the top of the head is 1.6 to 1 in the illustration of the Vitruvian Man by Leonardo da Vinci.

According to Vitruvius, proportion (or *analogia* or *eurythmia*), is one of six things of which architecture must consist, the others being order (or *ordinatione*), arrangement (or *dispositione*), symmetry, décor, and distribution (or *oeconomia*). Order is defined as the arrangement of the proportion, which results in symmetry, which consists in dimension, the organization of modules or units of measurement. Arrangement (the Greek *ideae*) is the assemblage of the modules to elegant effect; proportion gives grace to a work in the arrangement of the modules in their context. Alberti followed Vitruvius in his definition of *concinnitas* or beauty in *De re aedificatoria*: "It is the task and aim of *concinnitas* to compose parts that are quite separate from each other by their nature, according to some precise rule, so that they correspond to one another in appearance" (IX.5).

All proportion for Vitruvius is derived from the human body, which also contains order, arrangement, and symmetry. Proportion "consists in taking a fixed module, in each case, both for the parts of the building and for the whole, by which the method of symmetry is put into practice. For without symmetry and proportion no temple can have a regular plan; that is, it must have an exact proportion worked out after the fashion of the human body" (*De architectura* III.1.1). Proportion is achieved in architecture in the same way that proportion is achieved in the body, through the organization of modules resulting in symmetry.

> Therefore if Nature has planned the human body so that the members correspond in their proportions to its complete configuration, the ancients seem to have had reason in determining that in the execution of their works they should observe an exact adjustment of the several members to the general pattern of the plan.
>
> (III.1.4)

According to Alberti in *De re aedificatoria*, the proportions of a building correspond to what Alberti calls the "lineaments" of the building. Following Vitruvius, Alberti explained, "It is the function and duty of lineaments, then, to prescribe an appropriate place, exact numbers, a proper scale, and a graceful order for whole buildings and for each of their constituent parts" (I.1). Alberti described the building as "a form of body" (Prologue), but which consists of two things, "lineaments and matter, the one the product of thought, the other of Nature; the one requiring the mind and the power of reason, the other dependent on preparation and selection."

Alberti's definition of *concinnitas* is similar to the invocation by Alberti's acquaintance Nicolas Cusanus, in the papal curia in Rome, of the Platonic demiurge in *De docta ignorantia* (1440):

> In creating the world, God used arithmetic, geometry, music, and likewise astronomy. For through arithmetic God united things. Through geometry he shaped them . . . Through music he proportioned things in such way that there is not more earth in earth than water in water, air in air, and fire in fire.
>
> (II.13)[8]

In other words, nothing can be added, diminished, or altered for the worse.

In Book IX of *De re aedificatoria*, Alberti explained (IX.5):

> When you make judgments on beauty, you do not follow mere fancy, but the workings of a reasoning faculty that is inborn in the mind. . . . For within the form and figure of a building there resides some natural excellence and perfection that excites the mind and is immediately recognized by it. I myself believe that form, dignity, grace, and other such qualities depend on it, and as soon as anything is removed or altered, these qualities are themselves weakened and perish.

Beauty depends on the universal, archetypal Platonic idea, as in the lineament, where proportions in matter correspond to mathematical and geometrical proportions in the mind, and, following Socrates, beauty has the quality of *concinnitas*, that nothing can be altered for the worse. Beauty is found in the correct proportioning of the body as it is given by lineament and *concinnitas* and translated from nature to building through the Idea (IX.5):

> Beauty is a form of sympathy and consonance of the parts within a body, according to definite number, outline, and position, as dictated by *concinnitas*, the absolute and fundamental rule in Nature. This is the main object of the art of building, and the source of her dignity, charm, authority, and worth.

The beauty and harmony given by *concinnitas* and lineament through vision in nature and building is governed by the same underlying proportional systems and archetypal rule as the beauty and harmony in music. "The very same numbers that cause sounds to have that *concinnitas*, pleasing to the ears, can also fill the eyes and mind with wondrous delight" (IX.5). This is demonstrated in the application of the proportions of the Pythagorean harmonies to the façade of the Palazzo Rucellai—the octave, fifth, and fourth, the ratios described by Alberti in Book IX of *De re aedificatoria*. The ratios define relationships between consonant strings on a musical instrument that create contrasting sounds, which are then classified into sets of numbers. Alberti also described the *tonus*, or single note, the *diapason diapente*, or octave plus a half, and the *disdiapason*, or double octave. Mathematical proportions derived from musical harmonies are then translated into geometries in architecture, just like the children of the Demiurge in the *Timaeus* translate the mathematical proportions of the Divine idea into geometrical constructions. Thus "[f]rom musicians therefore who have already examined such numbers thoroughly, or from those objects in which Nature has displayed some evident and noble quality, the whole method of outlining is derived" (IX.5), as at the Palazzo Rucellai. The proportions of the Pythagorean harmonies can also be found on Alberti's façade for the nearby Church of Santa Maria Novella in Florence.

As reported by Iamblichus in *De vita pythagorica*, Pythagoras desired to invent an "instrumental aid for hearing,"[9] as the compass and ruler were used for sight. He happened to hear the sound of "hammers beating iron on an anvil" and recognized the combinations of the octave, fifth and fourth. He was able to repeat the consonances by hanging four strings from pegs attached to a wall with weights at the bottom. Striking the strings in different combinations reproduced the ratios of the octave, fifth and fourth. Pythagoras then transferred the strings from the wall to the lyre, which he called the "string-stretcher."

The proportions of the Pythagorean harmonies were also prescribed by Alberti in *De re aedificatoria* for the determination of plans of buildings, as "architects employ all these numbers in the most convenient manner possible" (IX.5). As with the area of the bay on the façade of the Palazzo Rucellai, dimensions of rooms in plans are modulated by the *sesquialtera* and the *sesquitertia* in order to ensure pleasing effect, beauty and *concinnitas*, where nothing may be altered for the worse, through harmonic proportions. "When working in three dimensions, we should combine the universal dimensions, as it were, of the body with numbers naturally harmonic in themselves, or ones selected from elsewhere by some sure and true method" (IX.6). The human body is the universal body, or the body of the universe, composed of the same proportional solids that are translated from the mathematical proportions of the divine, archetypal idea, the universal dimensions, in the creation of the universe. As the naturally harmonic ratios are used to determine the layouts of buildings, the *sesquialtera* and *sesquitertia* along with the *tonus*, *diapason*, and *disdiapason*, "[t]hese numbers which we have reviewed were not employed by architects randomly or indiscriminately, but according to a harmonic relationship" (IX.6).

Along with the set of simple harmonic ratios, Alberti prescribed a set of more complex mathematical relationships to determine proportional areas in buildings. These include roots and powers and three types of means: arithmetical, geometrical, and musical. "In establishing dimensions, there are certain natural relationships that cannot be defined as numbers, but that may be obtained through roots and powers" (IX.6). The square of a dimension defines an area, and the cube of a dimension defines a volume, as "[t]he cube is a projection of the square." The dimension that cannot be defined as a number is the diagonal or diameter derived from the area and projected in the volume.

Means are "methods of three-dimensional composition" and serve as "rules for the composition of outlines in three dimensions." The arithmetical mean is half the sum of two extremes, or the number equidistant between two extremes. This was the mean used by Alberti to justify the proportioning of the Doric, Ionic, and Corinthian columns as given by Vitruvius. The geometrical mean is the root of the product of two extremes. It is the mean used in the determination of the area in the plan of a building, being represented by the diagonal of the area; it is thus determined geometrically rather than mathematically, as "[t]his geometrical mean is very difficult to ascertain numerically, although it may be found very easily using lines."

The musical mean is the number between two extremes where "the proportion between the shortest and longest dimensions is the same as that between the shortest and the middle, and again the same as that between the middle one and the longest." The musical mean was defined by Plato in the *Timaeus* as the mean "exceeding one extreme and being exceeded by the other by the same fraction of the extremes" (36);[10] in other words the distance of the two extremes from the mean is the same fraction of their own quantity, as expressed in the equation (B – A) / A = (C – B) / C, B being the musical mean between A and C.[11] The mean proportional is seen as a means of determining musical consonances in the generation of series of plan sizes for buildings, beginning with the *sesquialtera* and the *sesquitertia*, and then determining more complex proportional relationships.

As noted by Rudolf Wittkower, Marsilio Ficino described the same three types of means, the arithmetical, geometrical, and musical, in the second book of the *Opera Omnia*:

> Divisions of that kind are triune, being arithmetic, geometry, and music. Arithmetic consists of equal numbers. Thus the median between three and seven is five, where the same binary number prevails below and above the median in equal proportion on both sides. Geometry is found in places of equality in reason, in which are both multiples and particulars: they can be seen to be clearly similar, as in three and nine, thus nine to seven and twenty [27], triple on both sides. As nine is close to six, so six is close to four. For the proportions are both one and a half [*sesquialtera*] . . . in three, four and six, the difference between six and four is binary, the difference between four and three is unitary, and further between six and three is a double ratio, thus between two and one is a double ratio. A similar proportion prevails here as well; the median exceeds and is exceeded by the same proportion.[12]

In the *Timaeus*, the musical mean (though not defined by Plato as musical) is described as an instrument used by God in the composition of the soul, in its subdivisions and mixture of the Same and Different. God divides the whole into a series of sections which are squares and cubes, the same squares and cubes used by Alberti in the determination of areas. A numerical sequence is thus produced from the series of squares and cubes: 1, 2, 3, 4, 9, 8, 27, that is, 2, 2 squared, 2 cubed, 3, 3 squared, and 3 cubed. The cube corresponds to the three-dimensional volume of geometrical matter. This is the so-called Platonic Lambda. The intervals between the squares and cubes are filled with the "musical" means, as described by Plato. The numerical series constitute the "fabric" of matter, which is then divided into the Same and Different, that is, the regular circular motion of the soul, which corresponds to the movement of the planets and the archetypal idea, and the irregular, contrary motion of the Different, which corresponds to the elliptic in the cosmos and unstable matter.

In *Timaeus* 36, "the motion of the outer circle he called the motion of the same, and the motion of the inner circle the motion of the other or different." The motions of the Same and Other "lie aslant" from each other, crossing on a diagonal. The motion of the Same includes the east to west rotation of the moon, sun and five planets (Mercury, Venus, Mars, Jupiter, and Saturn, in descending angular speeds), which were created "in order to distinguish and preserve the numbers of time" (*Timaeus* 38). Each heavenly body has independent rotations comprising the motions of the Other at an angle to the equator of the heavenly sphere and corresponding to the Zodiac.

While the outer motion of the Same is uniform and constant, the inner motion of the Other is "divided in six places and made seven unequal circles having their intervals in ratios of two and three, three of each" and set opposite the motion of the Same. Three sets of ratios of two and three describe the Lambda, the three ratios of two being 2:1, 4:2, and 8:4; and the three ratios of three being 3:1, 9:3, and 27:9, as odd and even numbers ascend in proportion from the unit or the number one, in describing the coincidence of opposites in the process of creation, following the Pythagorean Tetractys and the Egyptian Ennead. The ratios are either the distance of the successive radii of the bodies in Plato's *cosmos* or the differences between the radii.

The number series of the Platonic Lambda as described in the *Timaeus* is present in Alberti's design of the Church of San Sebastiano in Mantua late in his career. An examination of the dimensions of the building by Joseph Rykwert and Robert Tavernor revealed the presence of the number series 3, 9, 27 (a nine-square plan and a 27-square cube) and 2, 4, 8 (in the modules of the piers and the arms of the Greek cross).[13] The module is defined by Alberti in *De pictura*, his treatise on painting, as the *braccio*, which is a third of the height of a man, so it would be twice the width of a column.

Finally, Alberti described prescriptions for two other types of plans in *De re aedificatoria*, the round plan and the "many-sided" or polygonal plan. Obviously, "The round plan is defined by the circle" (VII.4). The round plan is the ideal plan for the temple, because, as Alberti explained, "It is obvious from all that is fashioned, produced, or created under her influence, that Nature delights primarily in the circle. Need I mention the earth, the stars, the animals, their nests, and so on, all of which she has made circular?" Polygonal plans are derived by subdividing the circle, in the same way that in the *Timaeus* the soul and the "pleats of matter" are derived by subdividing the divine whole. Alberti described the geometrical derivation of the polygonal plan (VII.4):

> For many-sided plans, the ancients would use six, eight, or even ten angles. The corners of all such plans must be circumscribed by a circle. Furthermore, they may be plotted exactly using the circle. For half the diameter of the circle will give the length of the sides of the hexagon. And if you draw a straight line from the center to bisect each of the sides of the hexagon, it is obvious how to construct a dodecagon. From a dodecagon it is obvious how to derive an octagon, or even a quadrangle.

The divisions of the circle in the composition of the polygons are translations of mathematical proportions into geometrical figures. The circle is the perfect eternal archetype from which all polygonal geometric figures of material reality are derived, the whole which is subdivided by mathematical and geometrical proportions. The inscription of polygonal figures in the circle is a representation of the process of the creation of material reality, like the Platonic Lambda in the *Timaeus*, in its finiteness and multiplicity, from the infinity and simplicity of God as an eternal archetype.

Alberti and Nicolas Cusanus certainly crossed paths as members of the papal curia in Rome. Cusanus was appointed Cardinal by Pope Nicholas V in Rome in 1450, the year that Alberti was writing *De re aedificatoria*. Cusanus and Alberti shared a close mutual friend, Paolo Toscanelli, and both dedicated works to him. Cusanus and Alberti can be seen to influence each other in certain works. Alberti's mathematical treatise *De Lunularum Quadratura* shows the influence of Cusanus, while Cusanus owned a copy of Alberti's *Elementa Picturae*,[14] and the *De Staticis Experimentis* of Cusanus shows the influence of Alberti's *De ludi matematici* and *De motibus ponderis* and contains an allusion to Vitruvius and ancient architecture as well.[15]

In *De docta ignorantia*, Cusanus explained the philosophical basis of the creation myth of the *Timaeus* as it unfolds through mathematical and geometrical proportions and is represented in the architectural theory of Alberti. Cusanus explained that the universe is one and whole and that its oneness is "contracted by plurality, so that it is oneness in plurality" (II.6). The singularity of the universe is derived from the singularity of the divine and absolute, the universal principle or *archê*, so the singularity of the universe is a secondary singularity, as in the singularity or harmony of *concinnitas*. Cusanus described the secondary singularity or oneness of the universe as tenfold, as all creation, or unfolding of the absolute oneness, is contained in the number ten, which is, as in the tetractys, the sum of the integers that comprise the Pythagorean harmonies and thus the number that governs the use of harmonic proportions in architectural composition. Therefore, the tenfold oneness of the universe enfolds the plurality of all contracted things. As the oneness of the universe is in all things as the contracted beginning of all—the oneness of the universe is the root of all things, and thus from this number arise squares and cubes as variations of the absolute oneness, as in the Platonic Lambda, and the prescriptions for the plans of Alberti. In prescribing such numerical proportions in the generations of plans in architecture, Alberti ensures that all composition in architecture contains the absolute oneness of the divine archetype in its multiplicity and variation, and thus *concinnitas*.

Cusanus compared the process of unfolding and the contracting of particulars in the universal, as the polygon in the circle, for example, to a geometrical progression from point to line to surface, as an allegory for the transformation of the archetypal idea to material form, or in Alberti's terms, lineament to matter. Particulars arise as a contractible universal that exists not in itself but in that which is actual, just as a point, a line, and a surface precede, in progressive order, the material object in which alone they exist actually. Thus all universals exist in the universe only in a contracted manner, as copies of an original singularity. The line and the surface exist in the material object as the universal exists in the particular. Nevertheless, by abstracting, the intellect makes them exist independently of things, as lineament. The universal is in the intellect as a result of the process of abstracting.

Plato conceived of the universe as being constructed in a geometrical progression from point to line to surface to solid, each corresponding to a level of the tetractys: one unit for a point, two units for a line, three units for a surface and four units for a solid. The atom for earth was a cube. In *De coniecturis* (On Conjecture, 1442), Cusanus described the transition from numerical to geometrical proportions in progression, in creation and unfolding, in the composition of matter. A unified body is perceptible as a combination of numerical figures,[16] he explained, as numbers are perceptible as a solid composition. The progression from the simplest unity is seen as a progression from the simplest point, to line, to surface, and to body. Unity is projected into line, surface, and body. The unity of the line is found in the surface and the body.[17]

In *De docta ignorantia*, Cusanus followed scripture (Wisdom 11:21) and the writings of Saint Augustine in proclaiming that "God, who created all things in number, weight, and measure, arranged the elements in an admirable order" (II.8). As for Plato and Ficino, God is architect of the world, and "[w]ho would not admire this Artisan, who with regard to the spheres, the stars, and the regions of the stars used such skill that there is . . . And He established the inter-relationship of parts so proportionally that in each thing the motion of the parts is in relation to the whole." The order of the universe, as seen in the geometrical and mathematical relationships between the stars and planets, was seen as a macrocosm of the order of the human soul.

In the *Opera Omnia*, Marsilio Ficino developed the idea that if geometrical variations are numerically controlled in relation to a singular form, the square or cube, then they will all contain the same universal principle. They all belong to one series or *scala* of values, and even provoke certain sequences of emotions and thoughts (2.1267),[18] like musical harmonies orchestrated numerically. Ficino identified the cube as the most singular and absolute of geometries, and thus the square and cube are the primary elements of a harmonious architecture and the cube is philosophically equated with the earth. Thus the cube as earth, the singular and primary geometry, contains all variations and possibilities within it, all absolute principles which are unfolded as ideas as the square is unfolded into an infinite variety of permutations through arithmetical proportions, as described by Alberti in *De re aedificatoria*, through the *sesquialtera*, *sesquitertia*, *diapason*, *disdiapason*, etc., all united by harmonic proportions as particular variations of a universal truth, as in the façade of the Palazzo Rucellai.

For Ficino the cube as earth is the "single work by the one God," and the cube is to the architect as the earth is to God. God is the "architect of the World" (*Opera omnia*, 2.1444), following Plato, and the ideas of things can only be found in the architect of the World. To the sequence of geometrical unfolding in point, line, plane, and solid, Ficino assigned particular thoughts or philosophical ideas: "as a solid [Earth] stands for action, as point stands for essence, line for being, and plane for virtue" (2.1447).[19] Numerical proportions are translated to geometrical proportions and then translated to human qualities. To the human qualities Ficino added all elements of knowledge: "the species, figures, quantities, magnitudes, and proportions of all things" (2.1446). All things in human thought are contained as particulars in the universal absolute of the cube, which becomes an "infinitely multiple frame of reference,"[20] as described by George Hersey, as the square and cube are the infinitely multiple frame of reference for the architecture of Alberti.

As all geometrical variations are contained in the one principal form, the cube, in the *Opera Omnia*, and as all particulars arise from the universal as all polygonal figures arise from the circle in the *De docta ignorantia* of Cusanus, so all particular points of light in vision arise from the light of the sun. According to Pseudo-Dionysius in the *Divine Names* (IV.4), "Light comes from the Good,"[21] that quality defined by Plato as the benevolent design of the Demiurge and the temperance and justice of the divine. As the circle of Cusanus participates in the polygonal figures but is at the same time inaccessible to them, as it has nothing in common with other figures in its inner simplicity and unity while it contains absolute eternity, which is the form of all forms folded perfectly in itself, so the Good according to Pseudo-Dionysius both participates in and remains inaccessible to all things: "The goodness of the transcendent God reaches from the highest and most perfect forms of being to the very lowest. And yet it remains above and beyond them all, superior to the highest and yet stretching out to the lowest" (IV.4). The Good, or the sun, provides everything with "measure, eternity, number, order"; the light from the sun is thus the instrument of God in the ordering of the universe through proportion, as God, for Cusanus, "who created all things in number, weight, and measure, arranged the elements in an admirable order," creating a *concinnitas* or beauty as defined by Alberti, "a harmony of all the parts . . . fitted together with such proportion and connection that nothing could be added, diminished, or altered for the worse."

According to Ficino in *De amore* (1469), the beauty of the body depends on three things: "Arrangement, Proportion, and Aspect. Arrangement means the distances between the parts, Proportion means quantity, and Aspect means shape and color. For in the first place it is necessary that all the parts of the body have their natural position" (V.6).[22] Vitruvius named arrangement, proportion, order, symmetry, décor, and distribution as the elements of architecture. Order is the arrangement of the proportion, resulting in symmetry, which consists in dimension, the organization of modules. Arrangement is the assemblage of the modules, and proportion gives grace to a work in the arrangement of the modules. Thus Ficino's formula for the beauty of the body is a condensed version of that of Vitruvius.

For Ficino in *De amore*, the qualities of arrangement, proportion, and aspect are not actually a part of the body, because they exist separately of an individual body and thus belong to the lineament of the body, or the lines, in Alberti's terms, rather than the matter. Ficino asked, "But who would call lines (which lack breadth and depth, which are necessary to the body) bodies?" (V.6). Arrangement entails spaces between parts rather than the parts themselves, and proportions are boundaries of quantities, which are "surfaces and lines and points," or points, lines, and planes, which Ficino defined as the qualities of essence, being, and virtue in the *Opera Omnia*. Thus, for Ficino in *De amore*, "From all these things it is clear that beauty is so alien to the mass of body that it never imparts itself to matter itself unless the matter has been prepared with the three incorporeal preparations which we have mentioned," which exist only in the mind of the artist or architect, as lineaments, in the terms of Alberti.

Through arrangement, proportion, and aspect, which are incorporeal qualities of the lineaments of matter or intelligences, which are copies of divine ideas and principles, "both the heavenly splendor will easily shine in a body which is like heaven, and that perfect Form of Man which the Soul possesses will turn out more clearly." Arrangement, proportion, and aspect are the perfect form that the soul possesses, the innate idea of the body in matter. The same formula can be applied to music: Arrangement is "an ascent from a low note to the octave, and thence a descent"; proportion is "a proper progression through third, fourth, fifth, and sixth intervals, and also full tones and half-tones"; and aspect is "the sonorous intensity of a clear note."

Piero della Francesca was principally known in the Renaissance as the author of treatises on mathematics and geometry. He is the author of the *Trattato d'abaco* (Abacus Treatise), *De prospectiva pingendi* (On Perspective in Painting), and the *Libellus de quinque corporibus regularibus* (Short Book on the Five Regular Solids). The first and last books are instruction manuals in applied mathematics.

His most famous works as a painter are the fresco cycle of the *Legend of the True Cross* (Figure 2.2) in the chapel of the main altar of the Church of San Francesco in Arezzo, painted for Luigi Bacci, and the *Flagellation of Christ*, which may have been painted for the Cappella del Perdono on the interior of the ducal palace of Urbino.[23] According to Giorgio Vasari, Piero studied arithmetic and geometry in his youth and went on to accomplish great results in both mathematics and painting ("maraviglioso frutto et in quelle et nella pittura").[24] Vasari also praised Piero for his mastery of the regular bodies of the Platonic geometries ("maestro raro nelle difficultà de' corpi regolari").

While he was working on the *Legend of the True Cross*, Piero traveled to Rome, in 1458–9. It is said that the architectonic quality of the picture—the architectural elements along with the foreshortening, the simplification of geometries, and the

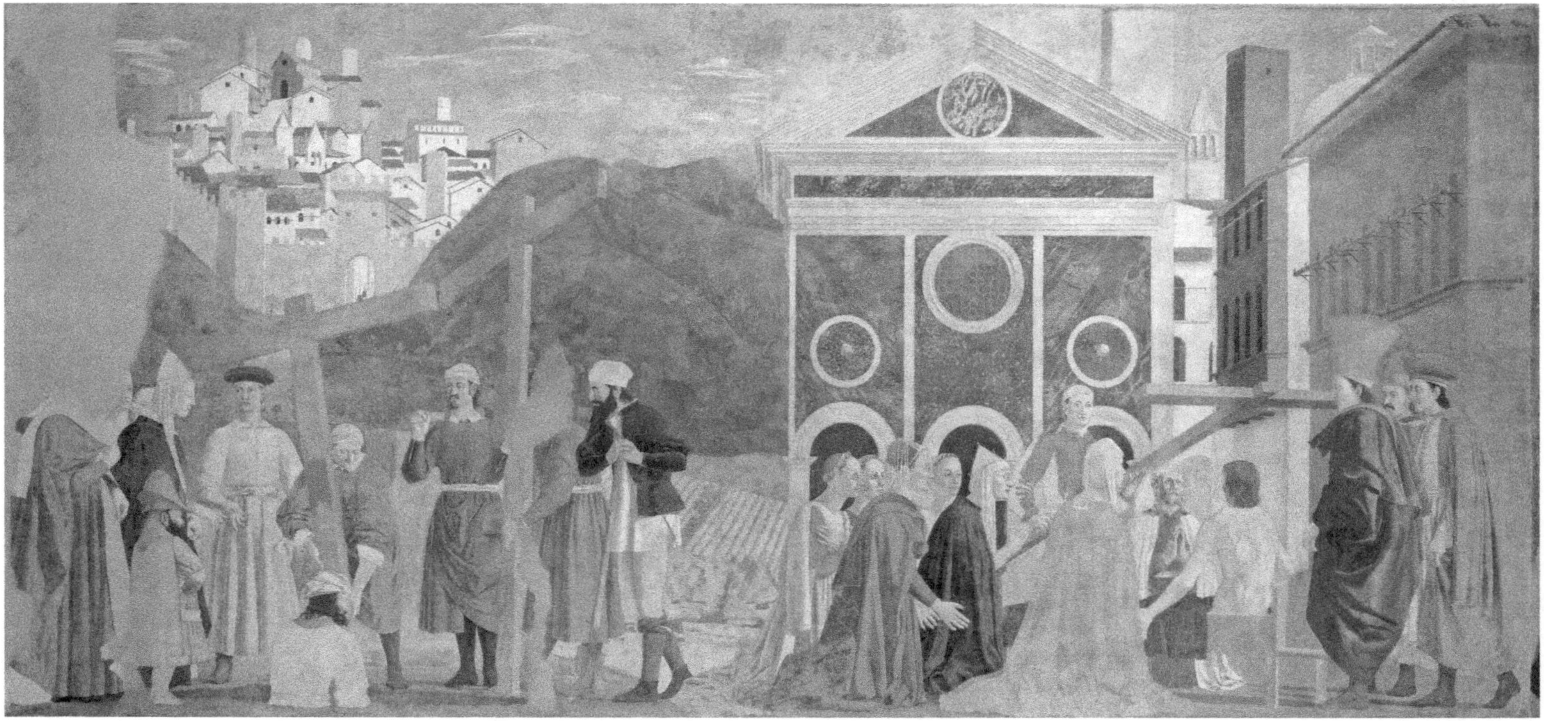

Figure 2.2 Piero della Francesca (1415/20–1492). *Legend of the True Cross: Finding of the Three Crosses and Verification of the True Cross*, c. 1452. San Francesco, Arezzo © 2016.

Photo credit: Scala, Florence—courtesy of the Ministero Beni e Att. Culturali.

continuity between foreground and background—were influenced by Piero's encounter in Rome with members of the humanist court of Pius II.[25] The picture represents an application of Piero's theories of perspective, as expressed in the treatises. The *Flagellation* dates to this time as well, and is also dominated by architectonic constructions, simplified geometries, and perspectival space. Both pictures display the measurement and scientific representation of solid geometries. Figures that Piero may have encountered in Rome include Cardinal Bessarion, Leon Battista Alberti, Nicolas Cusanus, Paolo Toscanelli, and Francesco del Borgo.

Leon Battista Alberti wrote a treatise on plane geometry, *De ludi mathematici*, and a treatise on plane and solid geometry, *Elementa Picturae*. Geometrical definitions are developed by Alberti for painting, beginning with the definitions in Euclid's *Elements of Geometry*. Geometrical constructions, regular and irregular polygons, are prescribed as the basis for constructing a picture. The use of geometry in painting is also prescribed in Alberti's treatise on painting, *De pictura* (1435), as he expressed:

> I want the painter, as far as he is able, to be learned in all the liberal arts, but I wish him above all to have a good knowledge of geometry. I agree with the ancient and famous painter Pamphilus, from whom young nobles first learned painting; for he used to say that no one could be a good painter who did not know geometry [in Pliny, *Historia naturalis*]. Our rudiments, from which the complete and perfect art of painting may be drawn, can easily be understood by a geometer, whereas I think that neither the rudiments nor any principles of painting can be understood by those who are ignorant of geometry.[26]

Piero was actually under the employ of Nicolas Cusanus in Rome in 1459, while Cusanus was governing Rome briefly in the absence of Pius II, Aeneas Silvius Piccolomini, who had gone to Mantua. Piero was employed to paint frescoes in the papal apartment and the Piccolomini room in the Vatican. For Nicolas Cusanus it was mathematics on which all knowledge must be based, as nothing beyond mathematics is certain, and mathematics is necessary for intellectual comprehension, as expressed in *De docta ignorantia* ("Nihil certi habemus in nostra scientia nisi nostrum mathematicam"; "Tuus intellectus sine numero nihil concipit").[27] Mathematics is translated into geometrical permutations in *De circuli quadratura* (1450), as a model for Platonic conceptions of the structure of being in divine emanation.

The *Trattato d'abaco* (1470) of Piero della Francesca contains a section on geometry, treating plane geometry and the measurement of polygonal figures, as in those described by Nicolas Cusanus in *De circuli quadratura* and the measurement of solid geometrical forms as they are inscribed into a sphere, as in the regular bodies described in the *Timaeus*.[28] The solid forms are analyzed mathematically, in terms of their proportional relationships to each other. Piero discusses in particular the five Platonic solids as they are inscribed in a sphere—the tetrahedron, cube, octahedron, icosahedron, and dodecahedron, along with irregular bodies. In the measurement of the solids, Piero uses principally the geometrical mean which became known as the "divine proportion," or the "golden section,"[29] that is, the division of a line so that the small section is in proportion to the large section as the large section is to the line. This golden mean corresponds to the mean described by Plato in *Timaeus* 36, in the construction of the soul by the Demiurge, in the construction of the subdivisions "each containing a mixture of Same and Different and Existence."

Plato described two means, "one exceeding one extreme and being exceeded by the other by the same fraction of the extremes, the other exceeding and being exceeded by the same numerical amount." The first mean corresponds to the golden mean used by Piero in the *Trattato d'abaco*. In the *Timaeus*, proportioning systems are applied to the solids in their subdivisions and reconfigurations, as "when larger bodies are broken up a number of small bodies are formed of the same constituents, taking on their appropriate figures; and when small bodies are broken up into their component triangles a single new larger figure may be formed as they are unified into a single solid" (54). Thus mathematical proportions are applied to regular and irregular bodies in the *Trattato d'abaco*.

The five regular bodies of the Platonic solids were the subject of Books XIII, XIV, and XV of Euclid's *Elements of Geometry*. Book XV in particular describes the role of the regular bodies in the construction of matter by the Demiurge of Plato's *Timaeus*. Piero's treatment of the solids in the *Trattato d'abaco* forms the basis for his continued examinations in both *De prospectiva pingendi* and the *Libellus de quinque corporibus regularibus*. *De prospectiva pingendi* (1480) was written for the ducal court at Urbino, late in Piero's career, and the autograph manuscript was placed in the library of Federigo da Montefeltro. The treatise is divided into three books. The first book is concerned with plane geometry, that is, points, lines, and surfaces. The second book is concerned with solid geometry, that is, cubic bodies. The third book is concerned with the proportioning of three-dimensional objects, including human heads and *torchi* or *mazzochi*, faceted geometric rings. The discussions on perspective focus on the construction of objects, many of which are parts of buildings, and as such the treatise was of particular interest to architects.

The treatise begins with the proclamation:

> Painting consists of three principal parts, which we call *disegno*, *commensuratio* and *colorare*. By *disegno* we mean profiles and contours which contain things. By commensuration we mean the profiles and contours proportionally positioned in their places. By *colorare* we mean the colors demonstrated in the thing—lights and darks according to how the lighting changes them.[30]

The triune division, as three manifestations of one entity, as in the Trinity, corresponds to Marsilio Ficino's criteria for the beauty of the body, as described in *De amore*: arrangement, proportion, and aspect. Arrangement corresponds to *disegno*, proportion to *commensuratio*, and aspect to *colorare*. Beauty in general "consists in a certain arrangement of all the parts, or, to use their own terms, in symmetry and proportion, together with a certain agreeableness of colors" (*De amore* V.3). Grace arises from harmony in beauty, which depends on line, proportion, and color, as "from the harmony of several virtues in soul there is a grace; from the harmony of several colors and lines in bodies a grace arises" (I.4).

Of the three parts of painting, Piero declared at the beginning of *De prospectiva pingendi*, only *commensuratio* would be discussed, or perspective, but "mixing in parts of *disegno*, without which it is impossible to demonstrate perspective."[31] Color would be left out, but the parts of painting would be discussed "that can be demonstrated with angled lines and proportions, that is, the points, lines, surfaces and bodies."[32] These classifications correspond to the definitions of Euclid's *Elements of Geometry*. Piero identified five elements that need to be considered in the perspectival construction of a painting: sight, or the eye; the form of the thing seen; the distance from the eye to the thing seen; the lines that connect the eye to the extremities or bordering lines of the thing seen; and the area between the eye and the thing seen.[33] These five elements need to be understood in order to understand perspectival construction.

The eye is defined as that in which are represented all of the things seen under different angles (*De prospectiva pingendi*, p. 64: "gli è quello in cui s'apresentano tucte le cose vedute socto diversi angoli . . ."). Objects appear as images in the eye depending on the angle of projection of the lines from the extremities of the objects to the eye; the larger the angle, the closer and larger the object ("cioè quando le cose vedute sono equalmente distante da l'ochio, la cosa magiore s'apresenta socto magiore angolo che la minore, et similmente, quando le cose sono equali et non sono a l'ochio equalmente distante, la più propinqua s'apresenta socto magiore angolo che non fa la più remota"). Objects in space occupy a hierarchy of being, or value, given by the variation in the relation to the angle of projection ("per le quail diversità se intende il degradare d'esse cose"). This is stated in the Eighth Theorem of Euclid's *Optica*.[34]

Sensible things, or objects in the sensible world, are therefore abstracted and transformed into images in the eye through mathematics and geometry. The images in the eye exist as copies of the sensible objects, and the objects become intelligible in the mind's eye, or objects of the intellect. The forms and proportions of sensible things are constructed in the mind from the idea of the things, or the intelligibles, which are translated to the sensible world through mathematics and geometry by way of perspectival construction, as it plays a role in vision. It is the form of the thing, according to Piero, rather than the thing itself, without which the intellect cannot judge nor can the eye comprehend the thing (*De prospectiva pingendi*, p. 64: "la forma de la cosa, perhò che senza quella l'inteletto non poria giudicare nè l'ochio comprendare essa cosa").

Following the definition of the elements of the painting, Piero proceeded in the treatise to discuss the elements of *commensuratio*, or perspective—in particular, in the first book, points, lines, and plane surfaces ("Intese le sopradecte cose, seguitaremo l'opera, facendo di questa parte dicta prospective tre libri. Nel primo diremo de puncti, de linee et superficie piane"). The point is defined as that which has no parts, something that is imaginative, according to geometers, a thing that is as small as the eye can comprehend and that does not contain quantity ("Puncto è la cui parte non è, secondo i geometri dicono essere immaginativo . . . Dirò adunque puncto essere una cosa tanto picholina quanto è posibile ad ochio comprendere . . . perchè nel puncto non è quantità").

Piero's *Libellus de quinque corporibus regularibus* was incorporated in the text of Fra Luca Pacioli's *De divina proportione* in 1509, translated into Italian, but without crediting Piero as the author.[35] Earlier, in 1494, Pacioli had included the solid geometry section of Piero's *Trattato d'abaco* in the *Summa arithmetica*, also without crediting Piero's authorship. The *Summa de arithmetica geometria proportioni et proportionalita*, Pacioli's first book, attempted to apply mathematics to the creation of works of art and architecture. Pacioli was familiar with both the *De architectura* of Vitruvius and the *De re aedificatoria* of Leon Battista Alberti. The *Summa* was dedicated to Guidobaldo da Montefeltro, Duke of Urbino, to whom the *Libellus* of Piero was dedicated. In the dedication, Pacioli explained that the study of mathematics, geometry, and proportion was necessary in particular in the construction of the Ducal Palace in Urbino, as well as the beauty of its ornament.

Fra Luca Pacioli's *De divina proportione* was dedicated to students of philosophy, perspective (optics), painting, sculpture, architecture, music, and mathematics ("philosophia, prospectiva, pictura, sculptura, architectura, musica, e alter mathematice").[36] Pacioli called all of the regular solids "divine" because of the necessity of employing the golden proportion in constructing the dodecahedron, the solid which subsumes the four regular bodies, called the *quinta essentia*, or Fifth Essence.[37] Thus the regular bodies take on a Platonic and mystical significance. Pacioli described all of the bodies in nature which are derived from the regular bodies as having forms in which "virtue is distilled."[38]

In Chapter Five of *De divina proportione*, entitled "Del condecente titulo del presente tractato" (On the governing title of the present treatise), Pacioli described the philosophical significance of the Platonic solids and their measurement. The purpose of the study of divine proportion is the intended contemplation of God ("intedemo a epso dio spectanti").[39] The proposition is divided into five parts, four of which concern the nature of God, the fifth of which concerns the nature of the Trinity. First, God is necessarily a unity and without differentiation.[40] Second, the same substance that is in the divine is found in the three persons of the Trinity—Father, Son and Holy Spirit; each person of the Trinity is a limit or boundary and contains the same invariant proportions.[41] Third, God exceeds all proportion, quantity and number. The proportion of God can only be expressed as occult, or irrational in mathematics.[42] Fourth, God exists in everything, as that part which is unchangeable, and can be apprehended by the intellect by the processes that will be demonstrated in the treatise.[43] Fifth, being confers celestial nature on us through the Fifth Essence, which mediates through the other four simple bodies, that is the four elements.

The triune hypostatic and numerical organization of the universe is the simplest manifestation of divine unity, in the same way that the triangle is the simplest manifestation unfolded from the circle, as the triangle is the most basic geometrical element in the construction of the Platonic solids, and, according to Aristotle in *De caelo*, the body comes to completion in the number three. According to Nicolas Cusanus in *De docta ignorantia*, the triangle is the simplest and most minimal polygonal figure. As every number is resolved in unity, so every polygonal figure is resolved in the triangle. Every polygonal figure is folded into the triangle and originates from it.[44]

All things in the sensible world contain an element of the divine given by a consistent and invariable proportion that is not a physical proportion but an intelligible proportion and can only be apprehended by intellect, as in the geometrical construction of being by Plato in the *Timaeus*. Such divine substance enters the sensible world by means of the Fifth Essence through the four simple bodies of the four elements in constant proportions, the proportions being intelligible copies of the divine.

The Fifth Essence is the fifth polygonal figure described in *Timaeus* 55, "a fifth construction, which the god used for embroidering the constellations on the whole heaven." The construction is the dodecahedron, corresponding to the 12 constellations, which approximates the spherical shape of the heavens as it approximates the spherical shape of the earth in the *Phaedo*. The fifth construction leads to the conclusion that there is a "single, divine world" (*Timaeus* 55), as opposed to five worlds. The Fifth Essence is the divine virtue, or the Good, mediated through the intelligible world, the solid bodies, into the sensible world of the elements. Such mediation is understood through mathematics and geometry by Plato, being the instruments of understanding in the soul.

Thus, in *Timaeus* 53–4, "So we must do our best to construct four types of perfect body and maintain that we have grasped their nature sufficiently for our purpose," that is, that they be comprehended in the intellect, as described by Pacioli, and demonstrated by mathematics and geometry. "Of the two basic triangles, then, the isosceles has only one variety"—thus it is the proportion of the divine substance, which is invariable, as described by Pacioli—"the scalene an infinite number. We must therefore choose, if we are to start according to our own principles, the most perfect of this infinite number. If anyone can tell us of a better choice of triangle for the construction of the four bodies, his criticism will be welcome; but for our part we propose to pass over all the rest and pick on a single type, that of which a pair compose an equilateral triangle" (*Timaeus* 54).

Of the four bodies constructed from the triangles, the tetrahedron for the atom of fire, the octahedron for the atom of air, the icosahedron for the atom of water, and the cube for the atom of earth,

> three are composed of the scalene [as well as the isosceles], but the fourth alone from the isosceles [the cube]. Hence all four cannot pass into each other on resolution, with a large number of smaller constituents forming a lesser number of bigger bodies and vice versa; this can only happen with three of them.

Thus, as described by Pacioli, the elements are bounded and contain the same invariant proportions, and that part of the sensible that is divine is unchangeable. Elements in the sensible world are subject to infinite variation, as geometrical solids are formed with consistent mathematical proportions.

Thus in the *Timaeus*,

> when larger bodies are broken up a number of small bodies are formed of the same constituents, taking on their appropriate figures; and when small bodies are broken up into their component triangles a single new larger figure may be formed as they are unified into a single solid.

Perhaps this is a reference to the dodecahedron, the fifth construction that the god used for embroidering the constellations on the whole heaven, and the Fifth Essence, which mediates the transformation of the substance of the divine through the intelligible solids and sensible elements. Thus, according to Pacioli in *De divina proportione*, being descends in nature through the elements, and as such being is formulated by sacred proportion; it is not possible to comprehend or demonstrate the formulation of being in the five regular bodies and the sphere in which they are inscribed without the use of proportion.[45]

Pacioli continued the mathematical and geometrical constructions of Piero and made explicit both their application to art and architecture and their philosophical and cosmological significance, as is demonstrated in the paintings of Piero, such as the *Flagellation* and the *Legend of the True Cross*, and was demonstrated in the architecture of Leon Battista Alberti. Mathematics and geometry as applied to architecture and painting continued as an important element of aesthetics until the beginning of the twentieth century, when they began to be less relevant to visual representation, resulting from the influence of science in the conception of the human mind in relation to the universe.

Notes

1 Leon Battista Alberti, *On the Art of Building in Ten Books (De re aedificatoria)*, trans. Joseph Rykwert, Neil Leach and Robert Tavernor (Cambridge, MA: MIT Press, 1988).
2 In *Opera Omnia*. See George L. Hersey, *Pythagorean Palaces, Magic and Architecture in the Italian Renaissance* (Ithaca, NY: Cornell University Press, 1976).
3 For a thorough discussion of the aesthetics of Alberti and Ficino, see the chapter "Alberti and Ficino," 99–148, in John Hendrix, *Platonic Architectonics: Platonic Philosophies and the Visual Arts* (New York: Peter Lang, 2004).
4 Arnaldo Della Torre, *Storia dell'Accademia Platonica di Firenze* (Florence: G. Carnesecchi and Sons Typography, 1902), 577: "Leon Battista Alberti, che il Ficino annovera fra coloro che nella sua adolescenza gli furono 'consuetudine familiares confabulators atque ultro citroque consiliorum disciplinarumque liberalium comunicatores.'" (Quoting Antonio Manetti, 1468; see Gaetano Milanesi, *Operette istoriche edite ed inedite di Antonio Manetti*, Florence: Successori Le Monnier, 1887, xix.)
5 Vitruvius, *On Architecture (De architectura), Books 1–5*, trans. Frank Granger (Cambridge, MA: Harvard University Press, 1931).
6 F. M. Cornford, *From Religion to Philosophy* (London: Edward Arnold, 1912), 205.
7 Ibid., 206.
8 Nicolaus Cusanus, *Nicholas of Cusa: On Learned Ignorance (De docta ignorantia)*, trans. Jasper Hopkins (Minneapolis: Arthur J. Banning Press, 1981).
9 Iamblichus, *On the Pythagorean Way of Life (De vita pythagorica)*, trans. John Dillon and Jackson Hershbell (Atlanta: Scholars Press, 1991), 139.
10 Plato, *Timaeus and Critias*, trans. Desmond Lee (London: Penguin Books, 1965).
11 See the discussion of the musical mean in Rudolf Wittkower, *Architectural Principles in the Age of Humanism* (New York: W. W. Norton, 1971).
12 Marsilio Ficino, *Opera Omnia* (Basel: Heinrich Petri, 1576), 2: 1454 f., quoted in Wittkower, *Architectural Principles in the Age of Humanism*, 111, n. 1:

> Item comparationem eiusmodi esse triplicem, scilicet arithmeticam, geometricam, harmonicam. Arithmeticam in numerii paritate consistere. Sic inter tria et septem medius est quinarius, numero eodem, scilicet binario alterum terminum superans, ab altero superatus, per proportionem utrinque bipartientem. Geometricam vero in rationis aequalitate sitam esse, in qua sunt multiplex atque superparticularis: quando videlicet ita comparamus, sicut se habent tria ad novem, ita novem ad septem atque viginti, nam utrobique tripla. Item quod est novenarius iuxta senarium, idem est senarius iuxta quaternarium. Nam et hic et ibi est proportion sesquialtera. . . . Sic enim ponas tria, quator, sex, differentia inter sex and quator est binaries: differentia inter quator et tria, unitas, sicut autem inter sex et tria dupla ratio est, ita inter duo et unum est ratio dupla. Viget hic altera quoque similitude, scilicet portionum: simili namque extremorum portione medius terminus excedit atque exceditur.

13 See the discussion in Joseph Rykwert and Robert Tavernor, "Church of San Sebastiano in Mantua," in *Leon Battista Alberti*, ed. Joseph Rykwert (London: Architectural Design, 1979), 90.
14 Karsten Harries, *Infinity and Perspective* (Cambridge, MA: MIT Press, 2001), 68.
15 Jasper Hopkins, *Nicholas of Cusa on Wisdom and Knowledge* (Minneapolis: Arthur J. Banning Press, 1996), 76: "In *De Staticis Experimentis* Nicholas alludes to several writers whose ideas he finds helpful, including Vitruvius." P. 509, n. 6: "In *De Staticis Experimentis* Cusa seems to have been influenced by ideas in circulation among his scholarly acquaintances—ideas expressed by Leon Battista Alberti in *De ludi matematici* and *De motibus ponderis* (lost)."
16 Nicolai de Cusa, *De coniecturis* (Hamburg: In Aedibus Felicis Meiner, 1972), 37: "Sensibilis corporalisve unitas est illa, quae millenario figuratur."
17 Ibid.:

> Solida atque compositissima est haec sensibilis unitas, uti ipse millenarius. Et ut harum unitatum conceptum subintres, eas concipe differentes, quasi prima sit unitas simplicissimi puncti, secunda simplicis lineae, tertia simplicis superficiei, quarta simplicis corporis. Scies post haec clarius unitatem puncti simplicissimi omne id esse, quod in lineali, superficiali atque corporali exstat unitate; sed unitas lineae est id omne, quod in superficiali et corporali est, atque superficialis pariformiter est id omne, quod in corporali.

18 For the references to the sections of the *Opera Omnia* see the discussion in Hersey, *Pythagorean Palaces*, 35–36.
19 Ibid., 36.
20 Ibid.
21 Pseudo-Dionysius, *The Complete Works*, trans. Colin Luibheid (New York: Paulist Press, 1987).
22 Marsilio Ficino, *Commentary on Plato's Symposium on Love* (*De amore*), trans. Sears Jayne (Dallas: Spring Publications, 1985).
23 See Marilyn Aronberg Lavin, *Piero della Francesca: The Flagellation* (New York: Viking Press, 1972), 51, 82.
24 Margaret Daly Davis, "Piero's Treatises: The Mathematics of Form," in *The Cambridge Companion to Piero della Francesca*, ed. Jeryldene M. Wood (Cambridge: Cambridge University Press, 2002), 134.
25 See for example Carlo Ginzburg, *The Enigma of Piero*, trans. Martin Ryle and Kate Soper (London: Verso, 2000), 18, 23, 27, 40.
26 Leon Battista Alberti, *On Painting (De pictura) and On Sculpture: The Latin texts of De pictura and De statua*, ed. and trans. Cecil Grayson (London: Phaidon, 1972), 88.
27 Maurizio Calvesi, *Piero della Francesca* (New York: Rizzoli, 1998), 73.
28 For a thorough discussion of the aesthetics of Piero della Francesca, and Luca Pacioli, see the chapter, "Piero della Francesca," 149–173, in John Hendrix, *Platonic Architectonics: Platonic Philosophies and the Visual Arts*.
29 Davis, "Piero's Treatises," 137.
30 Piero della Francesca, *De Prospectiva Pingendi* (Florence: Sansoni Editore, 1942), 63: "La pictura contiene in sè tre parti principali, quail diciamo essere disegno, commensuratione et colorare. Disegno intendiamo essere profile et contorni che nella cosa se contene. Commensuratio diciamo essere essi profile et contorni proportionalmente posti nei luoghi loro. Colorare intendiamo dare I colori commo nelle cose se dimostrano, chiari et uscuri secondo che i lumi li devariano."
31 Ibid., 63–64: "mescolandoci qualche parte de disegno, perciò che senza non se po dimostrare in opera essa prospectiva . . ."
32 Ibid., 64: "che con line angoli et proportioni se po dimostrare, dicendo de puncti, linee, superficie et de corpi."
33 Ibid.:

> La qual parte contiene in sè cinque parti: La prima è il vedere, cioè l'ochio; seconda è la forma de la cosa veduta; la terza è la distantia da l'ochio a la cosa veduta; la quarta è le linee che se partano da l'estremità de la cosa e vanno a l'ochio; la quinta è il termine che è intra l'ochio e la cosa veduta dove si intende ponere le cose.

34 Erwin Panofsky, *Perspective as Symbolic Form* (New York: Zone Books, 1991), 35.
35 Davis, "Piero's Treatises," 142.
36 Margaret Daly Davis, *Piero della Francesca's Mathematical Treatises: The "Trattato d'abaco" and "Libellus de quinque corporibus regularibus"* (Ravenna: Lungo editore, 1977), 43.
37 Ibid., 86.
38 Giusta Nicco Fasola, "Introduzione," in Piero della Francesca, *De Prospectiva Pingendi*, 20: "dai corpi regolari 'si distilla la virtù' in quelli da loro dipendenti."
39 Fra' Luca Pacioli, *De Divina Proportione* (Milan: Casa Scriptorium Editrice, 1988, Ristampa Anastatica dell'edizione del 1509), Cap. 5.
40 Ibid.: "La prima e che lei sia una sola e non più eno e possible di lei assegnare alter type de ne differentiae."
41 Ibid.:

> La seconda convenietia e dela sancta trinita. Cioe si commo in divinis una medessima sustatia sia frat re persone padre figlio e spirito sancto. Cosi una medessima proportione di questa sorte sempre conven se trovi frat re termini. E mai nei in piu ne in manco se po retrovare come se dira.

42 Ibid.:

> La terza conveniéta e che si commo idio propriamente non se po diffinire ne per parolle a noi intédere. Cosi questa nostra proportione non se po mai per numero intendibile asegnarene per quantita alcuna rationale exprimere ma ferro pre sia occulta e secreta e dali mathematici chiamata irrationale.

43 Ibid.:

> La quarta convenienta e che si commo idio mai non se po mutare e sia tutto in tutto e tutto in ogni parte cosi la presente nostra proportione semp in ogni quantita continua e discretato sienno grandio sienno picole sia una medesima e sempre invariabile e per verum modo se po mutare ne anco per intellecto altramate apprendere commo el nostro processo demostrara.

44 Nicolai de Cusa, *De docta ignorantia* (Lipsiae: Felicis Meiner, 1932), Book I, 40:

> Velis tamen circa hanc semper benedictam trinitatem advertere, quod ipsum maximum est trinum et non quaternum vel quinum et ultra. Et hoc certe est nota dignum; nam hoc repugnaret simplicitati et perfectioni maximi. Omnis enim figura polygonia pro simplicissimo elemento habet triangularem, et illa est minima figura polygonia, qua minor esse nequit. Probatum est autem minimum simpliciter cum maximo coincidere. Sicut igitur se habet unum in numeris, ita triangulis in figuris polygoniis.

45 Fra' Luca Pacioli, *De Divina Proportione*, Cap. 5:

> E per questi l'essere a cadauna altra cosa in natura. Cosi questa nostra sancta proportione l'esser formale da (secondo l'antico Platone in suo Timeo) a epso cielo atribuendoli la figura del corpo ditto Duodecedron, altra méte corpo de 12 pentagoni. El quale commo desotto se mostrara senza la nostra proportione non e possible poter le formare.

3 Design Method and Mathematics in Francesco di Giorgio's *Trattati*

Angeliki Pollali

During the 1950s, Rudolf Wittkower encapsulated his understanding of architectural proportion in the Italian Renaissance in the following statement: ". . . two different classes of proportion . . . derived from the Pythagoreo-Platonic world of ideas . . . the Middle Ages favoured Pythagoreo-Platonic geometry, while the Renaissance . . . preferred the numerical, i.e. the arithmetical side of that tradition."[1] He further contended that the Middle Ages favoured the perfect geometrical figures for both the plans and the elevations of buildings. Conversely, Renaissance artists adopted an empirical approach to measurement, in turn associated with a renewed interest in nature. As a result, the irrational proportions which often characterize geometrical figures were found to be inadequate. The Renaissance preference lay with integral numbers, or simple fractions. Consequently, "architects of the Renaissance and not of the Middle Ages, fully embraced Vitruvius' well known module system which was the only way of guaranteeing a rational numerical relationship throughout a whole building."[2] Commensurability of measurement, Wittkower concluded, was the "nodal point of Renaissance aesthetics."[3]

As Matthew Cohen has recently pointed out, Wittkower was the first scholar to engage in an historical study of Renaissance proportion, providing a "paradigm," which still underlies modern scholarship.[4] To a certain extent this would explain why architectural historians have traditionally scrutinized Italian Renaissance buildings for the modular system that underlies their design and have placed particular emphasis on the study of the orders. Cohen has also stated explicitly what intuitively appears to be a major shortcoming of Wittkower's thesis, namely that it consists of an "oversimplification of available evidence."[5] He has argued, instead, that geometry and numbers were employed equally in the architectural theory and practice of the Renaissance and the Middle Ages.[6]

While Wittkower's argument might be considered dated and overgeneralized, the writings of Francesco di Giorgio Martini invest it with a certain qualified validity. Francesco's treatise contains the only theoretical statement of a modular system in the fifteenth century. Unpublished during the Quattrocento, the treatise survives in a number of manuscripts dating from the 1470s to the late 1490s. The manuscripts are considered to fall into two major stages of preparation of the treatise. The first version, *Trattati I*, is found in the codex *Ashburnhamianus 361* in the Biblioteca Laurenziana in Florence and the codex *Saluzzianus 148* in the Biblioteca Reale in Turin. The second version of the treatise, *Trattati II*, is found in the codices *S.IV.4* in the Biblioteca Comunale in Siena and *Magliabechianus II.I.141*, in the Biblioteca Nazionale in Florence.[7] These reveal successive stages of preparation of the text, illustrating the trajectory of Francesco's architectural thinking.[8] The discussion of the modular system is found in codex *Magliabechianus II.I.141*, which represents the latest version of Francesco di Giorgio's text.[9] In this chapter, I will examine the nature of Francesco's modular system in relation to arithmetic and geometry, as well as its Vitruvian prototype. I will argue that Francesco's intent is to establish a correspondence between geometrical and numerical methods. In this process, he inventively applies the Vitruvian proportions of columns to the entire design of a building.

Francesco di Giorgio's modular system appears in *Magliabechianus II.I.141* in the context of both temples and private houses. The discussion of the temple is divided into three sections: exterior parts, such as columns, peristyle and steps; middle parts, such as cella, dome, doors and windows; interior parts, such as chapels, arcades and vaults. The modular system appears in the section concerning the middle parts; it follows the discussion of the configuration of the cella and precedes the discussion of the doors and windows. It is to be assumed, therefore, that the modular system refers to the main part

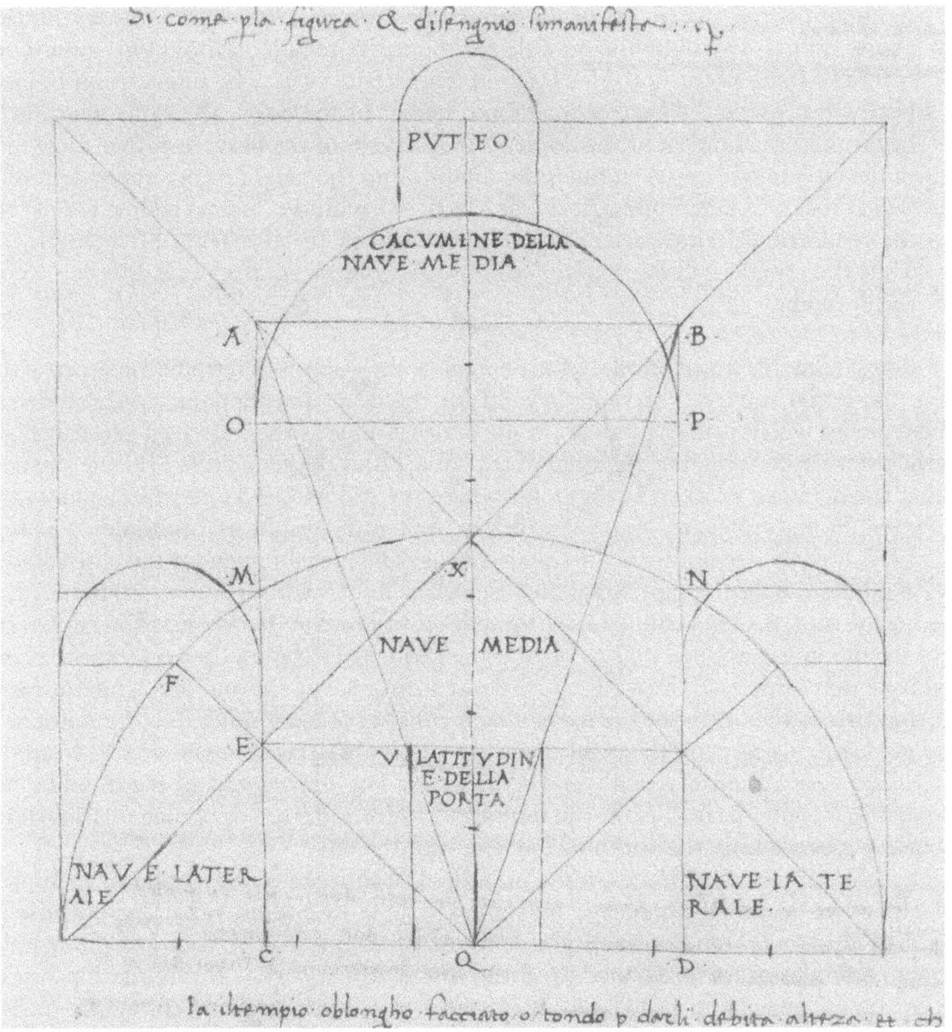

Figure 3.1 Codex II.I.141, folio 41 recto, detail. Biblioteca Nazionale Centrale, Florence.

of the building. Francesco provides three examples. The first diagram appears on folio 41 and it clearly concerns the elevation of a double-aisled basilica (Figure 3.1).

I will quote the passage at length to elucidate the exact procedure put forth by Francesco di Giorgio:

> first create a square of equal sides and then draw, from angle to angle, the two diagonals and divide the base in four equal parts. From the partition points C D draw two straight lines which will terminate at the diagonals forming a transverse line A B. Then draw a semicircle from the end points of the angles of the base, whose height [sic] will pass from the intersection point X of the diagonals and will meet [the lines C A and D B] at the points M N. Draw at these points [M, N] the transverse lines, and this will give you the correct [proportion] of the height to width of the lateral naves. Then take a line which will pass from the middle point of the greater and smaller square, and two more originating at the middle point of the base, which will intersect the diagonal lines at straight lines and will terminate at the end points of the semicircle. The part which remains within this portion, i.e. E F, will be the module for the entire temple. [Draw] two more lines from the same point Q to the squared height [sic] of B. The intersection point S [sic] will provide the correct width and height of the door. The same width is to be given to the lantern of the dome. The diameter of the base, or rather the width of the temple will be seven times the module and the height of the minor square A B C D will be 5½ modules. At the height of 4½ modules, draw the line O P. From the center of O P draw a semicircle; this will be the total height of the temple . . . Thus the temple will have *commensurate height and width*[10] as it can be seen in the drawing.[11]

It follows from this description that the height of the aisles is determined geometrically from the initial square. The same is valid for the door and the lantern of the dome. The segment which is the "module for the entire temple" is also derived geometrically as the distance between the diagonal of the square and the semicircle over the square's base. The distance is taken at the point where the radius Q F is perpendicular to the diagonal. Francesco claims that the width of the temple results as 7 modules. The segment E F, however, is not ⅐ of the square's base. The measurement of the entire width as 7 modules is only an approximation.[12] Similarly, if the width is considered to be 7 E F, the height of A B C D results as 5¼ modules and not 5½. It becomes thus apparent that the module is applied *a posteriori* to the geometrical figure. It is only for the central nave that Francesco draws a semicircle at a distance of 4½ modules from the base. So, apart from the height of the nave, which consists of a combination of a numerical and geometrical segment, the parts of the elevation are derived *geometrically* from the initial square through diagonals and semicircles. The dimensions of the building are not arithmetically derived from the module. In addition, the "modular relationship" seems to concern only the width to the height of the nave.

The following two examples found on folio 41v concern centrally planned temples, which are to be either circular or polygonal. The author claims that the purpose of the procedure is to achieve correspondence of the height to the width of the temple. These examples are less detailed, and the discussion focuses on the relationship of the width to the height of the building. In the first case, the starting configuration is a quadripartite square (Figure 3.2 upper diagram).[13]

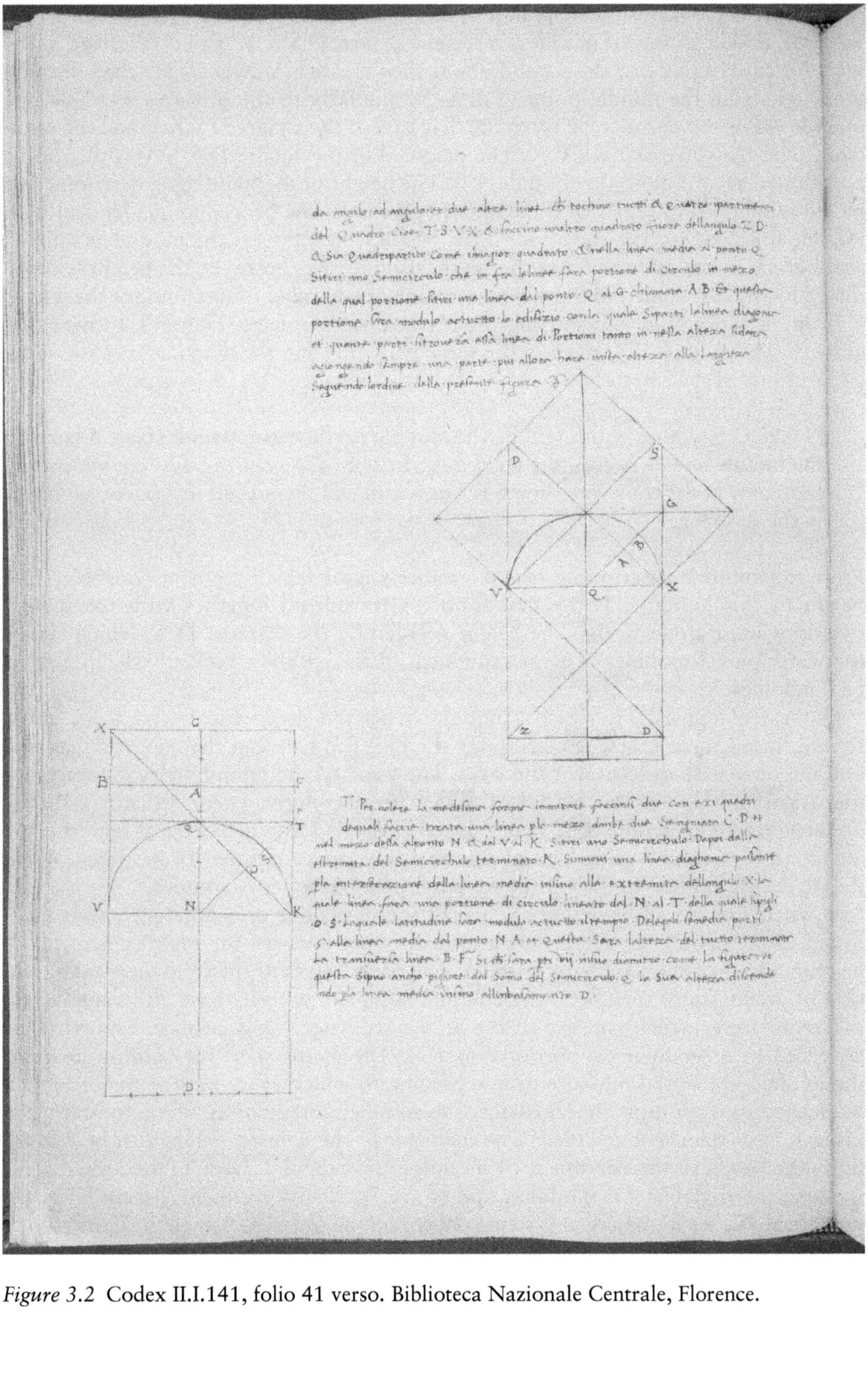

Figure 3.2 Codex II.I.141, folio 41 verso. Biblioteca Nazionale Centrale, Florence.

Francesco first draws the diagonals of the square and subsequently the lines, which unite the middle points of its sides. A second square, D S X V, is thus inscribed in the first. An equal square to the second one is then created: V X D Z. He then draws a semicircle from the middle point Q of V X. Similarly to the previous example, the module A B is the distance between the diagonal of the square D S X V and the semicircle over the square's base V X. The diagonal of the square D S X V is then to be partitioned according to the module A B. The height of the building is determined to be the number of modules of the diameter plus one. The procedure is identical with the one followed for the elevation of the basilica on folio 41: the module is applied *a posteriori* to the calculation of the height. In this case, however, the height is determined by the diagonal, and there is no mention of a number of modules for the width.

In the last example, the starting point is a double square (Figure 3.2 lower diagram).[14] The module is derived in the same way: through the diagonal and the semicircle of one of the squares.

> [T]ake O S, whose width will be the module of the entire temple. Give 5 parts to the middle line N A and this will be the height of the entire building, which will terminate at the transverse line B F. The width will then result in 7 parts, as shown in the drawing.[15]

From the author's description, it is not entirely clear which segment constitutes the height of this building. If the description is considered together with the respective drawing, it appears that the height is given by the segment D A, which equals the width plus 5 modules. The measurement of the width is explicitly mentioned to be 7 modules.

In these examples, Francesco's module is derived geometrically from an initial square. In all three cases, it constitutes the distance between the square's diagonal and the semicircle drawn over the base. The parts of the temple's elevation are not numerical multiples of this module, but they are also derived geometrically from the initial square. Finally, the total height is either defined as a geometrical segment, the diagonal, or a combination of a geometrical line and a number of modules. When the module is applied to the width, the latter is found to be 7 modules.

In Vitruvius' *De architectura*, the use of the module is of a different nature. It is found in the context of the orders and it does not concern the main dimensions—height/width—of the temple or its cella. The order refers to the following configuration of architectural members: the column, complete with its base and capital; and the entablature, which consists of the architrave, frieze, and cornice. The whole is governed by a modular proportional system. The diameter of the column provides the module, the initial measurement according to which every part of the order can be derived as a multiple. In Vitruvius, the complete modular system appears in the discussion of the Doric temple.[16] The diameter of the column is taken to be 2 modules. The height of the column is 14 modules; the capital 1 module; the architrave 1 module; the triglyphs 1½ modules; and so on. Vitruvius' modular system is clearly numerical. No such numerical system concerning the orders is present in Francesco di Giorgio's treatise. Even though the Doric entablature is discussed in the context of the temples in *Trattati II*, no proportions are given for its constituent parts.[17]

It has been suggested that Francesco's module is a means to derive integral number relationships from the main dimensions of the square: its side and diagonal.[18]

This interpretation would be in line with Wittkower's notion of Renaissance proportion. Francesco determines an arithmetical module to approximate incommensurable ratios; the architect's aim would be to abandon geometry and adopt a numerical system of proportions. As has become apparent, however, Francesco applies the numerical relationship to the geometrical drawings *a posteriori*. If anything, he is attempting to establish a *correspondence* between a geometrical and a numerical system. János Eisler has indeed pointed out that the main dimensions of the widths and heights of the nave and the aisles cannot be calculated by means of the length of the module.[19] Geometry remains Francesco's basic tool.[20] In fact, this entire section on the modular drawings in *Magliabechianus II.I.141* is introduced with the statement: "In order to illustrate some other geometrical proportions and measurements concerning the rectangular temples, start by drawing a square. . . ."[21]

Apart from the temples, the modular system also appears in the discussion of private architecture and in particular the rooms of the houses. Again this particular section appears only in *Magliabechianus II.I.141*.[22] Francesco states that even though he has discussed the width, length, and height of the *sale* and *triclini*, he will provide "some other different measurements,"[23] the reason being that "all the heights and proportions of the *sale* and *triclini* are calculated through simple numbers. These numbers cannot but have an irrational root . . . They can all, however, be determined through varied and proportionate lines."[24] As we shall see, the height is given by the diagonal in all cases. This would explain why the root of "simple numbers" is always an irrational number.

Francesco provides three examples.[25] First he takes a double square and derives geometrically a segment T R (Figure 3.3 upper diagram). The procedure is very similar to the one followed for the elevation of the double-aisled basilica on folio 41. The segment T R is equivalent to the segment E F of the basilica (Figure 3.1), and it consists of the distance between the square's diagonal and the semicircle drawn over the base. The height of the building is to equal the diagonal of the single square, which results as 5 T R. The width of the double square and thus of the building equals 7 T R. This segment, Francesco explains, is the module for the entire building. The second method consists in starting with a quadruple square as the basic unit (Figure 3.3 lower diagram). The height of the room is to equal the diagonal of the double square. Francesco divides the base of the quadruple square into eight parts and takes one part to be the module. He claims that the height of the double square subsequently results as 9 modules. The relationship 8 to 9 is only an approximation. In both cases, the methods to derive the height from the width are purely geometrical; the height is always the diagonal of the basic unit: a single or a double square.

The third method provides again a module for the design of the entire building (Figure 3.4).

In this case, however, Francesco does not specify any numbers. He only states that "The height of the major diagonal line is to equal the height of the entire edifice."[26] To conclude, in all three examples the height of the building is given by the diagonal of the initial square. As with the temples, Francesco is not attempting to avoid incommensurable ratios; the numerical method is applied *a posteriori* to a geometrical procedure.

If the modular system is essentially a geometrical method, why does Francesco attempt to make it *appear* numerical, especially since the height of the room is always

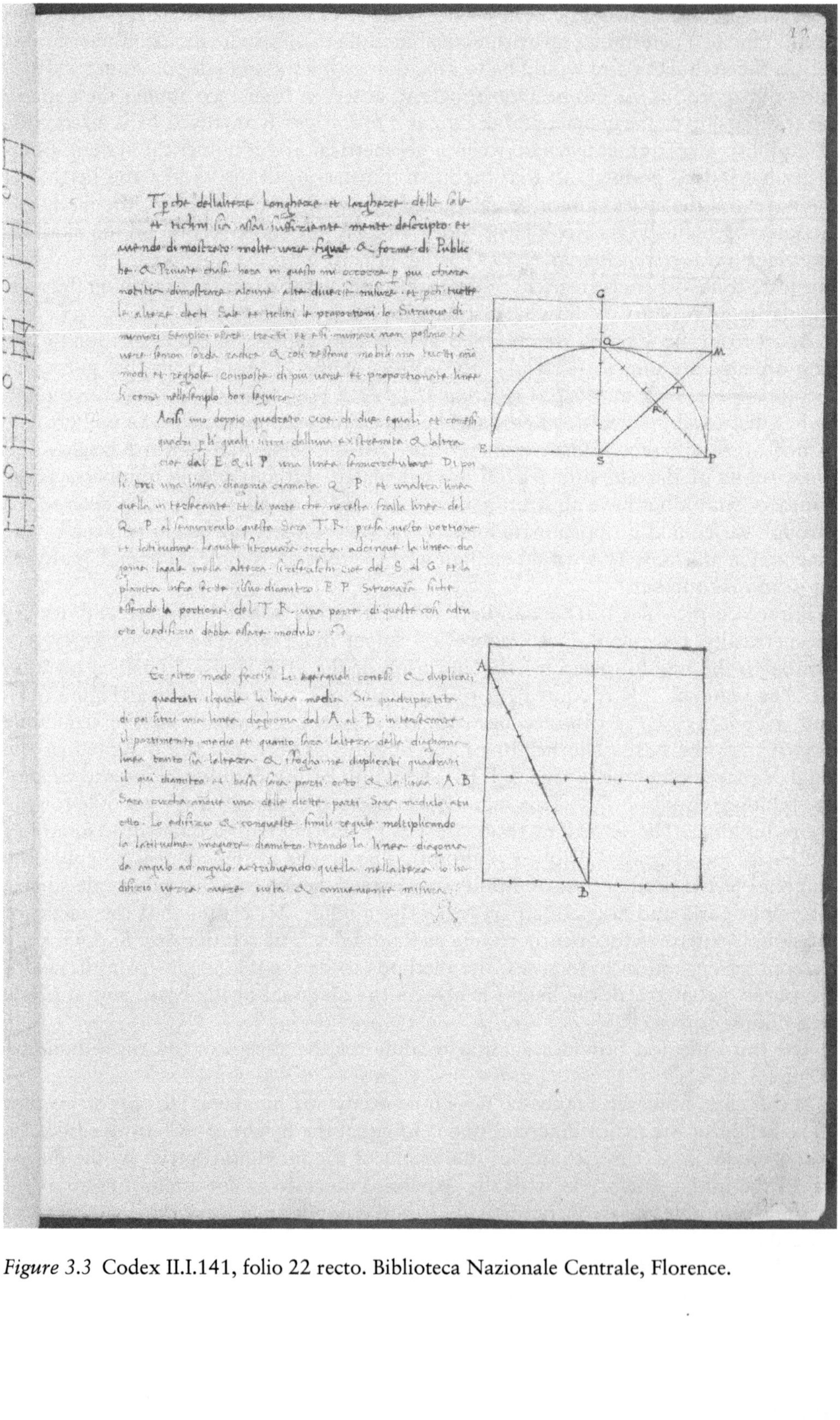

Figure 3.3 Codex II.I.141, folio 22 recto. Biblioteca Nazionale Centrale, Florence.

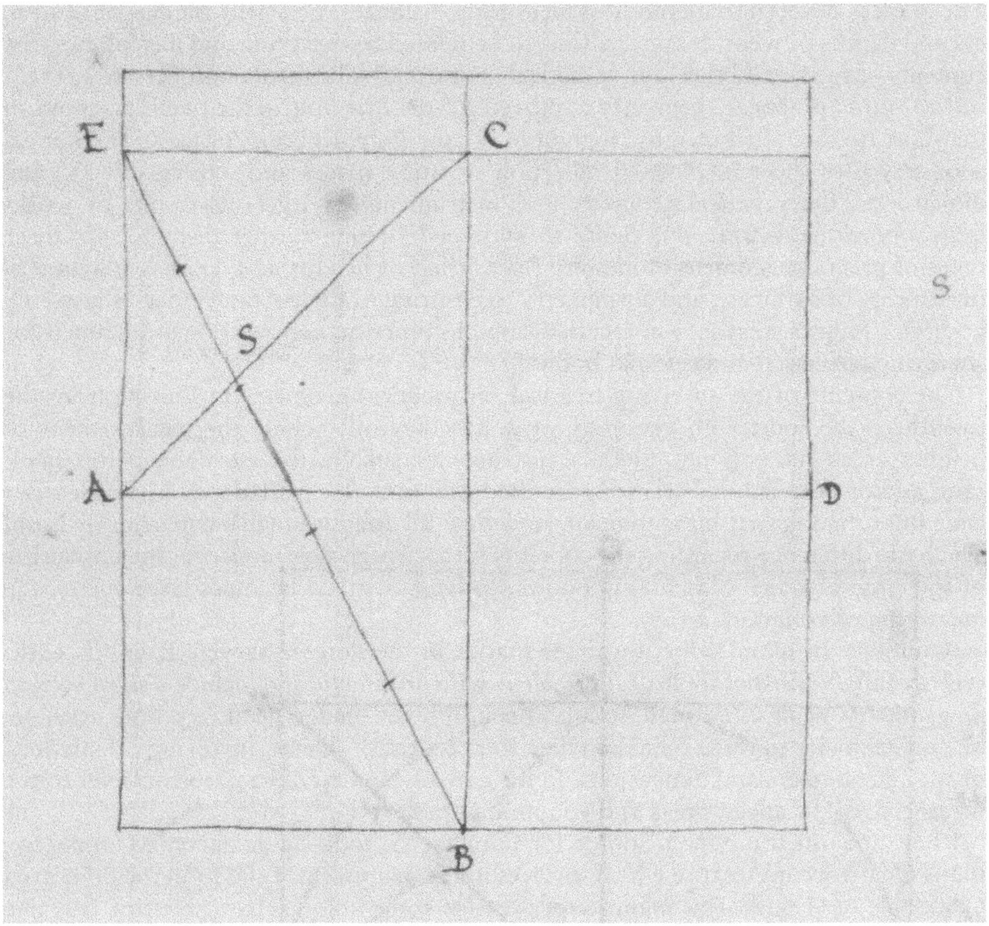

Figure 3.4 Codex II.I.141, folio 22 verso, detail. Biblioteca Nazionale Centrale, Florence.

given by the diagonal of the initial square? I believe that the answer lies with the numbers seven, eight, and nine. Reminiscent of the Vitruvian proportions of the columns, they reappear in relation to plans and elevations. These sections are included again only in *Magliabechianus II.I.141*, and they concern anthropomorphic plans and elevations. Before examining more closely these specifications, however, I will turn to Francesco's use and knowledge of geometry. I will propose that his knowledge of geometry relates to practical problems, in particular surveying methods, and not to theoretical considerations. His use of geometry as a design method appears thus to be related not to geometric axioms and propositions but to Quattrocento workshop practices and the representation of the human figure. It is precisely the human body which provides the link between Vitruvian columns and Francesco's modular system.[27]

In the first version of his treatise, Francesco devotes an entire book to geometry. Book IX in *Trattati I* relates to the following subjects: "Geometry and Ways of Measuring Distances, Heights and Depths."[28] Even though he begins with the definitions of various geometrical figures, Francesco includes no theorems, propositions, or proofs. The book is devoted to methods of measuring distances, primarily the height of towers and depths of wells. It appears thus to be related to surveying and metrology—the contents of medieval books on "practical geometry." The distinction between "theoretical" and "practical" geometry appeared for the first time in the twelfth century in the short treatise attributed to Hugh of St. Victor *Practica geometriae*:[29] "Theoretical geometry uses sheer intellectual reflection to study spaces and intervals of rational dimensions. But practical geometry uses instruments, and gets its results by working proportionally from one figure to another."[30] Hugh further distinguishes three types of practical geometry: altimetry (measuring of heights and depths), planimetry (measuring of surfaces), and cosmimetry (measuring of the circumference of heavenly bodies).[31] Hugh's treatise is in fact dedicated to practical geometry; it is a manual for surveying terrestrial and celestial bodies.

The contents of the surviving treatises on practical geometry in Europe from the twelfth to the fourteenth centuries vary, but they all include the measurement of heights, areas and volumes. Physical instruments, such as the astrolabe or the quadrant, and/or proportional relations are used as means of calculation.[32] These treatises constitute manuals of measurement, and they all follow Hugh's tripartite division. The basic difference regarding the contents is that Hugh's *cosmimetria*, the measuring of the circumference of heavenly bodies, in later treatises becomes stereometry, the measuring of volumes.

A number of manuscripts on mathematics in the *volgare* survive from fifteenth-century Italy.[33] Most of these deal primarily with arithmetic and include a short section on geometry, while a few manuscripts are entirely devoted to geometry. Both arithmetic and geometry in these treatises relate to practical problems, involving calculations of specific numbers and dimensions. In the case of geometry, the problems refer to the measuring of distances, areas and volumes.

I have identified the main source for Francesco's book on geometry in an anonymous Quattrocento treatise which survives in the manuscript *L.IV.18* in the Biblioteca Comunale of Siena.[34] This manuscript presents some unique characteristics. It is one of the very few extant texts in the *volgare* from Renaissance Italy which is exclusively dedicated to practical geometry. In addition, it is much more extended than similar treatises. Apart from the practical problems typical of such treatises, it also includes an extensive theoretical exposition: definitions of point, lines, areas, volumes. This is followed by 274 practical problems which relate to the calculation of areas and volumes of specific dimensions. The treatise concludes with 11 problems regarding measurements with the use of the quadrant.[35]

Let us consider the contents of Francesco's book on geometry from *Trattati I* in greater detail. The first part consists of the definitions of geometrical objects: point, line, angle, and various surfaces/shapes such as circle, triangle (and its variations), quadrangle, pentagon, hexagon, and the volumes of the column, pyramid, and sphere. The second part relates to measurements with the use of the quadrant: the height of a building, the shadow of the sun, the height of a tower, the depth of a well, and the dimensions of the earth. It follows the measurement of the width of a river through the use of similar triangles and the measurement of the height of a mill with the use of the astrolabe/quadrant. The book concludes with the calculation of the area of the circle and the inscription of geometrical figures in other figures: a circle in a square, a square in a circle, and a circle in a triangle. All calculations in the third part are practical problems; they relate to geometrical figures of specific dimensions.[36]

The first section of Francesco di Giorgio's book on geometry derives verbatim from codex *L.IV.18* and reproduces its initial definitions. The second section, on measurements with the use of the quadrant, also derives from the same codex and specifically folios 76 verso–82 recto. This section stands apart in *L.IV.18*, as the entire page before it, folio 76r, is left blank. Francesco's text reproduces this excerpt from *L.IV.18* with slight abbreviations. The measurement of the width of the river which follows in Francesco's text, however, does not appear in *L.IV.18*. Regarding the third section, only the calculation of the area of the circle is common to the two works. Francesco relates the first two out of the 12 ways to perform this calculation (*L.IV.18* folio 7 recto). Finally, Francesco's examples/problems of inscribing a circle in a square, a square in a circle and a circle in a triangle are different, i.e. relate to different dimensions, from those related in the Sienese manuscript. He might be deriving this material from another manuscript or he might be providing his own specifications. Francesco di Giorgio must have been familiar with similar Quattrocento treatises on practical geometry. In fact he begins his exposition with the definition of the "practice of geometry," which is divided into altimetry, planimetry, and stereometry and deals respectively with the "dimensions" of lines, surfaces, and volumes.[37] No such definition is present in codex *L.IV.18* but would have been present in other treatises.

Francesco's choice of material from codex *L.IV.18* is significant for the role that geometry plays in his treatise. He disregards the measurements of specific areas and volumes, which constitute the main body of codex *L.IV.18*, and he focuses on the measurements with the quadrant. Book IX in *Trattati I* does not concern typical problems of practical geometry, calculation of areas and volumes, but rather the surveying of heights of buildings, depths of wells, and widths of rivers. The initial definitions of the geometrical bodies are not relevant for the rest of the book; they might have been included to demonstrate the author's erudition on geometry. Finally, Book IX in *Trattati I* does not appear to give any indications regarding Francesco's design methods based on geometry.

Lon Shelby has identified three types of geometry present during the Middle Ages in Europe: theoretical, practical, and constructive.[38] The first involved axioms, theorems, and proofs and was based on the transmission of the texts by Euclid and Archimedes. The second related to practical problems of geometry and was based on the tradition of writings of the Roman professional surveyors—*agrimensores*. Constructive geometry, according to Shelby, was used by medieval craftsmen. It was primarily developed by masons and it can be reconstructed by the fifteenth-century booklets of craftsmen, such as Mathes Roriczer and Hanns Schmuttermayer.[39] This type of geometry did not involve geometric proofs or calculations to solve specific geometrical and arithmetical problems. It consisted of the physical manipulation of geometrical figures for the purpose of architectural design, such as the design of pinnacles. The basic geometrical instruments of the compass, the square, and the straightedge were used to manipulate simple geometrical forms. An example from *Geometria Deutsch*, attributed to Roriczer, illustrates this point:[40]

> In the first place, for an easy way to make a true set square, draw two lines that intersect perchance however you wish, and where the lines intersect mark an *e* there. Then set dividers with one leg on the point *e*, open them out however

> widely you please, and make a point on each line. There will be the letters *a b c*, which are all equidistant [from *e*]. Then draw a line from *a* to *b* and from *b* to *c*. Thus you will have a true set square, as in the example standing hereafter. Then if you eliminate the lines which one needs only for the setting out, you will have such a form as stands made hereafter.[41]

The procedure is very similar to the one employed by Francesco di Giorgio in his modular drawings. Even though Francesco makes no mention of the compass or the straightedge, it is apparent that his drawings are "constructed" with such instruments. In addition, in Francesco's drawings, it is the intersection of lines that coincides with characteristic points of the figure; the lines do not really serve the purpose of measurement.

Francesco's procedure of "constructive geometry" becomes more apparent in a number of passages, which again appear only in *Magliabechianus II.I.141*. As mentioned earlier, they concern anthropomorphic plans and elevations, and the integer numbers 7, 8, 9 appear in them. It is not coincidental, in my view, that these plans are discussed immediately after the modular proportions for the temple. The section opens with the origin of the partition—"partimento" of the temples. Francesco claims that the partition of the temples originated in the human body, which in turn can be divided into 9 or 7 heads. Only the partition of the human body into 9 heads is given in detail:

> The entire length of the face, from the extremity of the chin to the hairline, is one part. From the base of the throat to the extremity of the chest, it is another part. From there [the chest] to the groin we have two parts. From there to the kneecap we have two more parts. The [length of the] legs to the top of the foot [is to measure] two more parts. These make a total of eight. The height of the foot and the diameter of the throat make the ninth part.[42] This is the partition of the entire body. Finally divide the head in three equal parts.[43]

The partition is almost identical to the one given in Cennino Cennini's *Il libro dell'arte*.[44] The only difference between the two authors is that Cennini specifies the length of the throat and the height of the foot as one-third of the face. Francesco simply states that added together they make one part-head. His male figure results as 9 faces-heads, as opposed to the 8⅔ of Cennini. In his *Commentari*, Lorenzo Ghiberti provides a partition of the human body into 9 heads and an addition of throat and foot similar to Francesco di Giorgio: "from the earth to the fastening of the foot there is half a head, and another half head is found from the chin to the collarbone."[45] It seems, therefore, that the partition of the human body into 9 heads characterised Quattrocento workshop practice.

Francesco proceeds to obtain the plan of a "temple"—actually a basilica—from the partition of the human body into 9 heads (Figure 3.5 upper diagram). Considering the text together with the drawing on folio 42 v the procedure results as follows. First he draws a circle, whose radius begins at the lower part of the chest and terminates at the nose. The diameter of this circle determines the width of the basilica. He then draws the tangent lines to the circle all the way to the feet of the human figure: the height of the basilica is thus determined. Similarly to the modular elevation of the basilica on folio 41 (Figure 3.1), the base is divided into four parts. Circular chapels are added, taking the measurement from the head and specifically the distance from

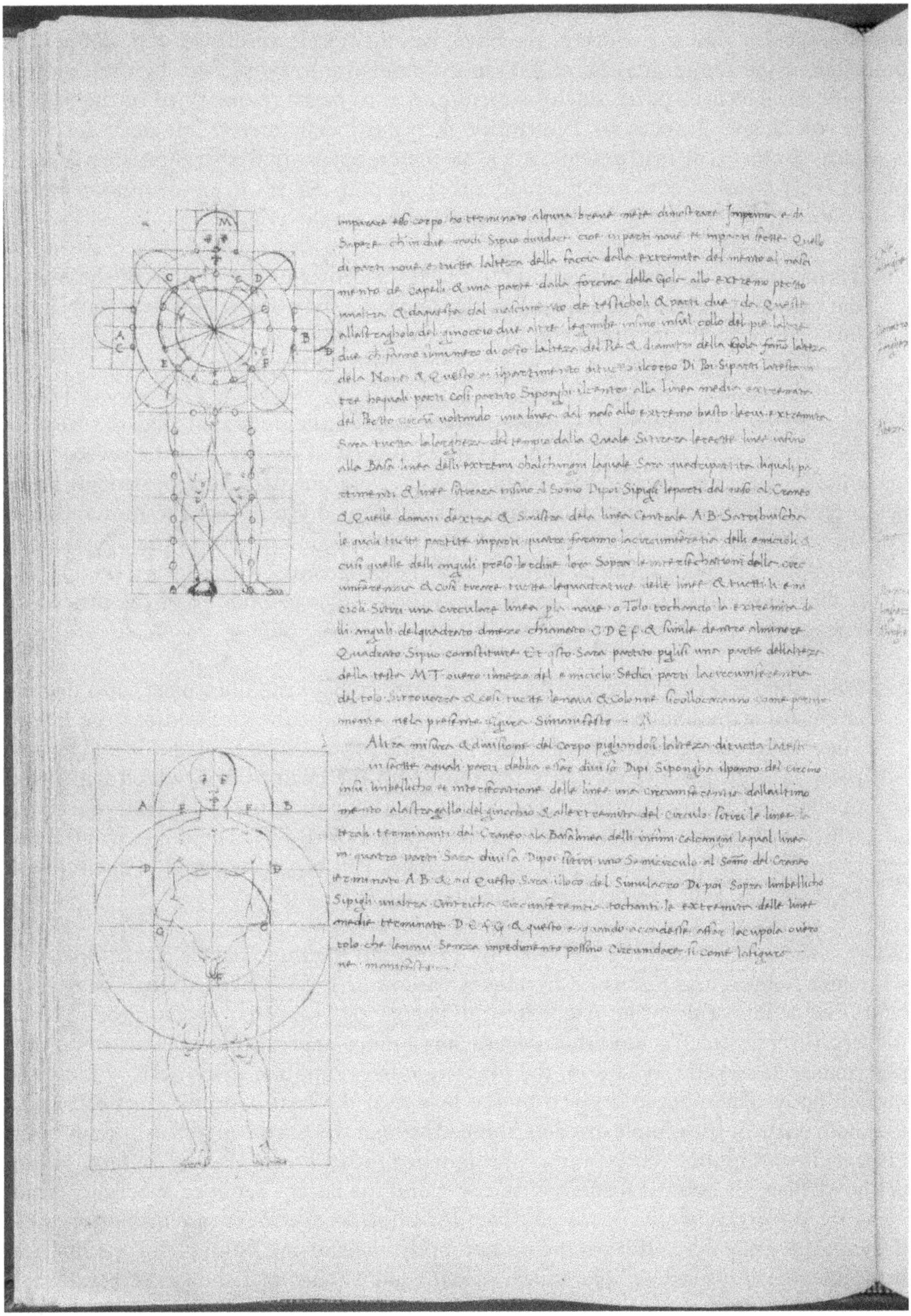

Figure 3.5 Codex II.I.141, folio 42 verso. Biblioteca Nazionale Centrale, Florence.

the nose to the top of the head. The square C D E F is drawn inside the initial circle. The side of this square is determined as two parts of the entire width of the basilica, and it coincides with the width of the nave. Another circle, inscribed in it, defines the diameter of the dome. It is here that the columns are to be placed, by dividing the circumference into 16 parts. Each segment/part is to measure one-third of the head.[46]

The wording of Francesco's description is of particular interest. He does not mention any geometrical instruments, such as the compass or the straightedge, but he makes every effort to convince the reader that the plan derives from the human figure. I am quoting the beginning of the passage to illustrate the point:

> Thus partitioned [the human body], place the center in the middle of the line at the base of the chest and draw a circular line from the nose to the lower part of the bust. The extremities of this line will give you the width of the temple. From this line draw straight lines all the way to the bottom of the heels.[47]

For the second partition of the human body, it is explicitly stated that the body is divided into 7 *equal* heads. Why does Francesco need to specify that the seven heads are equal? Does this signify that the nine heads were unequal? It appears that Francesco finds it difficult to accommodate two different divisions for the human body. For the seven head partition, he provides no analytical description of the proportions of each part. The plan result is simpler and the procedure itself, clearer (Figure 3.5 lower diagram). The use of the compass and the geometrical nature of the procedure become explicit in this passage:

> Place the center of the compass on the navel, where the lines meet, and draw a circle from the chin to the kneecap. At the edges of the circle, draw the lateral lines from the top of the head to the base line of the heels. Divide this line in four parts. Then draw a semicircle from the top of the head to A B, which will determine the position of the apse. Then draw another concentric circle from the navel, which will intersect the middle lines at the points D E F G. The dome will thus be constructed in such a way that no hindrances will be encountered in encircling it by the aisles, as it can be seen in the drawing.[48]

What is the exact procedure followed by Francesco for these plans? These drawings were first noticed and discussed by Henry Millon in 1958.[49] Even though the article bears the general title "The Architectural Theory of Francesco di Giorgio," Millon focuses on Francesco's modular system and, more specifically, the anthropomorphic plans. According to Millon, the plans represent modular grids derived from the human body. The module is given by the height of the head, and the proportions of the main parts of the temple are determined through the partition of the human body. His article exemplifies Wittkower's paradigm in a rather literal manner. In fact, Millon acknowledges his debt to Rudolf Wittkower and the latter's seminar, where the main ideas for the article were conceived. Even if Millon were correct in considering these drawings as grids derived from the human body, the starting point is not a numerical but a geometrical module. The initial module is a square and the grid constitutes a *geometrical* modular system. Considering the two different partitions of the human body into 7 and 9 heads, Millon proposes that "Francesco, like Procrustes, stretches or amputates . . . the human figure to conform to his abstractly conceived bed of modules."[50] Almost contradicting himself, therefore, Millon suggests that the human figure is *fitted* in the grid. A similar conclusion is reached by Günter Hellmann, who claims that Francesco first creates his plans geometrically and then inserts the human figure.[51] I believe that Francesco's procedure is more complex than either Millon or Hellmann would propose. The plans are not derived from the human body, using the height of the head as a module. It is clear from Francesco's drawings and textual descriptions that the plans are constructed geometrically; the human body seems to represent a pretext for their creation. Nor can I view this procedure as a simple placement of the human figure inside a geometrically constructed plan, as suggested by Millon and Hellmann. In the present drawings, there is a more active interaction between the human body and the plan created, especially regarding the employment of the circle. The circle appears to play a fundamental role in the determination of the design.

Similarly to the partition of the human body into nine heads, the use of the circle and square point to contemporary painting practice. In Francesco's partition of the human body, the face is divided into three parts. The unit is provided by the nose length. The same division is found in Cennini and Ghiberti. Cennini, however, refers also to the face's width: "From the point of the nose across the entire eye [give] one measure; from the end [point] of the eye to the ear, another measure, so that [the distance] from one ear to the other [measures] one length of the face."[52] The face, therefore, is as wide as it is long; it defines a square. In addition, Cennini suggests that the nose length is to be used as a measurement unit for the entire composition of a painting. This unit will be the "guide for the whole human figure, the houses and [the relationship of] one human figure to the other."[53] Unfortunately, Cennini does not relate the exact procedure. Analogous measurement units are present in the drawing of a human face set into a square by Villard de Honnecourt (folio 19v).[54] Each side of the square is divided into four parts. Once the vertical and horizontal lines are drawn, the square is divided into 16 units. The side of each unit equals one nose length. Finally, Ghiberti introduces his discussion of the partition of the human body by mentioning that the "circle of the painters" is responsible for the invention of forms as well as their measurements and symmetries.[55]

Originally a sculptor and a painter, Francesco di Giorgio was trained in a workshop. The association of *disegno* with the human body must have appeared natural to him. His "constructive geometry" is to be seen in connection with painting practices. As already mentioned, the partition into 9 heads appears in Cennini and must thus have characterised workshop practice. Yet, what about the partition into 7 heads, which Francesco seems to have difficulty accommodating? I believe that this constitutes a transformation of a Vitruvian notion. Vitruvius associates the male body with the seven diameter Doric column. Similarly, Francesco assigns 7 diameters to the Doric column. Concurrently, for him the human body is proportioned in terms of the head. By applying the number 7 to the human body, he creates a new partition of 7 heads.

This is corroborated by the last anthropomorphic drawing in *Magliabechianus II.I.141*. Francesco uses the 7-head partition of the human body to derive the anthropomorphic design of the façade of a building (Figure 3.6). The author relates that once the human body is divided into seven parts, one is to draw a middle line from the top of the head to the feet. Another straight line is drawn at the bottom of the feet, and it is divided into four parts. The width of the base is not specified in the text. As seen from the drawing, however, it coincides with the diameter

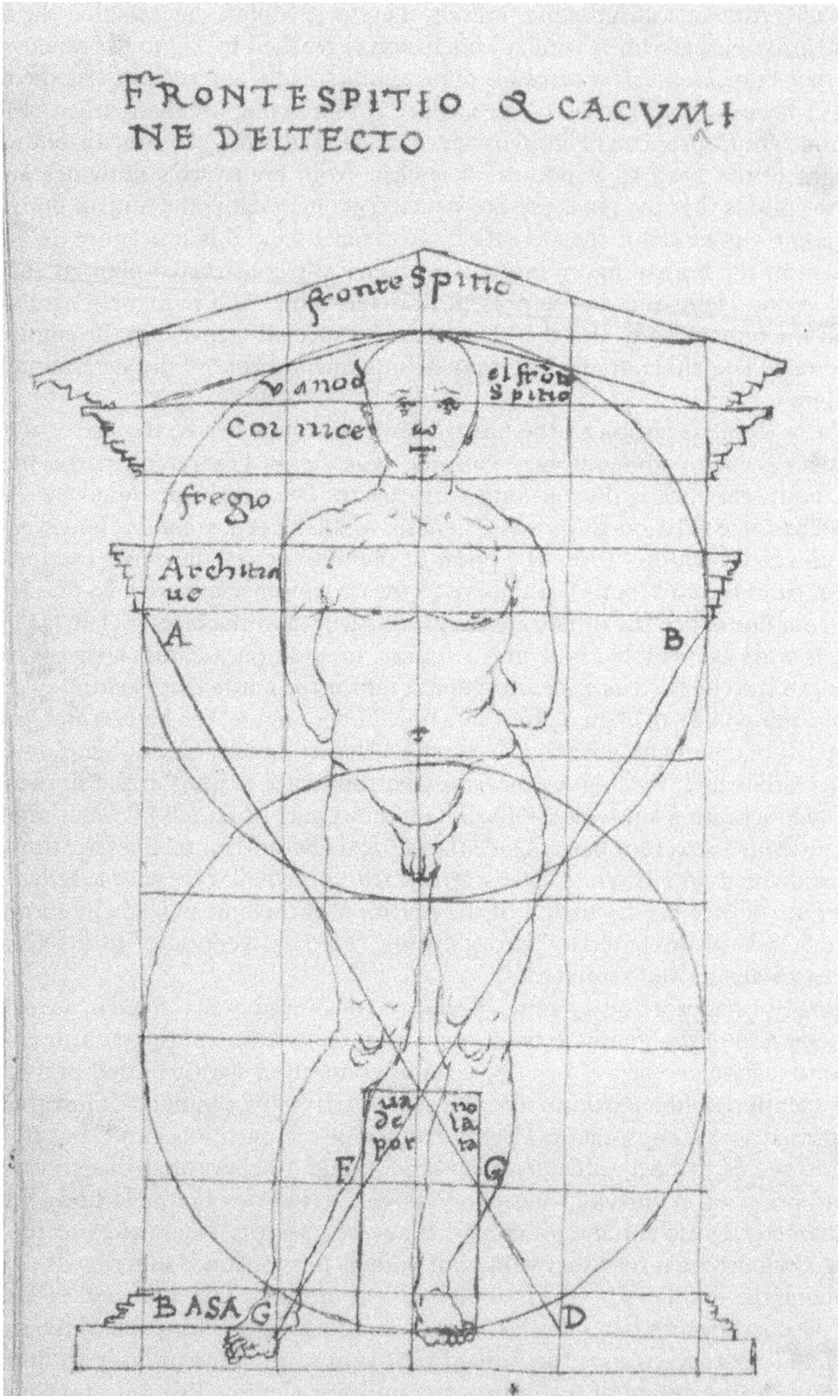

Figure 3.6 Codex II.I.141, folio 38 verso, detail. Biblioteca Nazionale Centrale, Florence.

of the two circles, which are derived as follows. The first circle runs from the top of the head to the groin. The second circle runs from the navel to the feet. The members of the entablature are given through the upper parts of the body. The distance from the chest to the throat is given to the epistyle; the distance from the throat to the chin defines the height of the frieze; the distance from the chin to the eyelids defines the height of the cornice; the rest of the head defines the height of the pediment. The entablature and its related members point to Vitruvius and a façade *all' antica*. Even more significantly, Francesco associates his 7-head anthropomorphic façade to a column. He explains that the height from the base to the first cornice is to be imagined as a column.[56]

If the division of the human body into 7 parts relates to the column, it is to be associated with the Vitruvian Doric column. Similarly, the division of the human body into 9 parts could thus be associated to the Vitruvian Ionic column of 9 diameters. Finally, Vitruvius prescribes 8 diameters for the Corinthian column. The numbers seven, eight, and nine that we have encountered in Francesco's modular drawings of the temples and rooms of private houses are thus of particular significance. They relate to the Vitruvian proportions of columns: 7 diameters for the Doric, 8 for the Corinthian and 9 for the Ionic.

The Vitruvian modular numerical system, referring to the columnar order, could not be easily used to derive the main dimensions, width to height, of a building. As it has emerged from Francesco's modular drawings, Quattrocento architectural practice relied on geometrical methods. Thus, commensurable ratios, to use Wittkower's terminology, could not be of much use for the Quattrocento theorist. Francesco's intention was not to abandon geometry but rather to establish a correspondence between geometrical and numerical methods. As a result, he inventively applies the proportions of Vitruvian columns to the design of the building as a whole. These considerations invest Wittkower's argument with a renewed, albeit qualified validity. Francesco is not transforming medieval geometry into Renaissance arithmetic; he only *appears* to do so. The very effort to establish a correspondence between geometrical and numerical methods in the last version of the *Trattati*, however, testifies to the importance that a numerical proportional system held for Francesco di Giorgio and Quattrocento architectural theory.

Notes

1 Rudolf Wittkower, "Systems of Proportion," *Architect's Yearbook* 5 (1953): 15. The first publication by Rudolf Wittkower on architectural proportion appeared in 1949 and consisted of a book which has proven to be extremely influential: *Architectural Principles in the Age of Humanism* (London: The Warburg Institute, 1949). The book received a revised edition in 1962 and two reprints: a 1952 reprint of the 1949 edition and a 1971 reprint of the 1962 edition. Wittkower's positions have also appeared in a number of articles: Idem., "International Congress on Proportion in the Arts," *Burlington Magazine* 94 (1952): 52–55; Idem., "Systems of Proportion," *Architect's Yearbook* 5 (1953): 9–18; Idem., "The Changing Concept of Proportion," *Daedalus* 89 (1960): 199–215. The summary presented here relies on Wittkower's 1953 article, as this publication presents a concise version of his main argument.

2 Wittkower, *Architect's Yearbook*, 17.

3 Ibid., 16.

4 Matthew A. Cohen, "Introduction: Two Kinds of Proportion," *Architectural Histories* (Special Collection, Objects of Belief: Proportional Systems in the History of Architecture)

2(1):21 (2014): 7–8. DOI: http://dx.doi.org/10.5334/ah.bv. For a more extended discussion on "Wittkower's paradigm," see Idem., *Beyond Beauty: Reexamining Architectural Proportion Through the Basilicas of San Lorenzo and Santo Spirito in Florence* (Venice: Marsilio, 2013), 36–51. Cohen makes a distinction between the historical and aesthetics premises of Wittkower's argument. I am focusing on the historical considerations and specifically the dichotomy between geometry and arithmetic.

5 Cohen, "Introduction," 5.

6 Cohen, *Beyond Beauty*, 36–51 and ch. 6.

7 They have been published by Corrado Maltese, ed., *Trattati di architectura ingegneria e arte militare*, 2 vols. (Milan: Il Polifilo, 1967). They will be hereafter cited as *Trattati*. In my designation of the first version of the treatise as *Trattati I* and the second version as *Trattati II*, I follow Maltese's edition. The first two manuscripts have been collated together to represent the first version and the other two the second version of the treatise. The dating of the manuscripts has been controversial. For a discussion of the major arguments, see *Trattati*, Introduction, xi–lxviii; Richard Betts, "On the Chronology of Francesco di Giorgio's Treatises: New Evidence from an Unpublished Manuscript," *Journal of the Society of Architectural Historians* 36 (1977): 3–14; Francesco Paolo Fiore, *Città e Macchine del '400: nei disegni di Francesco di Giorgio Martini* (Florence: Leo S. Olschki, 1978), 57–75; Carolyn Kolb, "The Francesco di Giorgio Material in the Zichy Codex," *Journal of the Society of Architectural Historians* 47 (1988): 132–159; Massimo Mussini, "La trattatistica di Francesco di Giorgio: un problema critico aperto," in *Francesco di Giorgio Architetto*, ed. Francesco Paolo Fiore and Manfredo Tafuri (Milan: Electa, 1993), 358–379; and Idem., *Francesco di Giorgio e Vitruvio: Le Traduzioni Del "De Architectura" nei Codici Zichy, Spencer 129 e Magliabechiano II.I.141*, 2 vols. (Florence: Leo S. Olschki, 2003), xiv–xxiii.

8 Two sixteenth-century manuscripts contain fragmentary sections of the treatise which do not coincide with any of the Quattrocento versions: *Codex Spencer 129* in the New York Public Library, which is thought to represent a draft between *Trattati I* and *Trattati II*, and the *Zichy Codex* in the Erwin Szabo Public Library in Budapest, which is thought to represent the earliest draft of Francesco's treatise. They have been published by Mussini, "La trattatistica di Francesco di Giorgio" 2003. A catalogue of the manuscripts relating to the treatise from the fifteenth to the nineteenth centuries is to be found in Gustina Scaglia, *Francesco di Giorgio: Checklist and History of Manuscripts and Drawings in Autographs and Copies from ca. 1470 to 1687 and Renewed Copies (1764–1839)* (Bethlehem, PA: Lehigh University Press; London and Toronto: Associated University Presses, 1992).

9 Catalogued by the Biblioteca Nazionale in Florence as codex *II.I.141*, the manuscript has been discussed in scholarship as codex *Magliabechianus II.I.141*, as it formed part of the Magliabechianus nucleus of the library. It is the latter designation that will be followed in this chapter.

10 The emphasis is mine.

11

> fatto prima uno quadrato d' equali lati nel quale da angulo ad angulo si tirino due linee diagonee, e la basa del quadrato divisa in quattro equali parti, e dal partimento C D si tiri due rette linee terminanti alle linee diagonee con una linea transversa A B. Di poi si tiri un semicirculo dalle estremità delli anguli della basa, passante la sua altezza all' intersecazione delle linee di<a>gonee X dove la linea del circulo passante interseca per M N, tirate in quel loco le transverse linee, sarà iusta altezza alla larghezza delle navi laterali; di poi si pigli una linea passante per lo mezzo del maggiore e minor quadrato, e due altre dal ponto medio della basa o passanti l' intersecazione delle rette linee e diagonee, e vadino a trovare la estremità della porzione del semicirculo, e quella parte che resta dentro alla porzione, cioè E F, sarà modulo a tutto il tempio; e due altre linee dal ditto ponto Q, e vadi insino alla quadrata altezza della B intersecando pe lu [*sic*] S: questa sarà la larghezza e altezza della porta. La qual medesima larghezza si dia al summo puteo overo lanterna del tolo. Perché il diamitro della basa overo latitudine di tutto il tempio si trovi parti sette del modulo E F, e l' altezza del minore quadrato A B C D saria parti 5 1/2, e a l' altezza di parti 4 1/2, si tiri la linea O P, in mezzo della quale si ponga

> il centro pigliando la circunferenzia dallo P. E questa sarà la somma altezza di tutto il tempio. . . . E così il tempio con ragione e l' altezza e le larghezze saranno commensurate, sì come per la figura e disegno si manifesta.

Trattati, 399, lines 29–34; 400, lines 1–15; and 401, lines 1–3. I have kept the translation of Francesco's text literal to bring forth the level of his mathematical erudition. All translations from Francesco's treatise in this chapter belong to the author.

12 This has been pointed out by Günter Hellmann, "Proportionsverfahren des Francesco di Giorgio Martini," in *Miscellanea Bibliothecae Hertzianae zu Ehren von Leo Bruhns, Franz Graf Wolff Metternich, Ludwig Schudt*, ed. Hanno Hahn (Munich: A. Scholl, 1961), 160, n. 10.

13 *Trattati*, 401, lines 4–17.

14 Ibid., lines 18–29.

15 Ibid., lines 24–28,

> si pigli O S, la quale latitudine sarà modulo a tutto il tempio. Delle quali se ne dia parti 5 alla linea media dal ponto N A, e questa sarà l' altezza del tutto terminata la transversa linea B F, si che farà parti VII in suo diamitro come la figura.

16 For the Ionic order, Vitruvius gives only the first member of the entablature, the architrave, in terms of the diameter. The height of both the frieze and the cornice is given in terms of the height of the architrave, whereas the height of the tympanum is given in terms of the length of the cornice.

17 *Trattati*, 389, lines 3–22.

18 Lawrence Lowic, "Francesco di Giorgio on the Design of Churches: The Use and Significance of Mathematics in the *Trattato*," *Architectura (Zeitschrift für Geschichte der Baukunst)* 12 (1982): 158.

19 János Eisler, "Remarks on Some Aspects of Francesco di Giorgio's Trattato," *Acta Historiae Artium (Academiae Scientiarum Hungaricae)* 18 (1972): 204.

20 The geometrical nature of Francesco's procedure has been acknowledged by every scholar who has written on the topic. The procedure has been primarily associated with medieval practice and specifically the quadrature technique. Hellmann, "Proportionsverfahren des Francesco di Giorgio Martini," was the first to associate Francesco's modular drawings to quadrature. He was followed by Eisler, "Remarks on Some Aspects of Francesco," 193–231, especially 202–204, and Richard Betts, "Structural Innovation and Structural Design in Renaissance Architecture," *Journal of the Society of Architectural Historians* 52 (1993): 5–25, especially 10–14. Eisler examines the modular drawing of the basilica on folio 41, whereas Betts focuses on the central modular drawings and specifically the upper drawing on folio 41v. Betts compares this drawing to the plans present in *Trattati I*, and he suggests that it represents a basilican plan rather than the elevation of a centrally planned church. He thus argues that the module constitutes the thickness of the walls which carry the vaults. Eisler, "Remarks on Some Aspects of Francesco," 203, considers the module to be "a part of the same radius of the in- and out-scribed circles belonging to a square which is drawn into the basic square in a way that its vertices fall upon the midpoint of the side of the basic square" while Betts, "Structural Innovation," 14, finds the module to be "equal to one-half the difference between the sides of the inner and outer squares in a two-step quadrature series." In part, this discrepancy relates to the variety of the scholarly definitions and applications of quadrature. The technique has been traditionally associated with the rotation of the square and the creation of a series of decreasing squares for the construction of Gothic pinnacles, as described in the fifteenth-century treatise by Mathes Roriczer. For a definition of *quadrature* and previous literature see Eisler, "Remarks on Some Aspects of Francesco," 202–204; note 24 and Betts, "Structural Innovation," 10–11; note 21. As I will argue in this chapter, the significance of the module does not lie on the actual segment it represents but on the module's connection to the numbers 7, 8, and 9. Lowic, "Francesco di Giorgio on the Design of Churches," has in turn associated Francesco's use of geometry to contemporary writings on mathematics and specifically the problem of the "squaring of the circle." He considers the square and circle as the basis of Francesco's church design and, as I have already mentioned, he views the module as a method to turn geometry into arithmetic.

21 *Trattati*, 399, lines 28–29, "E per volere dimostrare alcune altre geometriche proporzioni, commensurazioni de' tempi navali oblonghi, fatto prima uno quadrato . . ."

22 In the existing scholarly discussion of Francesco's modular system in *Trattati II*, the drawings of the houses have not received any attention either independently or in relation to those of the temples.

23 *Trattati*, 349, line 2, "alcune altre diverse misure."

24 Ibid., lines 2–6, "tutte le altezze <di> detti sale e ticlini le proporzioni lo<ro> si truova di numari semplici essare tratti, et essi numari non possano avere se non sorda radice . . . ma tutti hanno modi e regole composte di più varie e proporzionate linee."

25 Ibid., 349.

26 Ibid., lines 29–30, "l' altezza della maggior linea diagonia all' altezza di tutto lo edifizio attribuito."

27 The human body plays a significant role in Francesco's treatise, especially the first version of the *Trattati*. For a discussion of the use of human analogy in *Trattati I* and previous literature, see Angeliki Pollali, "Human Analogy in *Trattati I*: The *Ragione* of Modern Architecture," in *Reconstructing Francesco di Giorgio Architect*, ed. Berthold Hub and Angeliki Pollali (Frankfurt am Main: Peter Lang, 2011), 59–84.

28 "Geometria e Modi di Misurare Distanze, Altezze e Profondità," *Trattati*, 117. The title of the book, reflecting the contents, is given by the editor Corrado Maltese.

29 The distinction was first noticed by Roger Baron, "Sur l' introduction en Occident des termes 'geometria theorica et practica,'" *Revue d' histoire des sciences* 8 (1955): 298. For the Latin text of the treatise, see Roger Baron, ed., *Hvgonis de Sancto Victore Opera Propaedevtica: Practica geometriae, De grammatica, Epitome Dindimi in philosophiam* (Notre Dame, IN: University of Notre Dame, 1966), 15–64. For an English translation, see Frederick A. Homann, S.J. trans., *Practical Geometry: [Practica Geometriae]-Attributed to Hugh of St. Victor* (Milwaukee: Marquette University Press, 1991), 33–70.

30 Homann, *Practical Geometry*, 33–34.

31 Ibid., 34.

32 A survey of the contents of the major treatises is found in Stephen K. Victor, ed. and trans., *Practical Geometry in the Middle Ages: "Artis Cuiuslibet Consummatio" and "the Pratike de Geometrie"* (Philadelphia: The American Philosophical Society, 1979), 2–31. For the contents of *Quadrans Vetus* and *Geometrie Due Sunt Partes Principales* in particular, see Nan L. Hahn, ed., *Medieval Mensuration: "Quadrans Vetus" and "Geometrie Due Sunt Partes Principales . . ."* (Philadelphia: The American Philosophical Society, 1982), xli–lxxv.

33 A list can be found in Warren Van Egmond, *Practical Mathematics in the Italian Renaissance: A Catalog of Italian Abbacus Manuscripts and Printed Books to 1600* (Florence: Istituto e Museo di Storia della Scienza, 1980).

34 The manuscript has been published: Anonimo Fiorentino, *Trattato di Geometria Pratica: dal Codice L.IV.18 (sec.XV) della Biblioteca Comunale di Siena*, ed. Annalisa Simi (Siena: Università degli Studi di Siena, 1993). Lowic, "Francesco di Giorgio on the Design of Churches," 156–158, has pointed out that Francesco di Giorgio is the only fifteenth-century theorist who devotes in his treatise an entire book to mathematics. In an effort to identify the sources for this material, he has connected it generally to contemporary writings on mathematics, such as the *Trattato d' Arithmetica* of Paolo dell' Abbaco and the writings by Mariano Taccola. More recently but erroneously Elizabeth Merrill, "*Trattato* as Textbook: Francesco di Giorgio's Vision for the Renaissance Architect," *Architectural Histories* 1(1):20 (2013): 7–8, DOI: http://dx.doi.org/10.5334/ah.at, suggests that the contents of Francesco's book relate to the "*abaco* books," which she identifies as books used in elementary commercial schools. She claims, ibid., n. 30, that very few of these books survive, and she mentions codex Italiani IV, 35, 5570 in the Biblioteca Marciana as such an example. As I have mentioned in note 33, a number of manuscripts on arithmetical problems, including those for commercial purposes, survive from the fifteenth century, but Francesco's book relates to practical geometry and not arithmetic. On "abacus books" see Warren Van Egmond, "The Contributions of the Italian Renaissance to European Mathematics," *Symposia Mathematica* 27 (1986): 51–67. According to Van Egmond, "The Contributions of the Italian," 300 examples of abacus books survive in Italy in the *volgare*.

35 For the contents, see Anonimo Fiorentino, Introduction, especially pages: 4–18.

36 Lowic, "Francesco di Giorgio on the Design of Churches," 156–158, is the only scholar to date who has given some consideration to book IX from *Trattati I*. He divides the contents of the book in three groups: definitions of basic geometrical terms; methods of surveying and related instruments; selected mathematical problems, such as the squaring of the circle.

37 *Trattati*, 117, lines 11–19.

38 See Lon R. Shelby, "The Geometrical Knowledge of Mediaeval Master Masons," *Speculum* 47 (1972): 395–421, and Idem., "Geometry," in *The Seven Liberal Arts in the Middle Ages*, ed. David L. Wagner (Bloomington: Indiana University Press, 1983), 196–217. For a synopsis of the history of medieval geometry, see Homann, *Practical Geometry*, 2–6.

39 For the latest edition of these booklets, see Lon R. Shelby, ed., *Gothic Design Techniques: The Fifteenth-Century Design Booklets of Mathes Roriczer and Hanns Schmuttermayer* (Carbondale: Southern Illinois University Press, 1977).

40 For the attribution of *Geometria Deutsch* to Roriczer and previous literature, see Shelby, *Gothic Design Techniques*, 33–38.

41 *Geometria Deutsch*, No. 1 folio 1, fig. 24 in Shelby, *Gothic Design Techniques*, 114.

42 The "diameter" is to be understood here as the length of the throat. The use of the term diameter points to the employment of a compass for measuring/constructing the human figure.

43 *Trattati*, 403, lines 5–12, "tutta l' altezza della faccia, dalla estremità del mento al nascimento de' capelli, è una parte; dalla forcina della gola allo estremo petto un' altra, e da questa dal nascimento de' testicoli è parti due, e da queste all' astragolo del ginocchio due altre; le gambe insino in sul collo del piè l' altre due, che fanno il numero di otto; l' altezza del piè e diametro della gola fanno l' altezza della nona, e questo è il partimento di tutto il corpo. Di poi si parti la testa in tre equali parti."

44 Cennino Cennini, *Il libro dell' arte*, ed. Franco Brunello (Vicenza: N. Pozza, 1971), LXX: 82.

45 Lorenzo Ghiberti, *Commentari*, ed. Ottanio Morisani (Naples: R. Ricciardi, 1947), III. 44, 212, "da terra per insino alla chiavatura del piede è una mezza testa e una mezza dal mento alla forcella del petto."

46 *Trattati*, 403, lines 12–28.

47 Ibid., lines 12–16, "Così partito, si ponghi il centro alla linea media estremità del petto circumvoltando una linea dal naso allo estremo busto, le cui estremità sarà tutta la larghezza del tempio; dalla quale si tirarà le rette linee insino alla ba<s>sa linea delli estremi calcagni."

48 Ibid., 404, lines 1–10. As Maltese, n. 3, points out, the letters of the description do not correspond to the drawing.

> Dipoi si ponga il ponto del circino in su l' imbellico et intersecazione delle linee, una circunferenzia dall' ultimo mento a l' astragalo del ginocchio, e all' estremità del circulo si tiri le linee laterali terminanti dal craneo a la bas<s>a linea delli infimi calcagni, la qual linea in quattro parti sarà divisa. Dipoi si tiri uno semicirculo al sommo del craneo terminato A B, e a questo sarà il loco del simulacro. Di poi sopra l' imbellico si pigli un' altra cintrica circunferenzia toccanti le estremità delle linee medie terminate D E F G, e questo è quando accadesse a far la cupola overo tolo che le navi senza impedimento possino circumdare, sì come la figura ne manifesta.

49 Henry Millon, "The Architectural Theory of Francesco di Giorgio," *Art Bulletin* 40 (1958): 257–261.

50 Ibid., 258.

51 Hellmann, "Proportionsverfahren des Francesco di Giorgio Martini," 162–165.

52 Cennini, *Il libro dell' arte*, LXX: 82, "Dalla proda del naso per tutta la lunghezza dell' occhio, una di queste misure: dalla fine dell' occhio per fine all' orecchie, una di queste misure: dall' uno orecchio all' altro, un viso per lunghezza."

53 Ibid., XXX: 29–30, "guida di tutta la figura, de' casamenti, dall' una figura all' altra."

54 For the latest facsimile edition, see Carl F. Barnes, Jr., ed., *The Portfolio of Villard de Honnecourt (Paris, Bibliothèque nationale de France, Ms Fr 19093): A New Critical Edition and Color Facsimile* (Farnham: Ashgate, 2009).

55 Ghiberti, *Commentari*, III. 44, 211, "cerchio de' pittori."

56 *Trattati*, 394.

4 Mathematical and Proportion Theories in the Work of Leonardo da Vinci and Contemporary Artist/Engineers at the Turn of the Sixteenth Century

Matthew Landrus

Although there was a rise in books around 1500 that indirectly or directly assessed geometry and arithmetic, the means by which these disciplines were useful in the visual arts and engineering is difficult to prove. Written lessons of systematic sciences would not contribute in obvious ways to the systematic uses of those lessons for projects in the visual or mechanical arts. For example, evidence of the use of Euclidian geometry in machine design at this time is not well documented, though there is an example of this in Leonardo da Vinci's *Giant Crossbow* project.[1] Evidence of musical proportions in his *Last Supper* is generally agreed upon today, though proof of this is complicated.[2] Nonetheless, the use of Pythagorean theories in the visual arts at this time cannot be proven.[3] There was however an increase in the ways in which "mechanical artificers"—as Egnazio Danti would call them in the late sixteenth century—used mathematical and proportion theories for visual and mechanical projects.[4] Moreover, these practitioners were influenced by trends in humanistic discourses that valued engagements between moral philosophy and the practical arts. Although the uses of arithmetic and proportion theories by these mechanical artificers will be the focus of the following discussion, implicit in this chapter are the ways in which these projects would work or be of value because of their necessary agreements with the measures and rules of Nature.

From the period of a rise in publishing houses in the 1470s through the 1540s, algebra developed as a problem-solving technique, thereby orienting mathematical approaches away from the previous umbrella discipline of geometry.[5] Thus Euclid's readership predominantly addressed basic principles in the early sixteenth century rather than practical solutions. For example, large sections on Euclid in Luca Pacioli's books did not primarily target studio artists, though instead a much broader audience. Printed editions on engineering and mathematics replaced in popularity previous collectable manuscripts on the natural philosophy of these subjects, such as Leon Battista Alberti's rare *Ex ludis rerum mathematicarum*. By 1578—a century after Francesco di Giorgio's 1478 *Opusculum de architectura*—Egnazio Danti noted that students of mathematics, "may no longer go to the Philosopher's schools to learn it, since it has been banished from them, but the little which remains to us is limited to some practical aspects learned from the mechanical artificers."[6] The foundations of this shift away from the mathematical sciences in natural philosophy were during a period of practical developments in this field at the turn of the sixteenth century. At that time, artist/engineers in north Italy and south Germany continued a liberal arts tradition of studying mathematics, earning reputations as intellectual *uomini pratici* (practical men) in the pictorial, sculptural and technical arts.[7] This accompanied a rise at the end of the fifteenth century in humanist discussions of the visual and musical arts, such that those studies in morality that preoccupied the *studia humanitatis* for centuries were increasingly engaged more directly with lesser disciplines such as mathematics, astronomy, logic, natural philosophy, metaphysics, law and theology. Humanistic studies of poetry continued to rise in popularity at this time and would eclipse in academic importance the studies of musical, visual and other quadrivial arts in the sixteenth century, though at the turn at that century, a rational aesthetic involving systematic calculations in the visual arts and music appeared in the forms of painting, sculpture, engineering, architecture and academic treatises. Much of this happened within royal court and university environments that were part of the humanistic debates on moral epistemology.

The resulting interdisciplinary dialogues had in common an interest in harmonic relationships between constituent elements and ideas. These were considerations of universality among the arts, a kind of holistic decompartmentalization of approaches to the *studia humanitatis*. Precursors to this mixed engagement in the systematic arts and moral humanities happened to be the so-called practical men who also gained reputations for their skills in rhetoric and the *ars oratoria*. They include Lorenzo and Bonaccorso Ghiberti, Leon Battista Alberti, Filarete, Piero della Francesca and Francesco di Giorgio. At issue in the present study are the ways in which those kinds of approaches were manifest in systematic and mathematical contributions to the practical projects of artist/engineers around the turn of the sixteenth century and generally before Gerolamo Cardano's *Ars Magna* of 1545, which was an indication of a shift away from Neo-Platonic and Neo-Pythagorean concepts of universal truth in quadrivium and humanistic studies.

Whereas the original practical intentions for fifteenth-century treatises are not necessarily in question, what remain are questions on the extent to which their associated mathematical principles contributed to practical traditions in the liberal and technical arts toward the turn of the sixteenth century. Moreover, how influential were the popular manuscripts and incunabula of the works of Euclid, Vitruvius, Vegetius, Vigevano, Konrad Kyeser, Giovanni Fontana, Taccola, and Roberto Valturio on proportion theorists Fazio Cardano, Francino Gaffurius, Pomponius Gauricus and Luca Pacioli, and on the practical approaches of artist/engineers such as Andrea del Verrocchio, Francesco di Giorgio, Leonardo da Vinci, Mauro Codussi, Giovanni Antonio Amadeo, Donato Bramante, Bramantino (Bartolomeo Suardi), Michelangelo, Raphael and Giuliano da Sangallo? The works of Euclid, Archimedes and Vitruvius were standard resources for these individuals. Three of the most popular manuscripts of the fourteenth and fifteenth centuries were those of Vitruvius, Vegetius and Pliny the Elder, as part of a trend wherein manuscripts on the mechanical arts almost outpaced books on ancient historians into the fifteenth century. The first printed book in Verona, for example, was the first edition of Valturio's *De re militari* of 1472, and one of the earliest printed books in Utrecht was the 1473 Vegetius *Epitoma rei militaris*. The market for printed books on the *artes techinae* initially kept pace with books required for the *studia humanitatis*, so that in universities and royal courts around 1500 there had been an increasing reconsideration of the necessary connections and hierarchies among disciplines that were similarly interested in natural and divine paths toward universal truth. Moreover, in Florence, for example, the number of abacus treatises doubled in the latter half of the fifteenth century, and significant treatises on music appeared for the first time.[8] For the early modern "practical men," the universal form and function of Necessity was a matter of proper proportion, and by extension, the universal architecture of that Necessity was geometry. The following study will address the approaches of artist/engineers to proportional geometry in pictorial, mechanical and architectural projects, with particular interest in the work of Leonardo da Vinci.

Verrocchio

Until Leonardo was around 24 years of age, his primary professional influence was Verrocchio, a skilled painter, sculptor, goldsmith, bell-caster, cannon caster, geometer and musician, all of the skills that Leonardo supposedly mastered. A good example of this is Verrocchio's proportional measurement of a horse, which it appears Leonardo copied and may have owned. Leonardo referred to arithmetic as discontinuous quantity, and he called geometry continuous quantity, denoting the definitive points arrived at by the former and the continuous linear geometry arrived at by the latter. He favoured geometry and proportion over arithmetic and favoured visual solutions over numerical solutions, but he was also fascinated with arithmetical solutions to problems of all kinds. Leonardo used arithmetic and geometry especially for technical projects such as the Sforza Horse proposal for Ludovico Sforza around 1483–88, when Verrocchio worked on the equestrian monument of the *condottiero* Bartomomeo Colleoni in the Piazza San Marco of Venice. A better example of Verrocchio's influence is in the documentation regarding the gilded copper ball and cross at the top of the Florence Cathedral. In 1468–71, Leonardo would have seen the method of its construction. The ball—replaced after a lightning strike in 1601—was four *braccia* (arm lengths) wide, placed with the help of a massive scaffolding project and soldered in place with a burning mirror. This commission had been awarded to Verrocchio by the works department of the Duomo and was something that the cathedral's original architect, Filippo Brunelleschi, had been unable to complete. Much of the work at the Verrocchio studio was the result of teamwork, such that Paolo di Matteo cast the cross above the ball, Giovanni di Bartolomeo and Bartolomeo di Fruosino cast the knob that would link the ball to the lantern, and Luca di Piero, Salvestro di Pagholo di Stefano and Giovanni di Tomè hammered the ball into shape and attached it to the supporting internal armature. Like a number of Florentine artist/engineers, Leonardo made drawings of the scaffolding and crane used for the Duomo's construction, studying the methods of Filippo Brunelleschi. Calculations for Verrocchio's project included the estimates of proportions of metals in the solder, costs for materials and labor, weights of the materials to be lifted by the crane, and structural capabilities of the crane and scaffolding, as well as geometrical and measured assessments of the ball, button and other pieces. A mastery of Euclid was important for the design of the ball and its armature so that the inside angles and measurements were appropriate.

Leonardo is of particular interest because we have so much information about his personal engagement with ideas and intellectuals around 1500, especially with regard to the role of mathematics in the visual arts. Like other artist/engineers of this period, he was closely related to a notary, namely his father, and as a boy was given a basic education in arithmetic and vernacular reading and writing. This happened with the help of his father, mother and possibly a tutor and likely while he was apprenticed at the Verrocchio studio sometime between the ages of 10 and 16. He was taught addition, subtraction, multiplication, fractions and the rule of three. At around the age of 30 he sought an education in Euclid's *Elements* in Milan and Pavia with the help of Fazio Cardano (the father of Gerlamo Cardano), when he made numerous studies of Euclid in his notebooks.

Leonardo da Vinci

References to Leonardo's discussions of *scienza* (or science) rarely differentiate clearly between premodern concepts of this term and its modern associations. Modern familiarity with *scienza* in his work derives considerably from Francesco Melzi's compilation of the *paragone* of Leonardo's *Treatise on Painting*, wherein a decidedly mid-sixteenth century preoccupation with the term supports arguments about painting as a "science." As an example, the *Last Supper* exhibits a kind of science of visual and harmonic proportions by virtue of the arrangement of its tapestries, its composition, and its figures in sections of an octave, a fifth and a fourth.[9] Since the 1950s, approaches to Leonardo have often addressed his "art as science" or his "science as art," to use titles of each part (I & II) of Ludwig Heydenreich's 1954 monograph, *Leonardo da Vinci*. Moreover, to refer to Leonardo's work as "art" or '*Kunst*' is also a misleading sixteenth-century approach, particularly with respect to the German origin of the term (from *können*: capability and knowledge). It is nonetheless generally understood among academics today that Leonardo did not refer to his work as simply "art" or "science," though these terms are still used loosely to explain the artist's works. We know that he referred not to "art" but to specific kinds of the arts, and that his work was not scientific because he did not employ a modern, rigorous scientific method.

How therefore can we clearly or definitively associate his work with the general fields of art and science? Where in his studies and practices does he approach these disciplines? One specific link between medieval intuitions and proto-scientific approaches to so-called "art" and "science" is proportion theory, or, specifically, proportional geometry. Leonardo used this method of proportion to improve upon traditional technical and pictorial standards and practices. While in Milan, he devoted much of his time to understanding proportional geometry, studying with mathematicians Fazio Cardano and Luca Pacioli and with musician Franchino Gaffurio. Leonardo was an accomplished musician and was interested in Gaffurio's work and activities at the Milan Cathedral. He also knew of Gaffurio's publications, *Theorica musicae*, 1492, and *Practica musicae*, 1496, which were part of a recent trend to publish treatises on music. Three of the best examples of this trend time are Johannes Tinctoris's *Expositio manus* and *Proportionale musices*, c. 1480, and Florentius de Faxolis's *Liber musices*, c. 1495–96.[10] The latter was dedicated to Cardinal Ascanio Maria Sforza (1455–1505), Lodovico Sforza's brother. As a mathematical science of the *quadrivium*, music was regarded by Leonardo as an equal to the science of painting, which was itself the superior rhetorical mode to the other arts. Math, for Leonardo, was the universal governing principle in Nature, mastered with the help of mechanical processes of geometry and proportion, even when he would brag that a solution was "all mental" (not by virtue of mechanical shortcuts). What follows will be a discussion of the uses of proportional geometry as a way of determining form, balance, force and motion in the preparation for treatises. Primary attention will be given to solutions associated with the *Giant Crossbow*, a project for a treatise on military engineering that involved several distinct proportional methods. These recently discovered proportional innovations were the result of Leonardo's inventive approaches to technical and pictorial problems.

Proportional geometry, used to understand natural form, balance and force, also aided the visualization of problems of motion. Since Leonardo had no means to accurately calculate air or water friction, he often addressed the velocity of objects in air or water in a similar manner, producing proportional analogies for dynamic problems. He usually explained dynamics—the study of objects in motion—with medieval impetus theory, which did not account for air drag or viscous resistance. While making notes for treatises on painting and the nature of water in Manuscript A, he includes a note on impetus theory, along with a diagram, stating: "how admirable is your justice, Prime Mover! You have willed that no power lacks the order and qualities of its necessary effects" (folio 24r).[11] His illustration of the paths of a freely thrown ball and a bouncing ball has the balls stopping in different locations, though according to the impetus theory he quotes they are to reach the same distance, to the limit of the qualities of their necessary effects. Three years later, in the Codex Madrid, he compared the motion of a bouncing ball with the undulating movement of waves of water, addressing fluid mechanics and impetus dynamics as if similar pyramidal laws of the appearance of proportional oscillation governed them. He used this comparison to note that the bounce of a wave continues at a constant level much more than the bounce of a ball. The proportional model, or the constant in that case, was the form of the bounce, which helped isolate the differences in velocity, and thus the deduction that the property of water is more of a continuous quantity than the relatively discontinuous (or finite) impetus of the ball. Rather than quantify speed or friction in this case, Leonardo was primarily interested in the qualities of movement in air and water, using a proportional model as an analytical tool. Addressing the erosion of river banks, for example, he produced an illustration on Manuscript C of the flow of water past obstacles in the river, such that water flow is redirected by protrusions in the river bank toward the other side of the river downstream, eroding that point in the bank, and then reflected from there toward the other side further downstream. The neat crisscross of currents illustrates the proportional form, force and movement of water for one very specific case study about a possible form of riverbank erosion. Various other factors, such as alterations in channel depth and turbulence, are not considered in the study. Around 1505–8, Leonardo developed a greater interest in studying the properties of air and water turbulence, fascinated as ever by the visual appearances of these effects. Similar turbulent forms appear in both the Treatise on the Flight of Birds and on Windsor folio 12660v. He noticed that curls of turbulence develop as air rises from the outer edges of the downdraft of a bird's wing and as water rises around a pool in which water is poured. In both of these cases, he also illustrated these tight curls of turbulence at the inside of the bird's wings and the bubbling pool. There is a flow or an even stream of motion between inner and outer curls of turbulent air and water, a proportional similarity. Diagonal lines through air currents under the bird's wing seem to be the start of a diagram, possibly to map equal proportions of the strengths of currents around the curve of the lateral downdraft.

Leonardo used theories of proportional form, balance, force and movement in technical projects and treatises, including especially his development of the *Giant Crossbow*. As part of his extensive treatise programme, he illustrated the *Giant Crossbow* and other giant siege engines in preparation for an updated version of Roberto Valturio's *De re militari*, a popular treatise on military engineering best known around 1490 in the Italian and Latin editions published in Verona in February 1483. Manuscript versions had been available from as early as 1455, and the *editio princeps* was printed in Verona in 1472. Chief among Leonardo's contributions to the centuries old *De re militari* treatise tradition may at first appear to be the massive scale of his machines, whereas one could argue that the precision of the designs is most remarkable. Instead of using perspective constructions for the overhead views, he drew dimetric axonometric projections that have relatively equal measurements in the foreground and background. For the first time in the history of engineering, the proportions for these projections are accurate on at least two of the three dimensions, and thus an engineer should have enough information to build one of these projects according to the proportions represented on paper. In most cases, these designs are at a 30° 60° 90° angle, and they are dimetric because two of the directions of view are at the same scale. The *Giant Crossbow* is a good example of the use of proportional design because its dimensions are discussed in the text at the right side of the sheet. A comparison of written dimensions with measurements on the drawing show that the crossbow width and length are at the same scale of 108:1, whereas the components at the side are somewhat larger, at a scale of 80:1. In the statement at the right, Leonardo states that,

> This crossbow opens at its arms,
> that is where the rope is attached, 42 *braccia*,
> and is at its thickest, without its armature,
> 1 and 2 thirds *braccia*, and at its thinnest, 2/3rds of a *braccio*.
> It has an elevation [or draw] of 14 *braccia*. Its carriage
> is 2 *braccia* wide and 40 long and it carries
> 100 pounds of stone; and when it is
> moving, the carriage lowers itself and the
> crossbow directs itself along the length of the carriage.[12]

An arm length, such as *braccio a panno fiorentino* or *braccio mercantile milanesi*, in 1490 was approximately 58.4 or 59 centimetres. The width, where "the rope is attached," measures 22.67 cm, and would measure "42 *braccia*" at full scale, and thus the width is at a scale of 108:1. The length at the upper carriage measures 20 cm, representing approximately 37⅓ *braccia* of the crossbow. The lower carriage would extend an additional 2⅔ *braccia*, and this point will be discussed later. The total length of the carriage would be "40 [*braccia*] long" at full scale, and therefore its length in the drawing is at a scale of 108:1. The carriage would be "thickest, without its armature 1 and 2 thirds *braccia*, and at its thinnest, ⅔rds of a *braccio*," which is an estimation of the carriage's side measurements at full scale. On the drawing, the upper carriage is the "thickest" portion, measuring 1.17 cm, whereas the lower carriage is the "thinnest" portion, measuring .49 cm. These measurements are at a scale of 80:1. In fact the upper carriage is in two portions of ⅔ of a *braccio* (.49 cm) each, separated by a space of nearly a third of a *braccio*, illustrated as .19 cm. Through this open space would slide the beams that hold the trigger mechanism in place, as illustrated at the left of the drawing. This would reduce upward tension applied to the winding screw when the trigger nut is engaged. Conveniently, a third of a *braccio* is also the illustrated thickness of the wheel, felloe, hub, spoke, trigger braces, central side poles, side pole brace, u-bolt (at the back end), worm screw pole and some other items illustrated with parallel lines. Measurements of nearly all crossbow components on the drawing confirm that they proportionally fit one another, partially because they fit proportions that are measured in thirds. A measurement of the crossbow width at the back end of the carriage reveals an anomaly, such that the 1.62 cm width that Leonardo refers to as 2 *braccia* equals a slightly larger scale of 72:1. But this is the only example on the crossbow that is out of scale. Another mistake is the crossbowman, who, at a scale of 108:1, would be approximately 12 feet tall (3.65 m). Nonetheless, compared with leading engineering drawings of the time, including Francesco di Giorgio's illustrations for *De ingenis*, the proportional precision of Leonardo's *De re militari* series requires fewer estimates by an engineer.

His interest in third-part modular proportions was likely encouraged by the popularity of the medieval mercantile rule of three, wherein one would find the value of something by cross-multiplying unknown values with known values. For example: if seven pounds of barley cost nine *soldi*, how much would five pounds of barley cost? The unknown cost, divided by the proposed amount of five pounds, would equal the known cost of nine *soldi* divided by the amount of seven pounds. Five pounds multiplied by nine *soldi*, divided by seven pounds, would equal $6\frac{3}{7}$ *soldi*:

$$\frac{x}{5} = \frac{9}{7} \text{ cross-multiplied, yields: } \frac{45}{7} = 6\tfrac{3}{7} \text{ ths } \textit{soldi}$$

This is a proportional solution wherein three terms are given, and one has to find a fourth term that has the same ratio to the third term as the second term has to the first. For a water clock mechanism around 1510, Leonardo used the rule of three to determine the proportional differences between cylinder diameters that increased in size from one unit wide to 24 units, as these would fill with water and drop on each successive hour, triggering a flow of water that would activate a mechanical bell ringer.[13] Rather than follow the rule of three with mathematical precision, however, he added one-third to the diameter of each vessel, as his first calculation indicated that this would be relatively reliable. In practice he preferred geometrical and mechanical solutions to arithmetical problems. As a geometrical problem, the rule of three, or golden rule, adds a term to the method for finding the extreme and mean ratio, otherwise known as the golden section. In the *Elements*, Book VI, Euclid defined the latter ratio, stating that "a straight line is said to have been cut in extreme and mean ratio when, as the whole line is to the greater segment, so is the greater to the less."

Euclid also noted that, “a proportion in three terms is the least possible” (*Elements* V, definition 8).[14] Likely benefiting from geometry lessons provided by Fazio Cardano in Milan and Pavia during the 1480s, Leonardo used sensible mercantile and Euclidian proportional methods when considering the problems of scale and mechanical precision necessary for the giant siege engines of his *De re militari*. Siege engine components that are in third-part proportions can be consistently measured, or the scale changed, if only one or two general measurements are known for the entire structure.

Preparatory drawings along with dozens of related studies attest to the likelihood that Leonardo’s treatise program was more than just paper engineering.[15] Detailed studies of significant components of the crossbow appear on Codex Atlanticus (CA) folios 57v, 147av, 147bv, 148ar and 1048br. These orthographic projections show stages of design dominated by a concern for the proportions and structure of the massive armature on folios 57v and 147v, whereas folio 1048br has sketches of trigger as well as ratchet and pawl mechanisms, and folio 148ar is a precision demonstration of the characteristics and capabilities of the ratchet and pawl mechanism that would be a necessary feature of the winding screw of a giant crossbow. At the left of CA 147v, Leonardo isolates sketches of the winding screw mechanism and the lower carriage, illustrating ways in which the winding screw extends across the length of the upper carriage to allow one to move the trigger mechanism almost the entire 40 *braccia* length of the upper carriage and illustrating that the lower carriage would extend at least two or three *braccia* ahead of the of the armature, so that the entire machine would not tip forward. Since the crossbow’s length and width are at a scale of 108:1, the proportion of the 40 *braccia* carriage that is visible in the drawing is 37⅓ *braccia*, which means that the lower carriage would extend in front of the armature by 2⅔ *braccia*. In the drawing, one can see side poles under the armature that extend forward at an angle to meet this lower carriage extension.

To get precision representations of the appropriate proportional modules, Leonardo incised many features of the drawing with what appears to have been a metal stylus. Although metal stylus incisions also appear on other giant siege engines for his *De re militari* treatise—such as those on CA folios 141r, 145r, 181r and 182rb—the *Giant Crossbow* has the most extensive metal stylus preparatory marks in this group. Parallel incisions in the paper define most of the straight edges on the drawing. Curved incisions are apparent across the armature, likely applied by one end of a divider (a compass with two sharp tips) rotated around each arm, with the bottom end of the divider positioned at the right edge of the upper trigger mechanism and at the spokes of the rear left wheel. To see these stylus incisions one has to be in a relatively dark room and position the drawing at a horizontal angle to a strong source of light, so that this “raking light” reveals the shadows of the incisions. Guides for the metal stylus originally included a straightedge and possibly a metal or wooden template in the form of a 30° 60° 90° triangle. Parallel stylus incisions that extend horizontally across the upper sheet, between tips of the armature, were likely produced with the “crossed arc” method that is discussed by Cennino Cennini in his treatise *Il libro dell’arte*, a process that makes compositional lines that are at 90° angles to one another.[16] The Giant Crossbow drawing is the bottom half of folio 149r, the upper half of which contains studies of other crossbows and weapons. To cross arcs of a divider, Leonardo would have placed the bottom of the divider at the bottom centre of the lower folio, drawing the first arc with the upper portion of the divider across the upper portion of the sheet, and then he would have positioned the upper portion of the divider at the centre of the upper sheet, rotating the bottom portion of it in an arc across the upper portion of the lower sheet, thus creating an upward arc and a downward arc that cross each other at each side of the upper portion of the drawing. Between these crossed arcs, he drew a horizontal line with a metal stylus and straightedge, which is the upper parallel line on the paper. The two arcs, which are not visible on the sheet, sweep across the positions of the present locations of the armature and the ropes. Preparing to make the lower parallel horizontal line between armature ends, he could have repositioned a divider at the base of the page, rotating its upper portion across the second arc mentioned in the description. The horizontal line between second set of crossed arcs extends between the current locations of the tips of the crossbow, "where the rope is attached," as Leonardo notes at the right side of the sheet. After arranging the armature's location, it would appear that he drew the 60° parallel metal stylus incisions for the carriage, incised the carriage's horizontal features, incised portions of the armature, then the trigger mechanism, the worm screw mechanism at the right and finally the crossbowman. After making the preparatory incisions, Leonardo applied ink lines and wash to the page, finishing the drawing.

If there is very little that one can say is random about the *Giant Crossbow's* design—from its proportional components to its precision metal stylus incisions—what then can be said of its overall dimensions of 42 by 40 *braccia*? Why did he choose these numbers, and did he calculate the relationship between the drawing's measurements to this large scale? To answer the second question first, it seems unlikely that Leonardo would have chosen an odd scale of 108:1 for the width and length, given his consistent approaches to using relatively simple calculations. His choice of 108 is nonetheless a multiple of three and 36, a convenient guarantee that mechanical components designed in third-part proportions may be increased in scale without complex arithmetic. Assigning a width of 42 and a length of 40 is perhaps less obvious, unless one considers geometric methods for calculating the armature dimensions. As illustrated, the armature is spanned (drawn back) by ropes that are attached to a trigger mechanism three-quarters of the way back along the upper carriage. If one envisions the outer edge of this armature to be a portion of a large circle, one need only apply Euclidian and Archimedean principles to estimate lengths of the carriage and armature. The first clue is the 60° at which the carriage is placed on the folio. This is the angle one finds when inscribing within a circle an equilateral and equiangular hexagon, as noted in Euclid's *Elements* IV: 15. Any angle of the equiangular triangles that compose that hexagon will be one-third of two right angles, or 60°.[17] Euclid deduced in "Proposition fifteen" that the radius of the circle is equal to the side of the inscribed hexagon. Leonardo restates and illustrates Euclid's lessons regarding the inscription of a circle with a hexagon on Ms. A, folio 13v around 1490, close to the time of the *Giant Crossbow* and several treatise projects. In order to determine the outer measurement of the armature, Leonardo had to calculate the circumference of the circle in which it is inscribed. He appears to have understood he would have had to multiply the circle's diameter with $^{22}/_{7}$, which is Archimedes' approximation of the number used to calculate the circumference of a circle. To get the diameter, Leonardo seems to have assigned a number for the radius, selecting two-thirds of 60, which is 40. Another motivation for using this multiple of four may have been his memory of the geometry necessary for the Verrocchio studio to design the four *braccia* diameter copper ball for the lantern of the Florence Cathedral around 1468–71.[18] As for the circle around the armature, twice the radius of 40, multiplied by $^{22}/_{7}$ equals a $251^{9}/_{25}$ braccia circumference. If the armature's outer edge is a sixth of this circumference—which seems to have been a visualization of the geometrical problem, rather than a calculated approach—then a sixth of $251^{9}/_{25}$ is an armature length of $41^{17}/_{19}$, or approximately 42 *braccia*. Although this may seem to be wider than the armature span, this calculation is more than a coincidence, indicating that Leonardo's reference to the armature width may or may not have been a reference to the span. His reference to 42 *braccia* appears to have been a reference to the armature length. An engineer making these calculations would have been able to determine the lengths of wooden beams necessary for the project.

He also favoured proportional geometry as a way to calculate the impetus stored in a projectile before the *Giant Crossbow* launches it.[19] In the early 1490s—another possible period for the *Giant Crossbow* project—Leonardo illustrated in the volume now known as Codex Madrid I, on folio 51r, that equal amounts of draw tension applied to a crossbow string will not force the string to descend along the crossbow stock in equal proportions. He states that the "nature of this power shall be pyramidal . . . if I add 10 more pounds [to the string], the descent will not be equal to another ounce" of a *braccio* in descent along the stock.[20] An ounce is a twelfth of a *braccio*, or around five centimetres. Leonardo observed that with each addition of ten pounds to a crossbow string, its descent down the stock is somewhat less than in the previous descent, and that an illustration of the decreasing positions of descent of the strings will "almost appear to look like the foreshortenings of those who practice perspective."[21] To address the pyramidal nature of the problem, he devised a solution wherein the projectile's impetus would be relative to the angle of the crossbow string at the position of the nut of the trigger mechanism. He illustrates on folio 51r this inverse pyramidal proportion between the string angle and its power, noting that as the string angle decreases along the stock, its power increases: "should angle a shoot as far as one span, he would shoot the distance of 2 spans."[22] This theory of inverse pyramidal proportion significantly updates Leonardo's general assessment of the basic principles of impetus on MS A folios 30r and 35r around 1490. He applied a visual, geometrical solution to impetus theory, thereby providing a more reliable systematic approach with which to calculate impetus. Although he offered a proportional model, as opposed to a mere intuitive theory, this approach nonetheless lacked numerical estimates or quantifiable test results. In any event, had he published his treatise on mechanics—part of which is now in the Codex Madrid—perhaps his most influential contribution to the fifteenth-century treatise tradition would have been the illustrated proportional solutions to problems of form, balance, force and motion. Diagrammatic and systematic proportional solutions such as these were fundamental contributions to mid-sixteenth century scientific studies. As discussed in the present study, Leonardo developed these solutions with regard to fundamental governing principles such as pyramidal law, the centre of gravity, natural designs such as wings of nerve cord structures, models of motion and turbulence, dimetric orthographic projections, scale models, third-part proportions, the rule of three, and Euclidian and Archimedean geometry. Aiding him in this endeavour had been perspective theory, his anatomy treatise illustrations, studies of flight and flying machines, and engineering treatises. The unifying principle in all of this work had been proportion theories, used by Leonardo to improve on traditional methods that favoured intuition over systematic evidence of best practice.

Although it is not possible to identify his knowledge of best practice or tested results in many cases, he arranged proportional methods with which his audience could often test the accuracy of his assumptions. Some of the accuracy in the *Giant Crossbow* project is an example of this. For Leonardo, the art and *scienza* of technical illustration and natural philosophy was governed not by intuition but by proportional rules of form, balance, force and motion.

Giovanni Antonio Amadeo

In additional to Leonardo's innovative approaches to arithmetic and geometry in painting, Giovanni Antonio Amadeo applied similar interests—also between 1470 and 1500—to the facades of buildings. Like Verrocchio and Leonardo, he developed surfaces sculpturally, taking advantage of a pictorial trend since the early 1470s in the portrayal of *rilievo* or the illusion of deep relief and texture on surfaces.[23] Bramante developed this as a kind of sculptured wall motif, especially with Roman architectural features of his *Tempietto* in Rome. As a sculptor, engineer and architect, Amadeo applied the pictorial *rilievo* as a sculptured wall motif that used geometric forms and proportions, managing this with his first major commission—a funerary chapel in 1470 for Bartolomeo Colleoni in Bergamo. The virtuoso mathematics of the façade are proudly displayed with a rich tapestry of illusionistic cubes, the range of sculptures and contrasting colors, and the harmonic Pythagorean perfect intervals of a fourth (diatessaron 8:6 = 8 niches under a heptagon) a fifth (diapente third part divisions over paired divisions) and the octave (diapason of double units within single units). This is one of the most innovative facades of Renaissance architecture, a *bel composto* of systematic pictorial splendour, sculptural variety, and architectural *concinnitas* (indissoluble harmony). From 1481 until 1499, Amadeo directed another impressive project, the building of the Certosa di Pavia (Charterhouse of Pavia), taking over designs originally developed by with Giovanni Solari and developed by his son, Guiniforte Solari.[24] Pythagorean proportions appear to have been a governing principle of the façade, something Amadeo likely developed within 18 years, improving on Solari's designs.

Pietro Lombardo and Mauro Codussi

Another *bel composto* of pictorial, sculptural and architectural mathematics is the façade by Pietro Lombardo and Mauro Codussi for the Scuola Grande di San Marco (Confraternity of St. Mark). Here we see an effort to relate the building to its neighbour, the Basilica of St. Mark, though with the help of geometric expressions like perspective sequences of illusionistic hallways, deep and protruding Roman features that make sculptured wall motifs, and an emphasis on perfect harmonic intervals of ratios 1:2, 2:3 and 3:4. Cardussi's work on San Zaccaria and the Clock Tower of St. Mark also express his interest in mathematics.

Francesco di Giorgio Martini

Although trained as a painter and sculptor in Siena, Francesco di Giorgio was the leading military architect of Italy in the 1470s and 80s, responsible for 136 military fortresses.[25] He would also apply inventive and complex perspective construction techniques to interior spaces, such as the intarsia project of the Gubbio *Studiolo*, now located at the Metropolitan Museum.[26] He worked in Gubbio, Urbino, Naples and Milan on a wide range of civil and military projects and was also well known among intellectuals and engineers for his treatises on civil and military engineering, handwritten copies of which circulated widely in the 1480s and 90s. He was a major influence on Leonardo, whom he met in 1490 in Milan and Pavia. Leonardo improved on Francesco's technical illustration methods, developing precision axonometric illustrations that are diametric, with two proportional dimensions, rather than the trimetric illustrations by Francesco, wherein all three of the dimensions are not proportional. Francesco, Leonardo and their humanist colleagues shared an interest in structural forms and formulas that expressed the universal laws of Nature. Good engineering and good rhetoric were governed, they believed, in a geometric relationship between man and his surroundings, and the dialogues on these mechanical, mathematical and divine relationships were the subjects of their treatises. These writers conducted themselves as intellectuals, of course, conversant on structural engineering and the structure of the soul, on the mathematics of optical space and on the perfect Neo-Platonic and Neo-Pythagorean forms and numbers. As a student of mid-fifteenth century perspective ordering of pictorial space, Francesco continued to apply architectural principles to his paintings and sculptures. His *Coronation of the Virgin* includes a novel assessment of the heavens and structural intuitions of the supernatural elements. His works include great perspective depth and antique subjects such as mythological scenes in bronze bas-reliefs for doors of the Ducal Palace at Urbino. Those are some of the most complex perspectival scenes in low relief, with the mathematical tradition of Piero della Francesca's level of pictorial precision in Urbino.

Michelangelo

Much less inclined to reveal his tracks than Leonardo, Michelangelo deemphasized the roles of mathematics and proportion in his work. Vasari and Lomazzo quote Michelangelo's famous dictum regarding the "judgement of the eye," that "it was necessary to keep one's compass in one's eyes and not in the hand," and that "modern painters and sculptors ought to have proportion and measures right in their eyes" (Vasari), because "this science had been lost to the moderns, if compared to those marvellous statues of the classic artist, such as those of Phidias and Praxiteles located here in Rome" (Lomazzo).[27] As Vasari notes:

> I have some examples of his work, found in Florence and placed in my book of drawings; and these not only reveal the greatness of his mind but also show that when he wished to bring forth Minerva from the head of Jove, he had to use Vulcan's hammer: for he used to make his figures the sum of nine, ten, and even twelve 'heads'; in putting them together he achieved a certain overall harmony of grace, which nature does not present; and he said that one should have compasses in one's eyes, not in one's hands, because the hands execute but it is the eye which judges. He also used this method in architecture.[28]

This approach to a visible harmony of forms recalls Alberti's use of the harmonic term *concinnitas* to denote an indissoluble coherence of formal elements in architecture. Michelangelo entrusts primarily the "judgement of the eye" to such compositional matters in painting, sculpture and architecture, and this is in partial agreement with Vitruvius's recommendation for the training of the architect:

> He should be a man of letters, a skillful draughtsman, a mathematician, familiar with scientific inquiries, a diligent student of philosophy, acquainted with music; not ignorant of medicine, learned in the response of juris-consults, familiar with astronomy, and astronomical calculations. The reasons why this should be so are these . . . Mathematics again furnishes many resources to architecture. It teaches the laying out of buildings on their sites by use of set-squares, levels and alignments.[29]

Michelangelo's primary interest is in the "*concinni*" or agreeable, cohesive elements of a work, such as a larger head for a David sculpture, in agreement with its originally commissioned location at the roofline of the Florence Cathedral. This *concinnitas* was as important in music theory as it was in Albertian architectural theory, referring to an advanced branch of development in sound and form. Whereas initial sounds are continuous or definite, the relationships between the sounds are a bifurcation of the definite (discontinuous) branch into agreeable (*concinni*) and disagreeable (*inconcinni*) sounds; of these there is a range of agreeable sounds, from the grave to the acute.[30] Originally from Manuel Bryennius's fourteenth-century treatise *Harmonics*, this assessment of sounds was published by Franchino Gaffurio in his treatise on music theory, *Theorica musicae*, in 1492. He notes that the imagination (*phantasia*) converts these discontinuous sounds received by the ear into continuous sound. Leonardo referred to the definite, discontinuous quantities of mathematics, contrasting them with continuous quantities of geometry and line, both of which he considered to be properties of music and its equal, painting. Michelangelo used language similar to that of contemporary theorists like Gaffurio and the earlier Alberti and Vitruvius without leaving behind—for the most part—discussions and illustrations of his mathematical and proportional methods. An emphasis on "the judgement of the eye" over the skillful actions of the body must have been familiar to Michelangelo, who had some access in his formative years to the Florentine humanist discourses on the superiority of eye among the five senses, as discussed by Marsilio Ficino, Agnolo Poliziano and Girolamo Savonarola in the 1490s.[31] A handful of drawings nonetheless remain of Michelangelo's calculations with the compass, ruler and number. Two drawings—Casa Buonarroti folio 75Av and Archivio Buonarroti I, 155, folio 276Ar—on arithmetical estimates and measurements of a house and quarried marble blocks, respectively, suggest that he liked to think of numbers visually, that he did not necessarily estimate proportions strictly by eye but instead by the usual coordination of measure and proportion. As Vasari indicates, Michelangelo used interchangeable Italo-Byzantine 9⅓ face-length figures and Vitruvian ten face-length figures, "for he used to make his figures the sum of nine, ten, and even twelve heads."[32] He likely destroyed the majority of his measured drawings, whereas his finished projects show that he was very interested in the problems of measure, number and proportion, not to mention counting the fortune in ducats he earned from the Vatican. Although he likely had scholarly advice regarding the proportions of the Sistine Ceiling composition, he had final approval of that sophisticated proportional sequence. As he developed as an architect after 1515, "the judgement of the eye" would become—in many cases—secondary to his proportional calculations.

Raphael

Few artists took better advantage of the influences of their predecessors than Raphael, and almost all evidence of this is in the form of his finished work. For the inclusion of antique proportions in architecture and a broad depth of perspective field, Raphael would have witnessed Perugino's and Botticelli's paintings of temples within cityscapes in the Sistine Chapel while starting to work at the Vatican in 1508. He was in the process of applying more sophisticated architectural illusions—or *quadratura*—in the School of Athens and Disputa, the former with a model and plan of Bramante's Greek cross plan for St. Peter's, and the latter within an illusionistic apse of the old St. Peter's basilica. This followed Pope Julius II's interests in relating the intellectual content of the new architecture—and new universal order—to that of the older architecture and order. That room—the *Stanza della Segnatura*—is governed by Apollo and his *kythera* (harp), as he stands in the *School of Athens*. Also in the *Stanza*, Apollo sits on Parnassus, a fresco on the adjacent wall of the *School of Athens* and *Disputa*. Raphael's virtuoso use of perspective is also evident in the illusions of Old Testament scenes on the curved surfaces of the cramped spaces of Loggia of Pope Leo X, an approach that would be copied by followers such as Marcantonio Raimondi.[33]

Giuliano da Sangallo

Giuliano da Sangallo (c. 1443–1516) was a sculptor, military engineer and architect and one of the best authorities on the antique *symmetria* and *rhythmos*, thanks to his extensive studies, detailed measurements and proportional drawings of antique architecture and sculpture. He applied these approaches with mathematical precision in his work for the Medici.

Bramantino and Bramante

Bramantino's *Crucifixion* of 1510–20 updated Lucas Cranach the Elder's *Passion* of 1509 in a Lombardian architectural modular manner, with highly ordered, geometric, harmonious numerical ratios that seem to suggest his interest in geometrical rhetoric and decorum similar to that applied to architecture and similar to the Neo-Platonic humanistic debates on visual knowledge received through the superior sense organ of the eye and its internal compass (as per Michelangelo, etc.).[34] Bramante and Franchino Gaffurio were major influences on Bramantino's approaches to the *concinnitas* of pictorial composition, a harmonizing of pictorial elements that were only recently identified as a form of artificial *componimento* that can outdo Nature's *componimento*, or what we now call a painter's composition. As seen in radiographs of the painting, the comprehensive perspective construction is an architectural method for managing pictorial space, a descendent of Brunelleschi's initial approaches to artificial perspective, to legitimate mathematical space.[35]

The artist/engineers noted here developed their interests in these mathematical spaces and proportional Neo-Pythagorean expressions because of their interests in the universality of those spaces and expressions, as perceived optically and audibly, proportionally and harmoniously. Still, it is difficult to prove this practical engagement with engineering and mathematical books near the turn of the sixteenth century. As practitioners of arithmetic, and often mathematics, artist/engineers at this time apparently considered strategies that were both intuitive estimations of the proportional *quantita continua* of geometry and specific calculations of finite *quantita discontinua* of the optical point. Like their humanistic colleagues at court or the university, their approaches often involved ethical assessments of universal laws. Moreover, the doubling of the number of abacus treatises—from 26 manuscripts during the first half of the century to 60 manuscripts and six printed editions in the second half of the century—is an indication of the growing interests in the uses of arithmetic and mathematics in a broad range of disciplines.[36] This does not account for the increasing number of texts on practical geometry, tariffs, astrology and other related disciplines. Precision proportional form in a given visual art was believed to guarantee its proportional function as a work of art. This kind of holistic belief became less important into the sixteenth century, when other intellectual modes of technical and aesthetic assessment develop in popularity. The mathematics of the visual arts would cease to be as much about natural philosophy.

Notes

1 Matthew Landrus, *Leonardo da Vinci's Giant Crossbow* (Heidelberg: Springer, 2010), 80–94.
2 Thomas Brachert, "A Musical Canon of Proportion in Leonardo da Vinci's Last Supper," *Art Bulletin* 53 (December, 1971): 461–66. Matthew Landrus, "The Proportions of Leonardo da Vinci's Last Supper," *Raccolta Vinciana* 32 (2007): 43–100.
3 For evidence that Pythagorean theories were not used in the visual arts, see Anthony van der Schoot, *De ontstelling van Pythagoras* (Baarn, the Netherlands: Kok Agora, 1999). For an assessment of the reception of Pythagoras, see Christiane Joost-Gaugier, *Pythagoras and Renaissance Europe: Finding Heaven* (Cambridge: Cambridge University Press, 2009). For a history, see: Eli Maor, *The Pythagorean Theorem: A 4000-Year History* (Princeton: Princeton University Press, 2007).
4 Danti's proem to Euclid, Optics, transc. Heliodorus of Larissa, Ital. trans. Egnazio Danti, Florence, Bernardo Giunta, 1573. *La prospettiva di Euclide, nella quale si tratta di quelle cose, che per raggi diritti si veggono, et di quelle, che con raggi reflessi nelli specchi appariscono. Tradotta dal R.P.M. Egnatio Danti cosmografo del Seren. Gran Duca di Toscana. Con alcune sue annotationi de' luoghi più importanti. Insieme con la Prospettiva di Eliodoro Larisseo cavata della Libreria Vaticana, e tradotta dal medesimo nuovamente data in luce* (Firenze, Heirs of Bernardo Giunta, 1573).
5 Victor J. Katz and Karen Hunger Parshall, *Taming the Unknown: A History of Algebra from Antiquity to the Early Twentieth Century* (Princeton: Princeton University Press, 2014), 227–246.
6 Translation by Filippo Camerota, in his essay, "Teaching Euclid in a Practical Context: Linear Perspective and Practical Geometry," *Science & Education* 15 (2006): 323.
7 See for example, Long, *Openness, Secrecy, Authorship* as noted above.
8 Regarding mathematical treatises, see: Warren van Egmond, "The Commercial Revolution and Beginnings of Western Mathematics in Renaissance Florence, 1300–1500," PhD diss., Indiana University (1976), 590–596.
9 Landrus, "The Proportions of Leonardo," 43–100.
10 For assessments of the rise in music treatises, see: Claude V. Palisca, *Humanism in Italian Renaissance Musical Thought* (New Haven and London: Yale University Press, 1985), 161–225. F. Alberto Gallo, *Music in the Castle: Troubadours, Books, and Orators in Italian Courts of the Thirteenth, Fourteenth, and Fifteenth Centuries*, trans. Anna Herklotz and Kathryn Krug (Chicago: University of Chicago Press, 1995), 69–112.
11 J. Venerella, trans. and anno., *The Manuscripts of Leonardo da Vinci in the Institute de France: Manuscript A (Raccolta Vinciana)* (Milan: Castello Sforzesco, 1999), 72.
12 "Questa belestra • apre • nelle sue braccia / cioè dove s'ap[p]licca • la corda • b • 42 • • / ed è nel più • grosso sanza l'armadura sua / • b• uno e 2 terzi, e nel più sottile $^{2}/_{3}$ di b / ha di mo[n]tata b • 14 • il suo tinieri / è largo • b 2 • è lungho 40 e por[-] / ta lib[bre] • 100 di

pietra e quando è / in cam[m]ino il tenieri s'abbassa e la / balestra si diriz[z]a per lo lungo del tenieri."

13 See the designs on Windsor Royal Library folio 12688r and Codex Atlanticus folios 65v and 975v.

14 Thomas L. Heath, trans. and Dana Densmore, ed., *Euclid's Elements* (Santa Fe, NM: Green Lion Press, 2002), 99, 123.

15 Preparatory drawings include Codex Atlanticus fols. 57v, 147av, 147bv, 148ar/53ra, and 1048br, whereas related drawings include Ms. B fols. 5v-11v, 20r-21r, 23v-25r, 27r-v, 30v-33v, 35r-37r, 39v, 40v-56r, 57v-70r, 73v-93r, 94v-95r, 90r-100v, A.1, A.2, B.1, B.2, D, Codex Atlanticus fols. 7v, 27v, 50v, 62b, 72r, 90r, 112ar, 113r, 143r, 149ar, 149br, 153r, 154br, 155r, 175r, 175v, 754r, 1054r, 1054v, 1058r, 1063v, 1070r, 1071v, 1094r, Christ Church folio 20r&v, Louvre Inv. 2260, Codex Trivulzio p. 99 (64r), Uffizi no. 446ev, Windsor fols. 12647r, 12649r, 12469v, 12650r, 12650v, 12651r, 12652r, 12652v, 12653r.

16 Cennino d'Andrea Cennini, *Il Libro dell' Arte*, in *The Craftsman's Handbook*, trans. Daniel V. Thompson, Jr. (Yale 1933; reprint, New York: Dover Publications, 1960), 42–43.

17 I would like to thank Dr. J. V. Field for drawing my attention to this proposition in *Euclid's Elements*.

18 Leonardo referred to this project when designing burning mirrors around 1515, stating, "remember the soldering (*saldatura*) used to solder the ball for Santa Maria del Fiore," as noted in Manuscript G, 84v.

19 Franciscus de Marchia (fl. 1320) and John Buridan (c. 1295–c. 1358) formulated impetus theories as a way of explaining how an object keeps moving after it leaves the thing that throws it. This was first explained by the sixth-century Alexandrian John Philoponus, who corrected Aristotle's poor definition of impetus. One of Buridan's definitions of impetus theory that Leonardo appeared to know is that a falling body was believed to generate additional impetus embedded in that body: as impetus increases with the fall of the body, the body's velocity increases. A standard source for Marchia and Buridan is David C. Lindberg, *The Beginnings of Western Science*, 2nd ed. (Chicago: Chicago University Press, 1992), 302–303.

20 Ladislao Reti, trans., *Leonardo da Vinci: The Madrid Codices*, 5 vols. (New York: McGraw-Hill, 1974), 4: 102.

21 Ibid.

22 Ibid.

23 See for example the essays in Janice Shell and Liana Castelfranchi Vegas, eds., *Giovanni Antonio Amadeo. Scultura e Architettura del suo tempo* (Milan: Cisalpino 1993), and particularly chapters: Richard Schofield, "Amadeo's System," 125–156; and Pietro Cesare Marani, "L'Amadeo e Francesco di Giorgio Martini," 353–376; Jo Anne Giltin Bernstein, "Patronage, Autobiography, and Iconography: The Facade of the Colleoni Chapel," 157–174.

24 Ibid. Charles R. Morscheck, "Francesco Solari: Amadeo's Master?" in *Giovanni Antonio Amadeo: sculptore e architetto del suo tempo*, ed. Janice Shell and Liana Castelfranchi (Milan: Cisalpino, 1993) 103–124; and Janice Shell, "Amadeo, the Mantegazza, and the façade of the Certosa di Pavia," in *Giovanni Antonio Amadeo*,189–222.

25 See: Francesco Benelli, "Diversification of Knowledge: Military Architecture as a Political Tool in the Renaissance: The Case of Francesco di Giorgio," *Res* 57/58 (2010): 140–155. Allen S. Weller, *Francesco di Giorgio, Siena 1439–1501,* PhD diss., University of Chicago (Chicago: University of Chicago Press, 1943), 14–27. Gustina Scaglia, *Francesco di Giorgio: Checklist and History of Manuscripts and Drawings in Autographs and Copies from ca. 1470 to 1687 and Renewed Copies (1764–1839)* (Bethlehem, PA: Lehigh University Press; London and Toronto: Associated University Presses, 1992), 13–24. Pari Riahi, *Ars et Ingenium: The Embodiment of Imagination in Francesco di Giorgio Martini's Drawings* (London: Routledge, 2015), 11–45.

26 Spartaco Capannelli, ed., *Il Palazzo Ducale di Gubbio e Francesco di Giorgio Martini* (Gubbio: TMM, 2008), 227–294; Olga Raggio, *The Gubbio Studiolo and Its Conservation*, vol. 1, *Federico da Montefeltro's Palace at Gubbio and Its Studiolo*, vol. 2, Antoine M. Wilmering, *Italian Renaissance Intarsia and the Conservation of the Gubbio Studiolo* (New York: Metropolitan Museum of Art, 1999).

27 Giorgio Vasari, *Le Vite de' più eccellenti pittori, scultori, ed archittettori*, ed. Gaetano Milanesi, vol. 6 (Florence: Sansoni, 1878–85), 270. Giovanni Paolo Lomazzo, *Trattato dell'arte della pittura, scultura ed architettura* (Milan: Paolo Gottardo Pontio, 1584; Rome: Saverio Del Monte, 1844), 1: 45 and 2:165.
28 George Bull, trans., *Giorgio Vasari: Lives of the Artists*, vol. 3 (London: Folio Society, 1993), 210.
29 Vitruvius, *On Architecture* (*De architectura*), *Books I-V*, ed. and trans. Frank Granger (Cambridge, MA: Harvard University Press, 1931), 9 (Ten Books, 1: 1.3–4).
30 Ibid. Palisca, *Humanism in Italian Renaissance*, 209.
31 For the context regarding Michelangelo's approach to the 'judgment of the eye', see: Robert Clements, *Michelangelo's Theory of Art* (New York: New York University Press, 1961), 29–33.
32 Ibid. Vasari, *Lives*, 3: 210.
33 For a recent study of the Loggia and especially 'Raphael's Bible', see: Nicole Dacos, *The Loggia of Raphael: A Vatican Art Treasure*, trans. Josephine Bacon (New York: Musei Vaticani and Abbeville Press, 2008), 137–209.
34 See Luciano Patetta, ed., *Bramante e la sua cerchia: a Milano e in Lombardia 1480–1500* (Milan: Skira, 2001).
35 For a study of the restoration of this painting, see Pietro Marani, ed., *La Crocifissione di Bramantino: Storia e Restauro* (Florence: Centro Di, 1992).
36 Ibid. See Egmond, "The Commercial Revolution," 590–596.

Part II

Artists as Mathematicians

5 Dürer's *Underweysung der Messung* and the Geometric Construction of Alphabets

Rangsook Yoon

Introduction

Dürer's Mathematical Knowledge and the Revival of Euclidian Geometry in the Renaissance

Albrecht Dürer's theoretical writings have been claimed as the "birthplace of German scientific prose," to borrow Erwin Panofsky's term, and they have been extensively studied by art historians and historians of science alike, whether to explicate their sources of inspiration or to account for his scientific contributions and influences.[1] Dürer's published treatises, such as the *Course in the Art of Measurement with a Compass and Ruler* (*Underweysung der Messung mit dem Zirckel un[d] Richtscheyt*) and the *Four Books on Human Proportion* (*Vier Bücher von menschlicher Proportion*)—which were issued respectively in 1525 and 1528—as well as unpublished fragmentary notes related to them are indeed inexhaustible sources of information, not only about his own thoughts and ideas but also about his milieu and the Renaissance era, in which mathematical and geometric knowledge played an increasingly significant role. As the historian Paul Lawrence Rose writes, it is almost impossible to separate the classical basis of the mathematical renaissance from the general classical revival undertaken by the Italian humanists.[2]

It is abundantly clear that Dürer was well acquainted with writings of Euclid and Vitruvius, not to mention Italian mathematicians and theoreticians of his time. In 1507, during his second stay in Venice, he acquired the Latin edition of Euclid's *Opera*, published by Johannes Tacuinus in La Serenissima in 1505.[3] With the help of his best friend and patron, Willibald Pirckheimer (1470–1530), Dürer translated portions of Euclid's propositions, specifically those that pertained to perspective.[4] Dürer also wrote, "I took the matter to heart and studied Vitruvius who has written a little about the proportions of man," testifying to his familiarity with Vitruvius's *Ten Books on Architecture* (*De architectura*).[5] Furthermore, in 1523, he obtained ten books of unknown titles, which formerly belonged to Bernhard Walther, the Nuremberg mathematician and astronomer.[6] Thus it can be inferred that these books also had something to do with his scientific interest. It is documented that Walther possessed books in his library that Dürer would have been eager to obtain, notably a manuscript copy of Leon Battista Alberti's *De pictura* (1435–36).[7]

Extensive studies have been conducted on a range of Dürer's theorems and his use of the empirical applications of mathematical and geometrical problems, although not all aspects of his investigations have received an equal degree of scholarly attention. One noteworthy instance of relative neglect is a segment in the third chapter in Dürer's *Underweysung der Messung*, which deals with how to configure proper (*recht*) alphabet letters based upon geometric principles—or the art of measurement (*Kunst der Messung*). When examined by others, this was done largely to account for the sources that Dürer might have consulted, as well as his contributions to the development of typographic and calligraphic matters.[8] As can be seen in a 1976 article by Horst Heiderhoff, scholars have often emphasized the significance of this section of the *Underweysung der Messung* within the process of paving a way for mechanical "reproducibility" (to borrow a phrase by Walter Benjamin) and the standardization of alphabet letters as established according to Dürer's geometric rules.[9]

This chapter contextualizes Dürer's concern with the geometric construction of alphabets within the larger humanist culture of Renaissance Europe, where ancient and "modern," Northern European and Italian, science and art all seem to have converged relatively seamlessly. By examining how his methodical explanations about lettering processes are closely related to some of the better-known parts of the *Underweysung der Messung* regarding perspective, this chapter sheds light on some of the theoretical undercurrents. Specifically, I focus on his instructions to others regarding how to place lettering on high walls using perspectival techniques, which brings out the centrality of the privileged status of geometric knowledge regarding perspective in Dürer's time.

Body

The Content of Dürer's Underweysung der Messung and the Section on the Proper Geometric Construction of Alphabet Letters in the Third Book

Explicated throughout the *Underweysung der Messung* are the fundamental tenets of geometry and different sets of problems relating to the practical application of this math.[10] Dürer both deliberates and instructs others on all aspects of the art of design, ranging from perspectival drawing to constructing forms, based on the principles of Euclidian geometry. He regards the study of measurement as the real foundation for all types of art—a vital and necessary starting point for budding artists and also useful to painters as well as "goldsmiths, sculptors, stonemasons, and carpenters."[11] After the book's dedication to Pirckheimer, Dürer opens the theoretical tract by crediting Euclid as the authority and founder of geometry.[12] When Dürer employs the German term of measurement, *Messung*, he means geometry; the term "geometry" was not yet integrated into the German vocabulary at this time. When Joachim Camerarius translated Dürer's book into Latin in two parts in 1532 and 1535, he rendered it as *geometria*.[13]

The *Underweysung der Messung* consists of four chapters or books (*Büchlein*). The First Book elucidates principles of linear geometry. This is followed in the Second Book by an exposition of two-dimensional plane surfaces. The Third Book deals with applied geometry pertaining to solids, expounding on such matters as buildings and descriptions of various types of columns, capitals, bases, pedestals, sundials, and towers. This chapter ends with a detailed discussion on "proper" letter-design practices. Finally, the Fourth Book deals with solid and spherical geometries, and it is here that we find Dürer's famous dicta on regular polyhedra, stereometry, and perspective.

The somewhat lengthy section on how to create proportionally correct formations of alphabets is inserted like a digression, at first seemingly little pertinent to the exposition on architectural components and their constructions, which are the principal subjects in the rest of the Third Book. In this segment regarding the proper figuration of alphabet letters using Euclidean geometry, Dürer instructs the reader on how to construct each specific letter, from A to Z, in both Roman and Gothic types, which he calls *Lateinisch* and *Textur*, respectively. His prescriptive description of each letter is accompanied by two to five illustrations diagramming the letter, in which at least one of them is framed within a grid-like square. To form the Roman letter A (Figure 5.1), for instance, the encasing square is divided into four smaller squares by two bisecting lines, and each of these four squares is inscribed with circles and diagonals. According to the accompanying text, this diagram also serves as a template for constructing all other Roman capital letters, accomplished by using triangles, squares, and circles.

Dürer's Sources of Knowledge about the Proper Geometric Construction of Roman Alphabets

As has been amply discussed elsewhere by other scholars, Dürer did not invent the systematic application of geometry and the rules of proportion that underlie his descriptions of carefully crafted alphabet letters. Here, as in other discussions of theoretical matters, he was greatly influenced by the Italians of his day. Many prominent Quattrocento and Cinquecento art theoreticians and mathematicians in Italy wrote about how to form well-proportioned alphabets *all'antica*, under the assumption that ancient Roman inscriptions were shaped according to geometric and proportional rules. The revival of antiquity in the fifteenth century brought about a change of taste affecting all cultural activities, including writing, not only in terms of its rhetorical style but also in terms of its calligraphic and epigraphic styling.[14] With increasing antiquarian interest in Roman inscriptions, humanistic scripts reviving antique—that is, Roman—letters gradually replaced Gothic letters in manuscripts, as well as inscriptions added to paintings, sculptures, and architectural structures, and eventually type fonts used for printing presses.[15]

In his treatise *On Painting* (*De pictura*), written in 1435–36, Leon Battista Alberti (1404–72) briefly mentions the process of forming letters but does so without explicating the method in any detail.[16] A number of Italian contemporaries of Dürer wrote small manuals on how to construct Roman alphabets according to the geometrical principles "of the circle and the square" ("del tondo e del quadro"), based on observations of ancient Roman inscriptions that were considered the best exempla. These include: a manuscript by the calligrapher Felice Feliciano (1460), a friend of Andrea Mantegna;[17] a manual by Damiano da Moyle (or Damianus Moyllus) printed in Parma around 1480–85;[18] Sigismondo Fanti and his treatise published in Venice in 1514;[19] Francesco Torniello and his publication completed in Venice in 1517;[20] and most notably, a work by Luca Pacioli (c. 1445–c. 1514), who was a celebrated mathematician born in Borgo San Sepolcro and active in Bologna. Pacioli's text on alphabet construction was added to the eleventh chapter of his *De Divina proportione*, printed in Venice in 1509.[21] Despite certain differences in precise details, the methods established by these Italians all describe using a grid, starting with a square, and then emphasize proportionality when constructing individual letters.

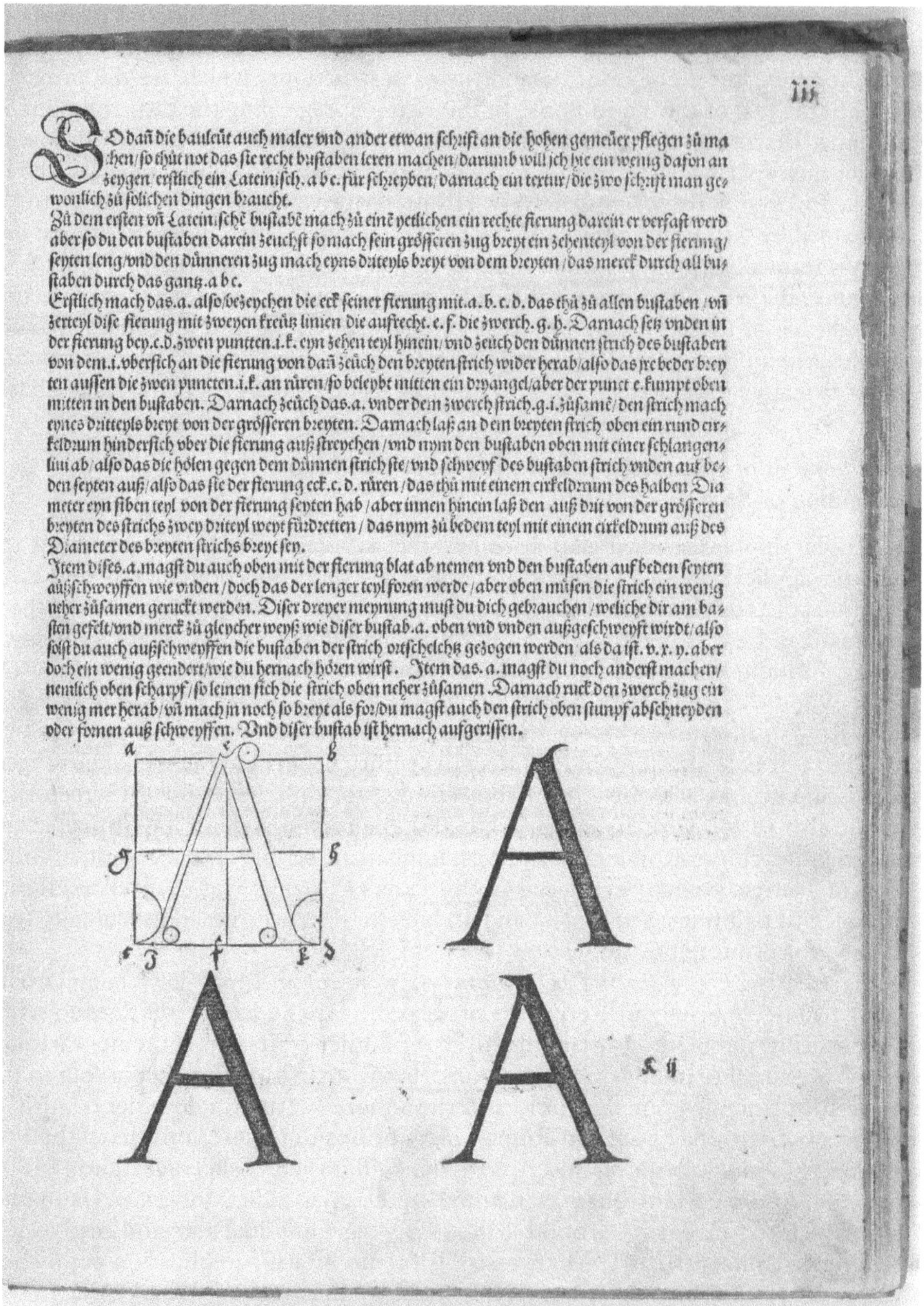
iii

SO dañ die bauleüt auch maler vnd ander etwan schrift an die hohen gemeüer pflegen zů machen/so thůt not das sie recht bustaben leren machen/darumb will ich hie ein wenig dafon anzeygen/erstlich ein Lateinisch.a b c. für schreyben/darnach ein textur/die zwo schrift man gewonlich zů solichen dingen braucht.

Zů dem ersten vñ Lateinischẽ bustabẽ mach zů einẽ yetlichen ein rechte fierung darein er verfast werd aber so du den bustaben darein zeuchst so mach sein grösseren zug breyt ein zehenteyl von der fierung/seyten leng/vnd den dünneren zug mach eyns driteyls breyt von dem breyten/das merck durch all bustaben durch das gantz.a b c.

Erstlich mach das.a. also/bezeychen die eck seiner fierung mit.a.b.c.d. das thů zů allen bustaben/vñ zerteyl dise fierung mit zweyen kreütz linien die aufrecht.e.f. die zwerch.g.h. Darnach setz vnden in der fierung bey.c.d.zwen puntten.i.k. eyn zehen teyl hinein/vnd zeüch den dünnen strich des bustaben von dem.i.obersich an die fierung von dañ zeüch den breyten strich wider herab/also das jre beder breyten aussen die zwen puncten.i.k. an rüren/so beleybt mitten ein dryangel/aber der punct e.kumpt oben mitten in den bustaben. Darnach zeüch das.a. vnder dem zwerch strich.g.i.zůsamê/den strich mach eynes dritteyls breyt von der grösseren breyten. Darnach laß an dem breyten strich oben ein rund cirkeldrum hindersich vber die fierung auß streychen/vnd nym den bustaben oben mit einer schlangenlini ab/also das die hölen gegen dem dünnen strich ste/vnd schweyf des bustaben strich vnden auf beden seyten auß/also das sie der fierung eck.c.d. rüren/das thů mit einem cirkeldrum des halben Diameter eyn siben teyl von der fierung seyten hab/aber innen hinein laß den auß drit von der grösseren breyten des strichs zwey driteyl weyt fürdretten/das nym zů bedem teyl mit einem cirkeldrum auß des Diameter des breyten strichs breyt sey.

Item dises.a.magst du auch oben mit der fierung blat ab nemen vnd den bustaben auf beden seyten außschweyffen wie vnden/doch das der lenger teyl foren werde/aber oben müsen die strich ein wenig neher zůsamen geruckt werden. Diser dreyer meynung must du dich gebrauchen/weliche dir am basten gefelt/vnd merck zů gleycher weyß wie diser bustab.a. oben vnd vnden außgeschweyft wirdt/also solst du auch außschweyffen die bustaben der strich ortschelchts gezogen werden als da ist.v.x.y.aber doch ein wenig geendert/wie du hernach hören wirst. Item das.a.magst du noch anderst machen/nemlich oben scharpf/so leinen sich die strich oben neher zůsamen. Darnach ruck den zwerch zug ein wenig mer herab/vñ mach jn noch so breyt als for/du magst auch den strich oben stumpf abschneyden oder fornen auß schweyffen. Vnd diser bustab ist hernach aufgerissen.

K ij

Figure 5.1 Albrecht Dürer. Constructing Roman Alphabet, 1525, folio *K2r*, woodcut, in the *Underweysung der Messung*. The George Khuner Collection, Gift of Mrs. George Khuner, 1981 (1981.1178.10), The Metropolitan Museum of Art, NY © The Metropolitan Museum of Art.

Photo credit: Art Resource, NY.

A closer look at, and a comparison between, Dürer's writing and similar Italian works, such as those by Pacioli and Feliciano, leave no doubt about the Nuremberg master's knowledge of the Italian works. As is generally acknowledged, one of the two principal sources of Dürer's theoretical dicta is an anonymous manuscript, *Ars literaria* (Cod. Lat. 451 in the Bayerische Staatsbibliothek, Munich), which once belonged to Hartmann Schedel (1440–1514), the author of the *Nuremberg Chronicle*.[22] The other acknowledged source is Pacioli's *De Divina proportione*. As is well known, Pacioli is a crucial figure in understanding Dürer's exposure to a range of theoretical ideas developing across the Alps at that time, as this Italian mathematician is the one most likely to have taught Dürer more about the secret art of perspective when the latter visited Bologna in 1506.[23] In 1507–8, in fact, Pacioli was preparing his own annotated edition of Euclid, and he might have been Dürer's point of contact with Leonardo da Vinci's ideas.[24] A summary view of Pacioli's dicta regarding alphabet letter formation is in order here to understand the Italian origins of Dürer's on the subject. In terms of proportion, Pacioli recommends 1:9 between the thickness of a letter's heavy limbs and its height, that is, one side of the square.[25] He also mentions that these proportional letters are in fashion on sculptures and monuments, and whether written or carved, their positioning in relation to the eye of the observer is critically important.[26] In addition, he stresses the proportions between individual letters, including their careful spacing on the same line and the distance between the lines.

Dürer's pronouncements on the subject of proper lettering echo Pacioli's in many respects, though not exclusively. Dürer states that lettering know-how is not just useful for calligraphers and typographers but also essential for stonecutters and architects, as they need to place inscriptions on the structures they erect. He writes that he prefers a proportion of 1:10, like most Italian theoreticians, but interestingly, he also advises the reader that his method is equally applicable for 1:9, thus clearly suggesting his acquaintance with Pacioli's work.[27] Despite its demonstration of the universal applicability of Euclidian geometry, the alphabet segment in itself may seem more in line with medieval "how to" manuals than with Renaissance articulations on the subject. However, as I will discuss, it is more complex and theoretically grounded than it may at first appear.

Alphabetic Letters on a High Wall as Metaphors for Vision

Though Dürer's method of lettering was not entirely original, he was certainly more than a mere synthesizer and disseminator of contemporary Italian opinions. Indeed, his adherence to conventional methods of forming proportionally correct letters can be assessed via approaches other than the originality of his ideas. This is especially true when we consider the logic of placing this particular dogmatic section that is focused on the proper geometric construction of alphabets in the Third Book. In spite of its seemingly digressive nature, with little relevance to the discussion of solid architectural components that precedes it, the segment is, in actuality, introduced according to the text's internal logic, and it follows the work's ample narrative structure.

Admittedly, Dürer had great difficulties in articulating and organizing his thoughts in his treatises. It should be recalled, however, that throughout the *Underweysung der Messung*, the functional use of applied geometry serves as an important link for the varying subjects found, and the math also facilitates points of digression into tangential subjects. The use of mathematics vis-à-vis architecture is thus an underlying theme utilized to illuminate such aspects as how to construct proper lettering in geometric terms, as found in the Third Book. In fact, this section on alphabet lettering forms an integral part of the book, not only due to Dürer's employment of a proportional system, using circles, squares, and division points, but first and foremost because of its fundamental subordination to general theoretical principles of perspectival geometry.

It is significant that the alphabet lettering section is immediately preceded by a page whose text essentially relates a corrective measure to counteract the rules of perspective. This information is provided with an accompanying woodcut (Figure 5.2), which illustrates an elevation of a high wall or tower inscribed with Roman capital letters (or majuscules) and adjoined by an Albertian diagram of vision. Dürer writes, "In order to be able to read the top line of what is written as well as the bottom one, it is necessary that the former be written in larger letters than the one below."[28] Elevated letters are not randomly enlarged, of course; instead, they are constructed with mathematical exactitude, as will be discussed in greater detail later. Reverberating Pacioli's statement that stresses the position of architectural letters in relation to the observer's eye while simultaneously articulating it with a didactic diagram of a perspectival triangle, step by step, Dürer explains how to determine the proper sizing of the letters on each line in relation to the other lines, as well as how to place them onto surfaces of tall structures through perspectival geometry, so that letters that are placed at different heights look both even and of equal size when viewed from the ground.

This description, which sits on a single page, is a crucial link and a segue in the Third Book. Here, Dürer explains how to solve the practical problem that an architect might face when inscriptions are placed on columns, towers, or high walls. Hence, this page that heralds the lengthy section of alphabet formation dealing with this practical problem regarding architectural structures is related—albeit loosely—to the preceding pages, with his discussion on the method of constructing a wall sundial.[29] In turn, the use of inscriptions, as a case in point accounting for optical distortions and ways to counteract them, allows Dürer to segue into his instructions about proper lettering in the following pages. In the past, more often than not, the section where Dürer painstakingly explains how to form the letters A through Z in the Third Book has been discussed in separation from this decisively theoretical page. However, in order to fully appreciate the theoretical ramifications of the alphabet lettering section, its close relationship with the immediate preceding page must be kept in mind. It is as though Dürer starts with a macroscopic glance, and then moves on to a microscopic analysis of each alphabet letter in the following pages.

Examining this diagram (Figure 5.2) closely, Roman capital letters are inscribed on a high wall of a tower and are represented as becoming progressively larger when moving from the bottom to the top, while at the same time, the spaces between the lines become wider towards the top. The full height of the tower is marked as *ab*. Dürer instructs the reader to place a standing person's viewpoint as far from the tower as possible, inscribing it as point *c*, and then to draw an arc by placing the legs of a compass on points *b* and *c*. The point of contact of this arc with the inclined line *ac* is then marked *e*. At this point, Dürer guides the reader to divide the arc *be* with the points that are spaced as far apart as the lines of the inscription (in this diagram, for the sake of clarity, the point *b* is double-marked), and to draw a web of straight lines, like rays, emanating from viewpoint *c* (conveniently illustrated with an eye in profile). These lines connect through all the points on the arc *be* to the vertical wall of the tower *ab*. Though not explicitly stated here, adjusting the

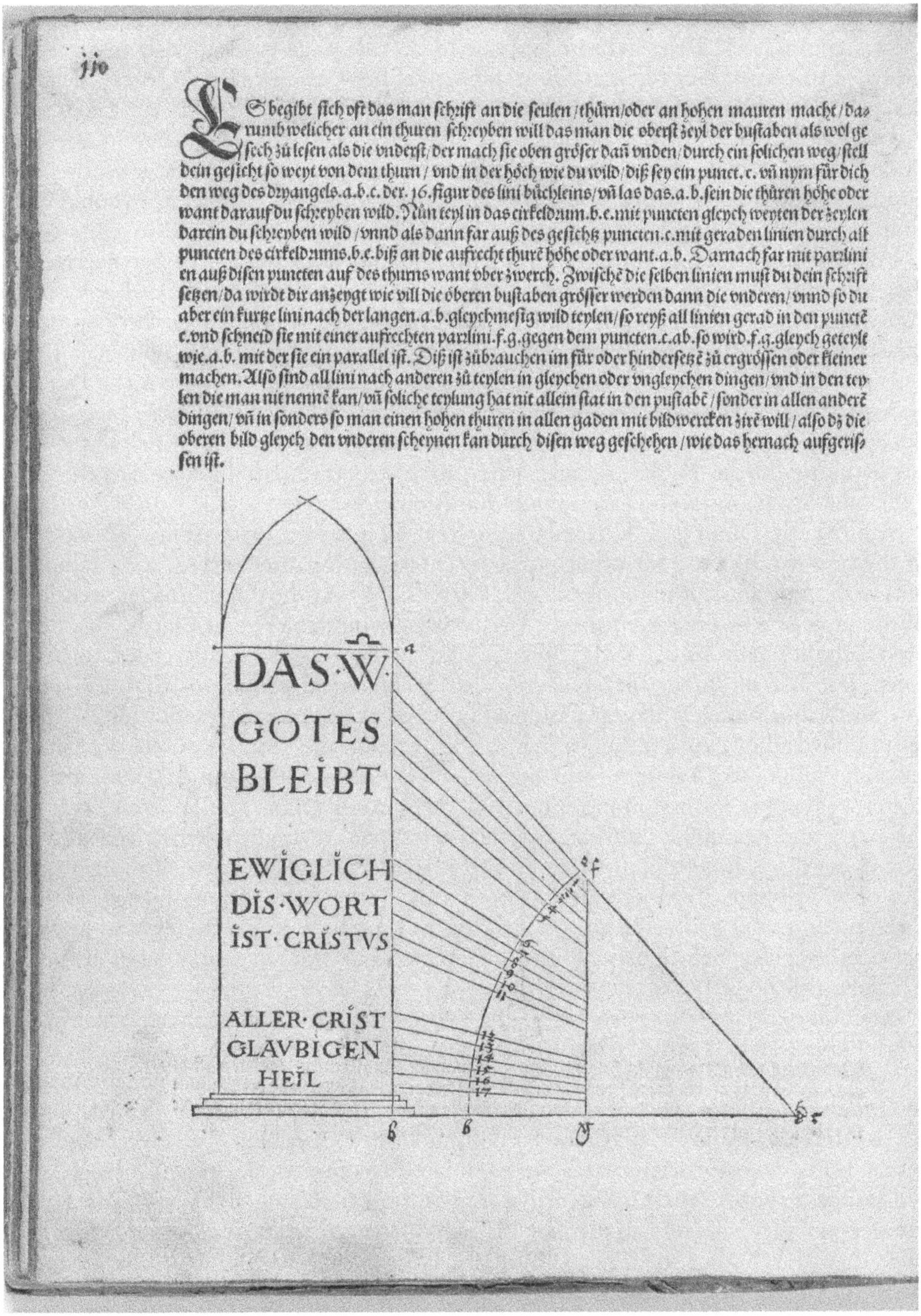

110

ES begibt ſich oft das man ſchrift an die ſeulen/thürn/oder an hohen mauren macht/darumb welicher an ein thuren ſchreyben will das man die oberſt zeyl der buſtaben als wol geſech zů leſen als die vnderſt/der mach ſie oben gröſer dañ vnden/durch ein ſolichen weg/ſtell dein geſicht ſo weyt von dem thurn/vnd in der höch wie du wild/diß ſey ein punct.c.vñ nym für dich den weg des dryangels.a.b.c.der.16.figur des lini büchleins/vñ las das.a.b.ſein die thüren höhe oder want darauf du ſchreyben wild. Nůn teyl in das cirkeldrum.b.c.mit puncten gleych weyten der zeylen darein du ſchreyben wild/vnnd als dann far auß des geſichts puncten.c.mit geraden linien durch all puncten des cirkeldrums.b.c.biß an die aufrecht thurē höhe oder want.a.b. Darnach far mit parlinien auß diſen puncten auf des thurns want vber zwerch. Zwiſchē die ſelben linien muſt du dein ſchrift ſetzen/da wirdt dir anzeygt wie vill die öberen buſtaben gröſſer werden dann die vnderen/vnnd ſo du aber ein kurtze lini nach der langen.a.b.gleychmeſig wild teylen/ſo reyß all linien gerad in den punctē c.vnd ſchneid ſie mit einer aufrechten parlini.f.g.gegen dem puncten.c.ab.ſo wird.f.g.gleych geteylt wie.a.b.mit der ſie ein parallel iſt. Diß iſt zůbrauchen im für oder hinderſetzē zů ergröſſen oder kleiner machen. Alſo ſind all lini nach anderen zů teylen in gleychen oder vngleychen dingen/vnd in den teylen die man nit nennē kan/vñ ſoliche teylung hat nit allein ſtat in den puſtabē/ſonder in allen anderē dingen/vñ in ſonders ſo man einen hohen thuren in allen gaden mit bildwercken zirē will/alſo dz die oberen bild gleych den vnderen ſcheynen kan durch diſen weg geſchehen/wie das hernach aufgeriſſen iſt.

Figure 5.2 Albrecht Dürer. Determination of the Size of Lettering on High Buildings, 1525, folio *K1v*, woodcut, in the *Underweysung der Messung*. The George Khuner Collection, Gift of Mrs. George Khuner, 1981 (1981.1178.10), The Metropolitan Museum of Art, NY © The Metropolitan Museum of Art.

Photo credit: Art Resource, NY.

sizes of the letters (or the height of each row of letters) is carried out by modifying the visual angles.[30] Thus, what is explained on this page is essentially a corrective measure to counteract the rules of perspective, now often referred to as "negative perspective."[31]

The method of applying perspectival geometry is explained earlier in the First Book—twice, in fact—when discussing how to construct and project spirals, then two additional times in the Third Book to demonstrate ways to construct both convex and round spiraling columns.[32] This explication is also closely associated to parts of the Fourth Book, where Dürer provides further theoretical principles of perspective and the different means of executing these tenets. He writes in the Fourth Book, "First you will have to determine the point of sight of the eye. Second, you will require the object, whether it is seen head on or in profile. Third, you require light [. . .]. The eye sees the object placed before it only by straight lines. [. . .] There must also be a proper distance between the eye and the object to be viewed."[33] These facts reveal, once again, the importance of perspectival geometry as one of the principal theoretical leitmotifs in the book, creating a narrative structure that interweaves theorems explained in various segments in the publication.

Significantly, Dürer ends his explanatory text on this page by writing, "This method of division can be used for other purposes, not only for inscriptions, for example, if you wish to decorate a high tower with artwork, so that the upper images seem to be of the same size as the lower ones."[34] What can be inferred here is that he considered alphabets not only as art forms, whose underlying geometric rules need to be put forth, but also as things that possess phenomenological properties that are subject to optical illusions, just like any other objects pertaining to representational art *and* nature. Instead of treating the geometric construction of letters as purely typographical and calligraphic matters, which exist unchanged on two-dimensional surfaces, Dürer weaves his discussion into the visual theory of his time, and he connects it with the larger and (seemingly) universal optical rules that are applicable to physical things occupying three-dimensional space. In other words, he uses the case of inscribing letters on a wall high above eye level as both a metaphor for vision and a case study of a representational problem, not to mention as a prelude to practicing and applying perspectival rules, before delving into such matters in more exacting detail at the end in the Fourth Book.

In summary, Dürer's approach to the geometric and mathematical construction of alphabet letters is multifarious. It is not a simple how-to exposition using circles, triangles, and squares alone, as it has so often been assessed; rather, it is integral to some of the main theorems of the *Underweysung der Messung*. In fact, his use of an Albertian diagram of vision in conjunction with a high wall decorated with inscriptions, whose varying sizes are calculated to counteract optical distortions at a distance, exposes Dürer's habit of seeing and rationalizing sight based on Euclidian geometry, which was shaped by his exhaustive investigations of an optical science then developing in Italy. As William Ivins proposed, the "rationalization of sight" is typical of Renaissance culture, enabling individuals like Dürer to see and understand natural phenomena as representational problems and to make inferences and anticipations based on Euclidian geometry.[35] It can be inferred that, for Dürer, geometry, or the art of measurement, is not a mere drawing tool to construct forms, but, more importantly, it is a rational instrument of cognition that reveals the order of nature.

Period Eye: Perspective Geometry and Constructive Geometry as Mental and Expository Models of Beauty

The way in which Dürer subordinates schematic forms of alphabets to the visual field at large—meaning three-dimensional space—is a case in point for applying what Bob Scribner calls a "logical system of representing the location of objects in space."[36] In large part, this was conditioned by his exploration of Euclidian geometry and visual models in optical science. Revealing the close connections between one's perception of the world, optical science drawing on Euclidian geometry, and representational art, the diagram by Dürer analyzed earlier, and indeed, the lengthy section on lettering in the *Underweysung der Messung* in general, reinforce the value and validity of the concept called "the period eye," put forth by Michael Baxandall.[37] In his *Patterns of Intention* (1987), Baxandall writes, "Living in a culture, growing up and learning to survive in it, involves us in a special perceptual training. It endows us with habits and skills of discrimination that affect the way we deal with the new data that sensation offers the mind."[38] Dürer was not born in a culture that preconditioned his way of seeing the world based upon Euclidean geometrical optics per se, but his arduous two-decades-long pursuit of discovering and mastering Italian art theory undoubtedly transformed and reshaped his habits of seeing. This is crystallized on the page in the *Underweysung der Messung*, displaying an Albertian diagram of vision, reflecting conceptual and perceptual attitudes that were more typical of contemporaneous Italians rather than his own German compatriots.

In the section on the geometric construction of alphabet letters, Dürer uses adjectives such as *recht* and *korrect*, rather than *hübsch* or *schön*, to describe what essentially can be characterized as canonical proportions of letters that were considered beautiful. The fact that beautiful letters were defined as something both measurable and apprehensible and thus explicable in terms of harmony, number, and geometry rather than simply intuitive and subjective, evinces the cultural significance of Euclidian geometry as a metaphor for beauty: a matrix for creating—and understanding—anything perceived to possess beauty and "perfection" in the Renaissance. As Heinrich Wölfflin put it in his *Art of Albrecht Dürer* in 1905, "Geometry ruled the spirit of that time. To be able to measure things meant to be able to understand them."[39] In Dürer's so-called "aesthetic discourse," which appears at the end of the Third Book in the *Four Books on Human Proportion*, Dürer writes, "For indeed, art is embedded in nature. He who can extract it has taken hold of it. Once captured, it will save you from many errors in your work. And a great deal of your work can be given certainty by geometry."[40] In other words, ancient mathematical knowledge was regarded as a type of container for the underlying secret of beauty. Simply put, Euclidian geometry served as the theoretical undercurrent of recognizing, reasoning, and elucidating what makes something beautiful.

Dürer was occupied—one might say, preoccupied—with the concept of beauty and explaining its underpinnings. This is testified to by his many references to the concept of beauty in his mature, later manuscripts, although his reflections on the subject dating from different time periods also demonstrate an evolution of his aesthetic philosophical beliefs. In one of his earliest pronouncements, for example, he writes, "Harmonious things are those which, when compared to others, are beautiful,"[41] although in a later, undated fragment, he confesses, "What is beauty, I know not. It depends on many things. And if we want to introduce beauty into our work, it is no easy task and we must search it out far and wide."[42]

Textur Type vis-à-vis Roman Type; the Deutch *vis-à-vis the* Welsch

Dürer's addition of the geometric construction of Textur type, in addition to Roman type, is important in this regard. It affirms the fact that he saw Textur type—what Italians saw as "barbaric" Gothic type—as an art form that was as beautiful as Roman type, thus divesting it of an Italian cultural bias. As a counterpart to what is essentially a judicial recapitulation of Italian theory regarding the construction of Roman alphabet letters, Dürer extends the same rational principles in ingenious and entirely new ways in order to uncover the geometric underpinnings of Textur type, thus explaining its inherent beauty.

The ramifications of his theorizing Textur type in mathematical and geometric terms are many. The desire to understand, measure, and intentionally (re-)create beauty in everything is tied to a historically constituted social practice. This reminds us of Dürer's recognition of how differing bodies can have different proportional balances and his admission of the coexistence of varying types of ideal beauty found within differing proportions. Furthermore, it also demonstrates that a seemingly innocuous exposition of underlying mathematical and rational rules of Textur lettering is not ideologically neutral. Indeed, it unveils a nationalistic sentiment of the "Deutsch" and their cultural struggles to maintain their identity vis-à-vis "the growing cult of the *Welsch*," as Baxandall puts it.[43] Dürer measures German art against the art of Renaissance Italy. The way in which Dürer pairs the Roman and Textur alphabets, as well as the ways in which he accounts for their proper proportions and formations via seemingly rational and universal geometry, in large measure reflects the various concerns of Humanist culture of his time.

Conclusion

This analysis of Dürer's section on alphabet lettering has demonstrated an important aspect of the author's more complex and theoretical ideas regarding perception. This section in the Third Book is more than a simple explanation of how to properly form alphabet letters. Instead, it is integral to the entire book, which collected and organized a rich and diversified amount of knowledge on the subject of the art of measurement. The importance of lettering and calligraphy, which held such great significance and a prestigious status as an art form in the culture of early modern Europe, seems to interest a disproportionately small number of scholars today. Such relative inattention to this section of Dürer's *Underweysung der Messung* regarding Roman and Textur alphabet letters at least partly reflects an increasing depreciation of handwriting and a general lack of interest in calligraphic and typographic matters in our own time. In addition, the prevailing scholarly interest in the multifaceted print culture of early modern Europe also accounts for the fact that this text segment has been largely ignored, other than regarding its contribution to the development of early modern typography, standardizing the latter through its provision of mechanically reproducible types—a justifiable assessment but, simultaneously, a simplistic and outdated one. As this chapter demonstrates, Dürer's segment on the geometric construction of alphabet letters in his *Underweysung der Messung* testifies to a much greater picture, namely, the period eye or cultural mode of knowing, seeing, and reasoning of his time, which is particular to this specific moment in time and place in the evolution of Cartesian epistemology. Dürer's moment is the juncture when anything and everything that was considered "right," meaning truthful and beautiful, was understood through the acquisition and application of mathematical knowledge. This is testified to by his use of the terms *recht* and *korrect* to refer to what we often describe today as "ideal."

In turn, Dürer's theoretical tract also allows us to better reflect on what is considered knowledge in our own time, as well as our tastes, our own period eye and epistemological condition, which are in sharp contrast to those of the Renaissance. Luis Radford characterizes historical epistemology "as the study of how what we take to be knowledge changes where knowledge is conceived as the conscious, self-creative activity of representing reality by means of models. These models constitute the a priori, but historically mutable, forms of human perception and cognition."[44] Dürer's section on lettering, like contemporaneous manuals by other writers of his day, shows us a peculiar early modern phenomenon: the privileged status of the gaze in conjunction with perspectival geometry. Everything, be it of nature or of human invention, is something that is fundamentally explicable through math, including the letters of an alphabet. From the viewpoint of his time, all can be systematically formulated into a body of knowledge through the objectively rational—the universal language of persuasion—that is, through the elegant rules of Euclid's geometry.

Notes

1 Erwin Panofsky, *The Life and Art of Albrecht Dürer*, 8th ed. (Princeton: Princeton University Press, 1995), 244.
2 Paul Lawrence Rose, *The Italian Renaissance of Mathematics: Studies on Humanists and Mathematicians from Petrarch to Galileo* (Geneva: Librairie Droz, 1975), 2.
3 Panofksy, *Life and Art*, 117–118; Martin Kemp, *The Science of Art: Optical Themes in Western Art from Brunelleschi to Seurat* (New Haven and London: Yale University Press, 1990), 55. This copy once owned by Dürer is now in the Herzog August Bibliothek, Wolfenbüttel; on its title page, Dürer inscribed: "I bought this book at Venice for one ducat in the year 1507. Albrecht Dürer."
4 See Walter Strauss, *The Painter's Manual: A Manual of Measurement of Lines, Areas, and Solids by Means of Compass and Ruler Assembled by Albrecht Dürer for the Use of All Lovers of Art with Appropriate Illustrations Arranged to be Printed in the Year MDXXV* (New York: Abaris Books, 1977), 27.
5 Hans Rupprich, *Dürer: Schriftlicher Nachlass*, 3 vols. (Berlin: Deutscher Verein für Kunstwissenschaft, 1956–69), 1:103.
6 Strauss, *Painter's Manual*, 12–15; Jeanne Peiffer, "Le style mathématique de Dürer et sa conception de la géométrie," in *History of Mathematics: States of the Art: Flores Quadrivii, Studies in Honor of Christoph J. Scriba*, ed. Christoph J. Scriba and Joseph Warren Dauben (San Diego, CA: Academic Press, 1996), 49–61. It is also known that Willibald Pirckheimer purchased many manuscripts from the libraries of Regiomontanus (d. 1476) and Bernhard Walther (d. 1504) pertaining to ancient mathematics.
7 See Jeffrey Chipps Smith, *Dürer* (London and New York: Phaidon, 2012), 362.
8 For instance, Strauss, *Painter's Manual*, 21–24. On Dürer's sources and the past scholarship about them, see the section in this chapter, "Dürer's Sources of Knowledge about the Proper Geometric Construction of Roman Alphabets."
9 Horst Heiderhoff, "Dürers Beitrag zur technischen Reproduzierbarkeit der Schrift," *Imprimatur* 8 (1976): 299–304.
10 About the *Underweysung der Messung*, see Rainer Schoch, Matthias Mende, and Anna Scherbaum, eds., *Albrecht Dürer: Das druckgraphische Werk*, vol. 3, *Buchillustrationen* (Munich, Berlin, London, and New York: Prestel Verlag, 2004), 168–278, especially 168–172.
11 Strauss, *Painter's Manual*, 9, 37.
12 Ibid., 12.
13 Smith, *Dürer*, 360.

14 For more on this subject, see John Sparrow, *Visible Words: A Study of Inscriptions in and as Books and Works of Art*, 2nd ed. (Cambridge and New York: Cambridge University Press, 1969, reprinted 2011), especially Chapters II ("The Inscription in Renaissance Works of Art") and III ("The Inscription as a Literary Form").

15 Regarding the development of Humanist scripts in Renaissance Italy, see James Wardrop, *The Script of Humanism: Some Aspects of Humanistic Script, 1460–1560* (Oxford: Clarendon Press, 1963).

16 For more on Alberti, see Giovanni Mardersteig, "Leon Battista Alberti e la rinascita del carattere lapidario romano nel Quattrocento," *Italia Medioevale e Umanistica* 2 (1959): 285–307.

17 Regarding Felice Feliciano's *Alfabeto* (Codex Vaticanus Latinus 6852), see Giovanni Mardersteig, ed., *The Alphabet of Francesco Torniello da Novara* (Verona: Officina Bodoni, 1971), 31–45; Mardersteig, "Leon Battista Alberti," 298–304; and Lucia A. Ciapponi, "A Fragmentary Treatise on Epigraphic Alphabets by Fra Giocondo da Verona," *Renaissance Quarterly* 32 (1979): 18–40, especially 19.

18 About Damiano da Moyle's *Alfabeto*, see Stanley Morison, *A Newly Discovered Treatise on Classical Letter Design, Printed at Parma by Damianus Moyllus circa 1480*, reproduced in facsimile with an introduction by Stanley Morison (Paris: Pegasus Press, 1927); Giovanni Medri, "Le opere calligrafiche a stampa. IV. Damiano Moile," *All'insegna del libro* I (1928): 172–179; Mardersteig, "Leon Battista Alberti," 304–305; and Mardersteig, *Alphabet of Francesco Torniello*, 60–61.

19 Sigismondo Fanti, *Theorica et practica [. . .] de modo scribendi* (Venice: Joannes Rubeus Vercellensis, 1514).

20 For more on Francesco Torniello da Novaria's *Opera del modo de fare le littere maiuscole antique* (Milan: Gotardo da Ponte, 1517), see Mardersteig, *Alphabet of Francesco Torniello*.

21 For a discussion of Luca Pacioli's *De Divina Proportione*, published by Paganino Paganini, see R. Emmett Taylor, *No Royal Road: Luca Pacioli and His Times* (Chapel Hill: The University of North Carolina Press, 1942), Chapter 15, 252–287; Giovanni Medri, "Le opere calligrafiche a stampa. I. Luca Pacioli e il trattato della Divina Proporzione," *All'insegna del libro* I (1928), 19–27; Mardersteig, "Leon Battista Alberti," 307; and Mardersteig, *Alphabet of Francesco Torniello*, X–XI.

22 About the *Ars literaria*, see Béatrice Hernad and Karl Dachs, eds., *Die Graphiksammlung des Humanisten Hartmann Schedel* (Munich: Prestel Verlag, 1990), 304–305; Strauss, *Painter's Manual*, 21–23; and Georg Dehio, "Zur Geschichte der Buchstabenreform in der Renaissance," *Repertorium für Kunstwissenschaft* 4 (1881): 269–279. Dehio believes the author of the method explained in this manuscript to be Leonardo da Vinci. Dehio cites Geoffroy Tory de Bourges; in his *Champ Fleury* (Paris, 1529), Tory accuses Luca Pacioli of having plagiarized Leonardo.

23 About Dürer's letter to Willibald Pirckheimer, written from Venice on October 13, 1506, see Rupprich, *Schriftlicher Nachlass*, 1: 58; 2: 78.

24 A copy of Pacioli's *Summa di Arithmetica, geometria, proportioni et proportionalita* (Venice, 1494) was in Pirckheimer's library. See Strauss, *Painter's Manual*, 13, 15.

25 Mardersteig, *Alphabet of Francesco Torniello*, 35–36, 57–60. Feliciano recommended the ratio 1:10 in his treatise. An analysis of Alberti's inscriptions, such as those on the frieze of the Holy Sepulcher in the Rucellai Chapel in Florence (1467) reveals that Alberti preferred a ratio of 1:12. In his *Alfabeto*, Damiano da Moyle also recommended 1:12.

26 See Ciapponi, "Fragmentary Treatise," 133–138.

27 Strauss, *Painter's Manual*, 23.

28 Ibid., 259.

29 Ibid., 257.

30 Kirsti Andersen, "The Mathematical Treatment of Anamorphoses from Piero della Francesca to Niceron," in *History of Mathematics: States of the Art: Flores Quadrivii, Studies in Honor of Christoph J. Scriba*, ed. Joseph W. Dauben, Menso Folkerts, Eberhard Knobloch and Hans Wussing (San Diego: Academic Press, 1996), 3–28, especially 4–5.

31 See Andersen, "Mathematical Treatment of Anamorphoses," 5. In *The Oxford Companion to Western Art*, Harold Osborne defines "negative perspective" as: "A term used to

describe the application of lines of sight to the adjustment of proportions in large-scale decorations, paintings, or statuary in order to counteract the perspective effect of the more remote parts." He uses Dürer's example, where "all the letters will appear equal to an observer placed at the viewpoint because they have been designed to subtend equal angles at the eye." See Harold Osborne, "Perspective," in *The Oxford Companion to Western Art. Oxford Art Online* (Oxford University Press), http://www.oxfordartonline.com/subscriber/article/opr/t118/e2024.

32 Strauss, *Painter's Manual*, 50–51, 64–65, 203–206, 209–214.

33 Ibid., 371; Rupprich, *Schriftlicher Nachlass*, 373; and Panofsky, *Life and Art*, 42–43. As Strauss mentions, this is a verbatim translation from Piero della Francesca's manuscript, *De prospectiva pingendi.*

34 Ibid., 259.

35 William M. Ivins, Jr., *On the Rationalization of Sight: With an Examination of Three Renaissance Texts on Perspective* (New York: Metropolitan Museum of Art, 1938), 7. Also see Kemp, *The Science of Art*; Samuel Y. Edgerton, *The Renaissance Rediscovery of Linear Perspective* (New York: Basic Books, Inc., 1974); Idem., *The Mirror, the Window, and the Telescope: How Renaissance Linear Perspective Changed Our Vision of the Universe* (Ithaca and London: Cornell University Press, 2009); and Karsten Harries, *Infinity and Perspective* (Cambridge, MA: MIT Press, 2001).

36 Bob Scribner, "Ways of Seeing in the Age of Dürer," in *Dürer and His Culture*, ed. Dagmar Eichberger and Charles Zika (Cambridge: Cambridge University Press, 1998), 94; Kemp, *The Science of Art*, 53–61.

37 Michael Baxandall, *Painting and Experience in Fifteenth Century Italy* (Oxford: Oxford University Press, 1972), 29–38.

38 Michael Baxandall, *Patterns of Intention* (New Haven, CT and London: Yale University Press, 1985), 107.

39 Heinrich Wölfflin, *The Art of Albrecht Dürer*, trans. Alastair and Heide Grieve and ed. Kurt Gerstenberg (New York: Phaidon, 1971), 283.

40 Strauss, *Painter's Manual*, 12; Rupprich, *Schriftlicher Nachlass*, 3: 295.

41 See Rupprich, *Schriftlicher Nachlass*, 2: 100.

42 Strauss, *Painter's Manual*, 12; Rupprich, *Schriftlicher Nachlass*, 2: 119.

43 Michael Baxandall, *The Limewood Sculptors of Renaissance Germany* (New Haven and London: Yale University Press, 1980), 135–142.

44 Luis Radford, "On the Epistemological Limits of Language: Mathematical Knowledge and Social Practice During the Renaissance," *Educational Studies in Mathematics* 52 (2003): 123–150, especially 141. Also see, Lorraine Daston, "Historical Epistemology," in *Questions of Evidence: Proof, Practice, and Persuasion across the Disciplines*, ed. James K. Chandler, Arnold Ira Davidson and Harry D. Harootunian (Chicago: University of Chicago Press, 1994), 282–289, especially 282.

6 Circling the Square

The Meaningful Use of Φ and Π in the Paintings of Piero della Francesca

Perry Brooks

In Memory of James Beck and David Rosand[1]

The physical, metaphysical, and esthetic associations of the irrational number known successively in mathematical history as "mean and extreme ratio," "the golden section," and φ ([1+√5]/2, or 1.618. . .)—based on the division of a line into two parts, such that the ratio of the whole to the greater part equals the ratio of the greater to the lesser part—fascinated certain commentators of the first half of the twentieth century, notably Matila Ghyka in writings such as his 1931 book *Le nombre d'or*.[2] The four diagrams of Figure 6.1 show ways of generating the proportion and some of its properties.

Diagram I is, in essence, the construction offered by Euclid in the *Elements* Book II Proposition 11. Diagram II shows a variation of the demonstration of *Elements* II.11 developed by Hero of Alexandria, a mathematician of the first century A.D., and transmitted to the West through medieval Arabic sources.[3] (Because of its simplicity, Hero's construction became the most common one in modern accounts of the golden section.) Diagram III reveals its pervasive presence in the equilateral/equiangular pentagon, whose construction was the earliest use to which Euclid applied the ratio (*Elements* IV.10–14, and XIII.7–8). The pentagon is constructed from an isosceles triangle whose two base angles are each twice its remaining angle. (In terms of the angular degrees—absent in Euclid, where everything is non-arithmetized—this is the triangle whose two base angles are 72° and whose peak angle is 36°.) For artists and those not concerned with rational proofs, the pentagon would likely provide the most mnemonically trenchant image of the golden section ratio.[4] Diagram IV shows another significant geometrical figure—the right triangle whose hypotenuse and base are in the golden section ratio (φ:1), and whose upright measures √φ—which, although not discussed in ancient sources (and apparently not before Johannes Kepler [1571–1630] in surviving writings),[5] can be derived by using Euclid's method of determining the geometrical mean between two lengths (*Elements* VI.13) when the relevant lengths are in the golden section ratio (φ:1). √φ is the geometric mean between φ and 1, and the resulting right triangle is the only one whose sides are in continuous proportion. The angle between the segments in the golden ratio measures 51°50'.

Ghyka and like-minded observers were prone to find φ everywhere and to produce self-generating analytic schemata that often seemed to lose touch with the natural objects or works of art whose beauty they purported to explain. (Among the more well-known and rather comical examples of such "harmonic analyses" are a dour

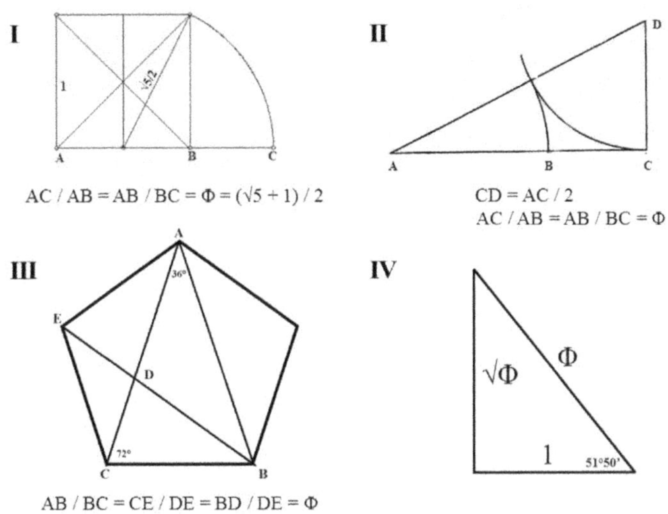

Figure 6.1 Diagram I: Generation of φ from a square, via a compass-swing based on the diagonal from the midpoint of a side to a corner of the square. Diagram II: Division by φ, via two compass-swings, of the side of a right triangle, where the side to be divided and its adjacent side form a right angle and are in the ratio 2:1. Diagram III: Angular measures and φ-ratios in the regular pentagon. Diagram IV: The right triangle with sides of φ, √φ, and 1, and the significant angle of 51°50'.

two-dimensional mechanical mask abstracted by Ghyka from a photo of—and presumably pinpointing the allure of—the comely face of tennis celebrity Helen Wills and a complex cat's cradle of lines and angles superimposed by Charles Funck-Hellet—at compounded levels of removal from its aesthetic stimulus—upon a reproduction of Cesare da Sesto's copy of Leonardo's lost *Leda and the Swan*.)[6] At mid-century Rudolf Wittkower warned that "more often than not, manipulating a pair of dividers has tempted scholars to exchange the thorny path of exact analysis for the easy road to amateurish exercises."[7] Wittkower, in his 1949 book *Architectural Principles in the Age of Humanism*, downplayed the use of irrational ratios like φ in Renaissance art in favor of the simple ratios like 1:2, 2:3, and 3:4, corresponding to the musical intervals of the octave, the fifth, and the fourth.[8] Yet when Wittkower teamed in 1953 with the painter B.A.R. Carter, an expert in the mathematics of perspective, to analyze the perspective of Piero della Francesca's *Flagellation of Christ*, they found pervasive evidence of modules interrelated by that other fascinating irrational number π (3.14 . . .)—the ratio of the circumference to the diameter of a circle—essential to finding a circle and square with equal areas.

Toward the end of the century, the historian of mathematics Roger Herz-Fischler produced a detailed historical account of φ in his 1987 book *A Mathematical History of Mean and Extreme Ratio*, beaming onto the subject a more realist light that tended to debunk metaphysical associations and artistic relevance of the ratio.[9] Herz-Fischler's hard-mindedness was a welcome approach but seemed out of tune with the meanings that might be found in φ by Renaissance mathematicians like Luca Pacioli, whose 1509 work *Divina proportione* was the first complete book devoted to the proportion.[10] The "neo-Platonic mooning" that the historian of science Lynn Thorndike frequently belittled in early modern science can be a fascinating source of meaning for the art historian, and in suggesting that Piero's work is not totally exempt from such interests, I am going against the current of the recent writings of two important Piero scholars, Judith Field and James Banker.[11]

Ancient sources are mostly silent in regard to the possible symbolic meaning of "extreme and mean ratio," which in early Greek mathematics was literally "unspeakable" because of its irrationality. (Pythagorean lore recounted the death by drowning of the first person to divulge the secret of surds.)[12] Yet a sacred aura seems implied by the fact that it was essential for constructing the pentagon, which in turn constituted a building block for the dodecahedron, the structural form of the heavenly Quintessence in the cosmology of Plato's *Timaeus*.[13] Within the dry rationalism of mathematical texts, the earliest emotional outbreak is that of Johannes Campanus, the thirteenth-century translator of Euclid's *Elements*. Discussing the golden section relationships between the dodecahedron and icosahedron, he declares,

> Wondrous, therefore, is the power of the line divided according to mean and extreme ratio; since very many things worthy of the admiration of philosophers come together there, this foundation or pre-eminence proceeds from the invariable nature of higher foundations, so that, with a certain irrational harmony, it rationally unites so many solids diverse in size, in number of sides, even in shape.[14]

Piero della Francesca used "mean and extreme ratio" in solving problems in his *Trattato d'abaco* and *Libellus de Quinque Corporibus Regularibus* and followed the pioneering example of the twelfth-century mathematician Leonardo Fibonacci in treating geometrical problems with a thoroughly algebraic approach and translating proportional relations into specific numerical quantities.[15] Like almost all of the mathematical treatments of the proportion, Piero is totally silent with regard to reasons for fascination with the proportion. This silence would be broken by Piero's admirer, "plagiarizer," and fellow townsman, the Franciscan friar Luca Pacioli—famously portrayed by Jacopo de' Barbari with a dodecahedron—who praises the proportion in all-out symbolic and allegorical terms.[16]

Pacioli extolled the proportion of φ as being like God in sharing the following five properties:[17] 1) Unity: there is one & only one. 2) Trinity: as there is one divine substance in three persons, so there is one proportion which must be found within three terms. 3) Incommensurability: as God cannot be defined or understood through words, so the divine proportion cannot be expressed through rational numbers. (The incommensurability that had been a stumbling block for the Greeks is here explicitly identified as a divine virtue for perhaps the very first time in mathematical history: "commo Idio propriamente non se po dìffinire nè per parolle a noi intendere, così questa nostra proportione non se po mai per numero intendibile asegnare, nè per quantità alcuna rationale exprimere, ma sempre fia occulta e secreta e da li mathematici chiamata irrationale."[18]) 4) Immutability: as God is unchanging all in all, and all in every part, so the divine proportion is invariable whether manifest in large or small, continuous or discrete quantities. 5) Source of being: As God gives form to the Celestial Virtue or Fifth Essence, and through this to the four earthly elements, so the divine proportion gives form—through the 12 pentagons of which it consists—to the dodecahedron, the form of Heaven, according to Plato.

Pacioli's approach to π, our other number under inspection, is more practical and optimistic, since he believed that this number (known today to be both irrational and transcendental) would someday receive a rational solution.[19] A more profound thinker, the fifteenth-century philosopher Cardinal Nicholas of Cusa (Nicolaus Cusanus), once shared this optimism but came to be less confident of a solution in the search to square the circle (at least on earth), yet the pursuit was both an object of practical research and of metaphorical contemplation on the relationship of the human intellect to divine truth. The determination of the number we call π engaged such mathematicians as Cusanus, Paolo Toscanelli, and Regiomontanus in exchange and argumentation, as part of a fifteenth-century revival of interest in the work of Archimedes. (Among James Banker's spectacular discoveries, a manuscript of Archimedes's works in Piero's hand in the Biblioteca Riccardiana, Florence, confirms the artist's part in this Achimedean renaissance.)[20] Archimedes had arrived at his well-known approximation $3^{10}/_{71} < \pi < 3^{1}/_{7}$ by using the "method of exhaustion," where ever-more-complex regular polygons alternately circumscribe, and then are inscribed within, a circle; the circle's circumference can be ever-more-precisely approximated in relation to the difference between that of the slightly larger circumscribing polygon and the slightly smaller inscribed polygon, whose measurements can be calculated through rational means. (Archimedes himself finally became exhausted with 96-sided polygons!)[21] Transferring this method to theology, Cusanus believed that through awareness of an "inapprehensible precision of the truth," the human intellect could approach the divine with ever greater propinquity but not gain equality except through a dissolution into it, as an infinitely sided polygon would dissolve into the circle.[22] Elsewhere, Cusanus explained the mystery of Christ's dual nature as the infinite polygon of his humanity melding into the circle of his divinity: "quasi utsi polygonia circulo inscripta natura esset humana & circulus divina."[23] Prompting his readers to transcend physical with immaterial vision, Cusanus wrote in one of his late works: "Those who have sought the squaring of the circle have presupposed coincidence in equality of the circle and square, which is certainly not possible in sensible things. For there is no given square which is not unequal to any given circle in matter. This equality, then, which they have presupposed, they have not seen with carnal eyes, but with mental ones."[24]

Cusanus' meditations are hardly translatable into pictures, yet to an intellectual and mathematically minded visual artist like Piero della Francesca, these ideas might imply the virtue of intentionality toward precision of proportional statement, along with an awareness of the inevitable frustration of that desire due to the inherent materiality of painting versus the immateriality of mathematical being. Euclid's first two definitions in the *Elements* are "A point is that which has no part" and "A line is a breadthless length."[25] The nonvisuality of this realm, as Leon Battista Alberti and Piero both knew, could be a stumbling block to the painters whose endeavor they wanted to bring into Euclid's domain. In a not easily translatable phrase, Alberti famously placed painting under the patronage of a "*pinguiore*" or "*più grassa*" Minerva—a fatter Minerva.[26] Piero della Francesca redefined the mathematician's dimensionless point as "the smallest thing the eye can see" in his treatise *De Prospectiva Pingendi*:

> Puncto è la cui parte non è, secondo i geumetri dicono essere inmaginativo . . . Et perchè questi non sono aparenti se non è a l'intellecto et io dico tractare de prospectiva con dimostrationi le quali voglio sieno comprese da l'ochio, perhò

> è necessario dare altra difinitione. Dirò adunqua puncto essere una cosa tanto picholina quanto è posibile ad ochio comprendere.[27]

Measurement of the fictive base molding of Piero's fresco of Mary Magdalene in the Arezzo Duomo (Figure 6.2) shows that in the most precise and nonimagistic passages of painting, the painter's mark measures a half-centimeter, defeating exact numerical quantification and referring us back to a visual rather than mathematical sense of the miniscule. Piero, we might add, almost surely knew his fellow townsman from Sansepolcro, Luca Pacioli, and perhaps knew Cusanus, who served as "vice-pope" in Rome in 1459 during the absence of Pope Pius II at the Council of Mantua while Piero painted

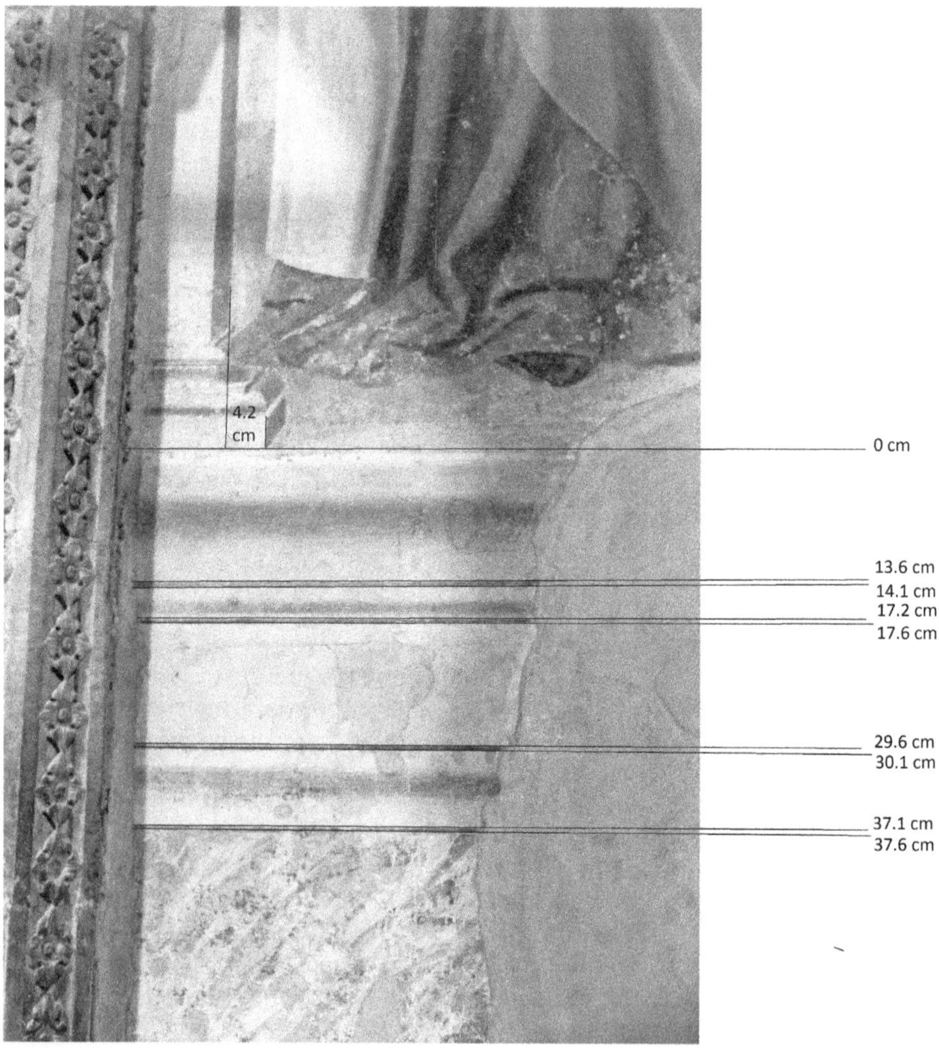

Figure 6.2 Piero della Francesca. *Mary Magdalene*. Duomo, Arezzo. Detail of base with measurements.

Photo credit: Perry Brooks.

frescoes in the Vatican Palace.[28] Thus it does not seem farfetched to find metaphorical use of φ and π in Piero's paintings, despite the absence of such in his writings; I would find it more surprising if Piero did not share this interest characteristic of his age.

The painting whose scale and finesse of execution most invite precise measurement is the *Flagellation* (Galleria Nazionale delle Marche, Urbino), the subject of the groundbreaking 1953 study by Wittkower and Carter mentioned previously, which attempted to construct a meticulously measured floor plan for the space of the image (Figure 6.3). Despite the clear sense of measured order in the painting, Piero introduced an ambiguity into this order (and an opportunity for conjecture on the part of interpreters) by apparently cutting the picture plane (the intersection of the visual pyramid) not precisely at one of the seams of the "terracotta" brick-colored tiles of the pavement but at a distance measuring six-and-a-fraction of the "terracotta" tiles from the closest white "marble" divider in the representation of the pavement. This gave the opportunity for interpreters, wishing to simplify things, to "move out" (toward the viewer) or to "move in" (toward the paintings depicted characters) in order to derive the picture's measurements. Carter (to whom Wittkower gives credit for the measurements) "moved out," placing the ideal intersection at a distance of seven tiles in front of the white divider. While applications of the golden section (φ) to analyses of Piero's work have been legion,[29] Carter seems to have been the first to detect a symbolic use of π. Carter identified the basic module of the architecture as a 1.85″ unit that measures, at the picture plane, one-half of the brick-colored tiles and one-third of the larger white marble pavers. (The proposed ratio of the width of white marble to red tile is thus 3:2, or 1.5) When multiplied by π, the 1.85″ module produces a new "mystic" unit of 5.8″. The distance of the ideal viewer's eye to the picture plane measures ten "mystic" units, the distance from the picture plane to the center of the circle in which Christ stands measures 20 "mystic" units (at the scale of the picture), and all the other figures are placed at multiples of the unit as well. As Wittkower commented (295): "This seems to be more than chance: Piero may have chosen this curious relation between the module scale and the 'mystic' scale to symbolize the interweaving of this worldly space with that belonging to the Kingdom of Christ." Carter attempted to ground the basic 1.85″ module in fifteenth-century measurements by referring it to a ten-unit line printed twice in the margins of Pacioli's *Divina proportione*, without apparently reading Pacioli's text. (Carter measured Pacioli's unit as 0.74″, and the 1.85″ module as 2.5 such units.) Pacioli's text explains that the line's purpose is relative, provided to give the reader a concrete idea of the height of his friend Leonardo da Vinci's ongoing Sforza equestrian monument, which is 12 braccia high and 37⅘ times the line in the margin. Carter's 1.85″ module becomes the highly contingent figure of 2½ of ⅒ of 1/37⅘ of 12 *braccia Milanesi*![30] Had Carter referred to Quattrocento units—as he later would do in his analysis of Piero's *Baptism* appended to Marilyn Aronberg Lavin's monograph on the painting—he might have discovered another instance of the use of π: Carter found that the height of a depicted male standing forwardmost at the picture plane would measure 22″, and—referring this to the 6-foot or 72″ Anglo-Saxon ideal—identified the scale of the picture to the scale of the world as 22 to 72, or 1 to 3.27. It went unnoticed that if referred to the ideal Renaissance male height of three Florentine *braccia*—equal to 175 cm, or 69″—the scale of the picture to the scale of the world is 22 to 69, or 1 to 3.14; in other words, *we* relate to the picture as π to one!

Twenty years after the Wittkower-Carter study, Warman Welliver published a methodologically suggestive article also employing π in the analysis.[31] Welliver "moved in" to place the ideal picture plane at a distance of six brick-colored tiles from the white paver (Figure 6.4). Welliver sought to root his measurements in fifteenth-century units

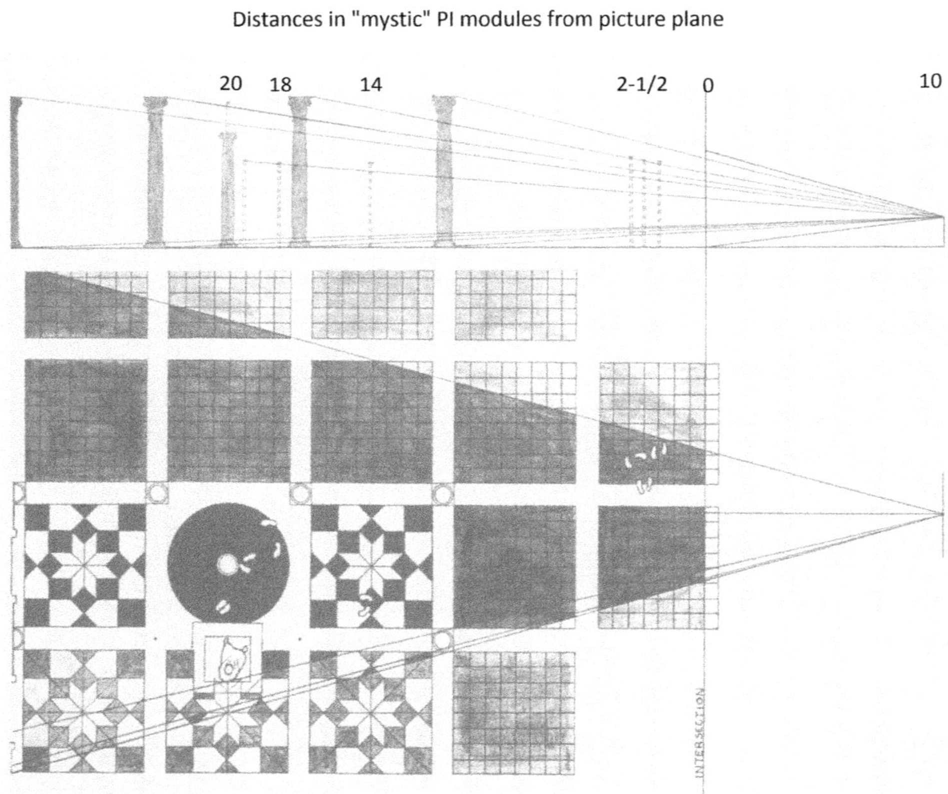

Figure 6.3 Piero della Francesca. *Flagellation*. Galleria Nazionale delle Marche, Urbino. Hypothetical Floor Plan as drawn by B.A.R. Carter (with indication of *Π*-module distances provided by present author). *Journal of the Warburg and Courtauld Institutes* 16 (1953), pl. 44.

Photo credit: B.A.R. Carter Estate.

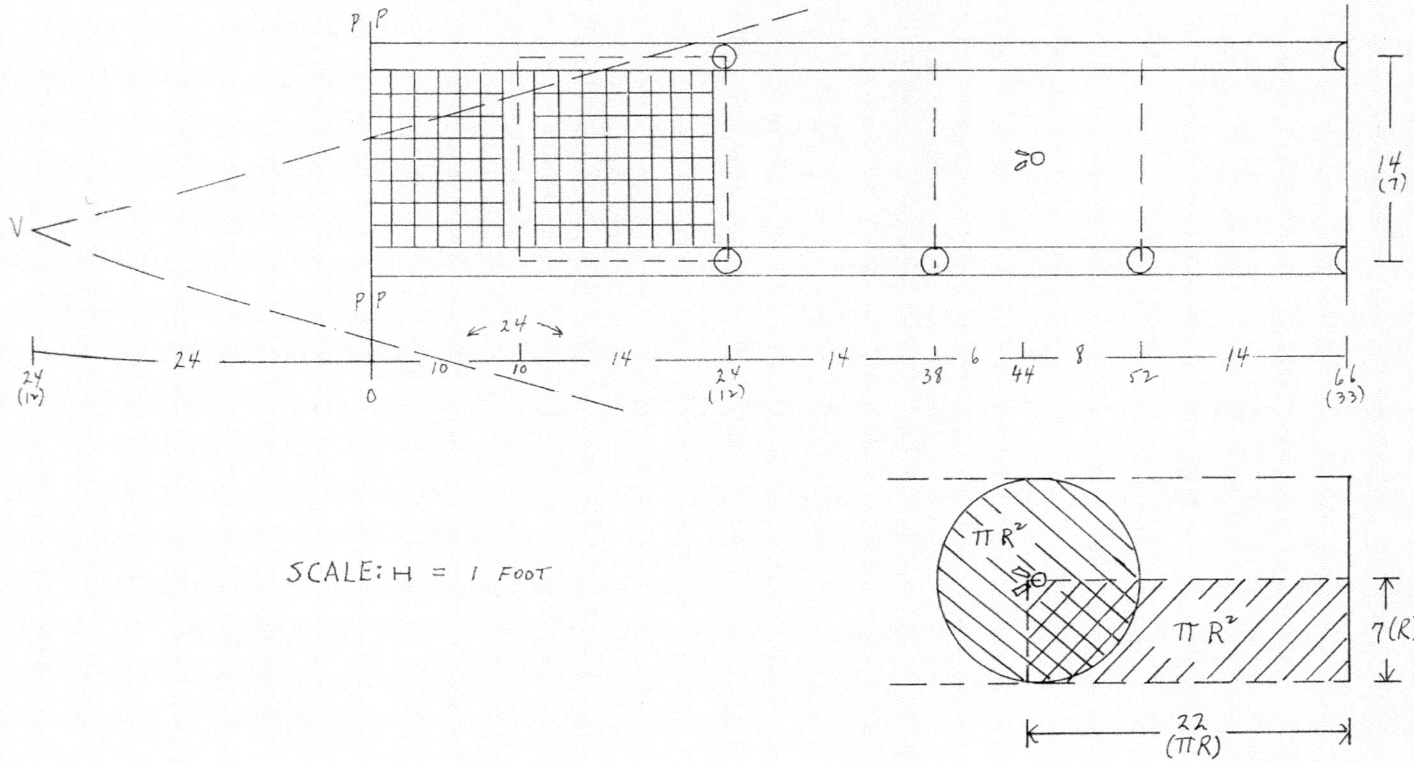

Figure 6.4 Piero della Francesca. *Flagellation*. Galleria Nazionale delle Marche, Urbino. Hypothetical Floor Plan of left side according to Welliver. The ratio of the distance from the picture plane to the back wall compared to the distance of the Column of the Flagellation to the back wall equals π ($^{66}/_{21}$, or $3^{1}/_{7}$). *Art Quarterly* 36 (1973), fig. 12.

Photo credit: Lucy Welliver Scanlon.

like the Florentine *braccio* (or cubit) and *piede* (or foot, equal to a half-*braccio*), to relate them to Renaissance beliefs that the ideal body measures 3 *braccia* or 6 *piedi*, and to read depicted characters as inviting measurement through their bodily display and gestures. (One of Welliver's examples is the figure of St. John the Baptist in Domenico Veneziano's *St. Lucy Altarpiece* [Uffizi, Florence], who offers the profile of his foot as a measure, both for his own height and for the architecture, and points to the Christ Child in a demonstrative gesture that is, simultaneously, a gauge for the rate of foreshortening of the distance between columns behind him.) For Piero's *Flagellation* Welliver took the foot of one of the foreground bystanders as module. He measured the red tiles as 1½ feet wide and the white marble dividers as 2 feet wide—thus recalibrating the proportions of white divider to red tile from Carter's ratio of 3:2 (or 1.5) to a ratio of 4:3 (or 1.33). (Measurements taken in pixels from digital reproductions suggest a ratio of the width of the white marble dividers to the red tiles as 1.36–1.37, i.e., somewhere between 3:2 and 4:3, but closer to Welliver's figure than to Carter's; such measurements from reproductions might involve inaccuracies due to any convex curvature of the wooden panel.) Welliver interpreted the left-side space of the scene as based on square units of 14 × 14 feet, measured from the central axis of the white marble paving dividers. He measured the picture plane as 66 feet from the back wall and the column of the Flagellation as centered 21 feet from the back wall; $^{66}/_{21}$ being equal to $^{22}/_{7}$, the ratio is π. Christ can be interpreted as standing a foot in front of the column, or 22 feet from the back wall. If one calculates the area of a circle centered on the column and inscribed within the larger 14-foot square, the measure will be π times the radius squared, or $^{22}/_{7}$ times 7^2, or 154; Christ also stands at the corner of a rectangle measured from the axis of the column, the axis of the white paver, and the back wall whose dimensions are 22 by 7, and whose area is 154. Welliver concludes (19): "Thus the moral of the geometry of Piero's floor plan appears to be that Christ's tormentors, in binding Him to this column, have unwittingly brought Him to precisely the point where one of the great mysteries of classical mathematics, the rectangling of the circle, would testify to His divine nature." It should be noted that the circle is an invisible one, though the inattentive reader might identify it as the visible one in the dark paving on which Christ stands, whose radius is 6 feet. Despite Welliver's starting point in the 6-foot ideal height that does plausibly apply to Domenico Veneziano's stocky St. John, Welliver does not comment on the fact that Christ is more than a head taller than the 6 feet measured by the radius of the pavement circle.

Both of these articles involve elements of simplification and conjecture, and they cannot both be valid *in toto*, yet both fuel the interpretive imagination. Inspiration preceded minute mathematical analysis when they influenced my looking, as a younger art history graduate student, at some of Piero's works. (Coming of age in the 1970s, I was also spurred, perhaps subconsciously, by some of the art I first encountered at the time, such as the wall drawings of Sol Lewitt, prompting musings about an art that could be both richly sensuous and yet somehow existent in an invisible and "mentally portable" realm.) Welliver's hypothesis of the presence of beckoning "gaugers" among a painting's depicted characters proved particularly fruitful in analyzing Piero's *Resurrection*, where significant measure configures the surface design without resorting to the hypothetical virtual spaces that seem problematic in the two inspirational articles. (My analysis was subsequently confirmed by measurement *in situ*.)[32] Piero's painting, as has often been noticed, is a visual rethinking of an iconic fourteenth-century altarpiece in his hometown, now attributed to Niccolò di Segna.[33] Piero's cogent restructuring utilizes both φ and π to order, coordinate, and accentuate elements of the image, and as Wittkower and Carter and Welliver suggested, endow his picture with the mark of the numinous.

Christ's triumphal tread on the top of the tomb is both measured and measuring. The reader—invited to divide the width between the columns at the level of the top of the tomb according to Hero's construction (Figure 6.1, Diagram II) on a reproduction of the fresco—will find that the "φ-cut" marks the axis of Christ's left foot. (When the direction of division is reversed, the φ division defines the axis of the sleeping soldier directly at Christ's right, assuring his significance in the work.) The smaller section of this division is equal to the height, on the picture plane, of the sarcophagus from the architectural base.[34] If the axis, from architectural base to tomb-top, of the sleeping soldier at Christ's right is divided according to the golden section, the division falls at a nodal point of the figure, pinning the center of his breast.

Christ's tread—measured with respect to the picture's width—simultaneously measures significant lengths on the top of the tomb. To his right, the length is equal to the height that his body rises above that tomb; to his left, the shorter length is in approximate φ-ratio to the length to his right. The lance-bearer to Christ's left joins in the revelation of measure by marking off the same longer length on the lance above his hands. The lance-bearer signals the presence of φ by the angle at which he tilts his lance; measuring 72° with the tomb-top, this is the angle of the diagonals that connect the points of a pentagon and crisscross in φ-ratios, as seen in Figure 6.1, Diagram III.[35] (Extension of the contour of Christ's upper left arm to meet the horizontal of the tomb-top reveals another significant φ-related angle: the nearly 52° angle between the base and hypotenuse of the $\varphi - \sqrt{\varphi} - 1$ triangle seen in Diagram IV.) Below his hands, the lance-bearer marks off a length equal to the height of the top molding of the sarcophagus (and approximately one half the smaller φ division from center of the sleeping soldier's breast to the tomb-top). Reexamination of the φ-related proportions of the lower part of the fresco suggests that the artist used a 5-by-13 grid to approximate them (Figure 6.5). Five and thirteen are members of the Fibonacci series of numbers (1, 2, 3, 5, 8, 13 . . .), where each new number is the sum of the previous two and where the ratio of each number to the one before comes ever more proximate to φ; this series had entered twelfth-century mathematics as a solution to the problem of determining the reproduction rate of rabbits, with no mention of a connection to the golden section.[36] According to Leonard Curchin and Roger Herz-Fischler, the earliest surviving written connection of the series to φ is an anonymous early sixteenth-century annotation in a copy (located in the Bibliothèque Nationale, Paris) of Pacioli's 1509 edition of Campanus' translation of Euclid's *Elements*; the painting offers an earlier *visual* text.[37]

The height measured off on the top of the lance is one half the height of space seen behind the fictive framing entablature; it forms the radius of a circle whose area closely approximates that of a square whose side is the horizontal dimension measured between the columns at the top of the tomb, proportionally related through π (Figure 6.6). (The crossing on Christ's banner marks the height of the implied

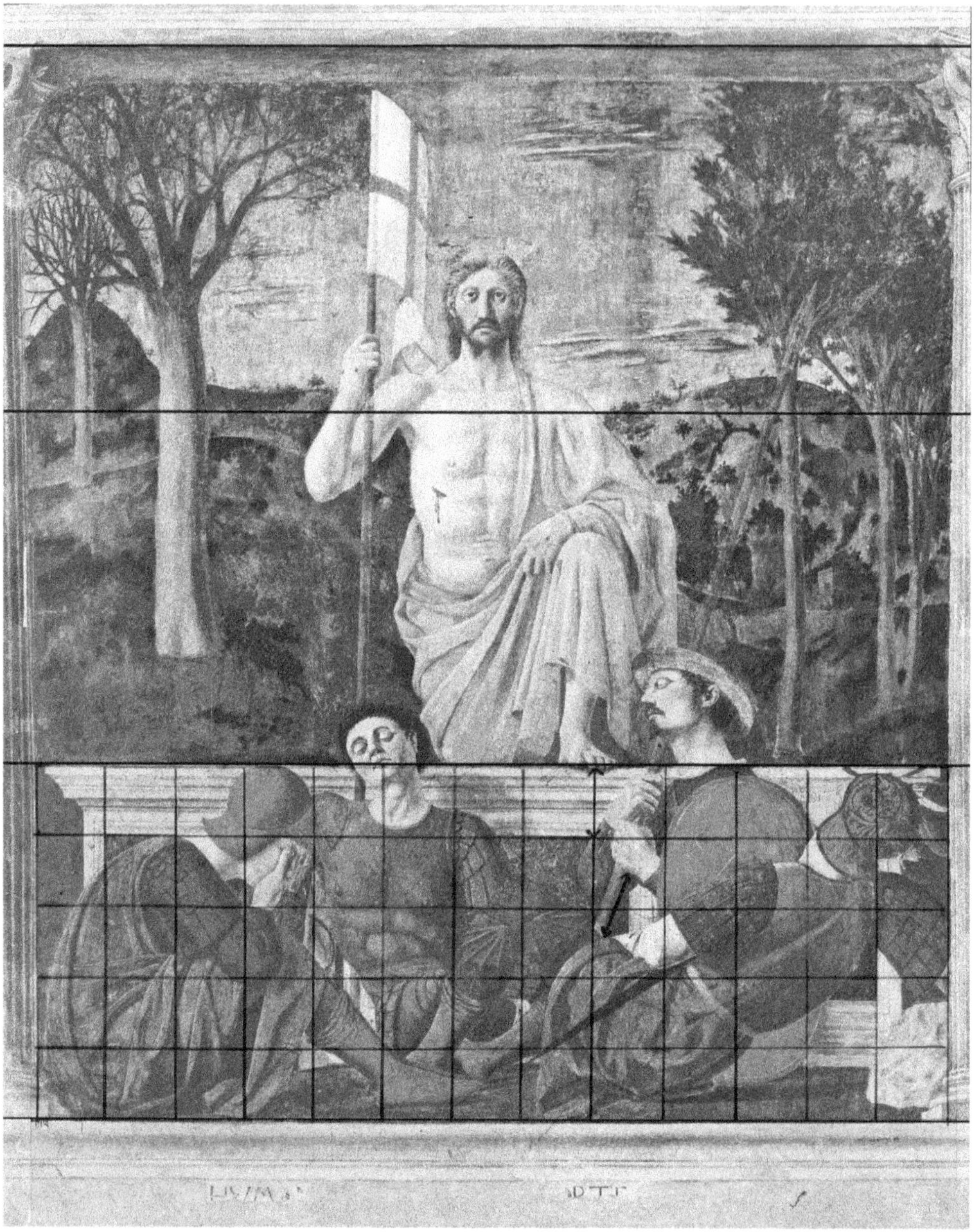

Figure 6.5 Piero della Francesca. *Resurrection*. Museo Civico, Sansepolcro. Modular schema to be consulted in conjunction with the Appendix. The module measures approximately 15.5 cm and is marked off on the part of the lance visible beneath the lance-bearer's hands. The height of the fictive entablature on the viewer's side is three times the distance from the fictive architectural base the top of the tomb. Author's analysis based on Millard Meiss, *The Great Age of Fresco* (New York: George Braziller, 1970), 141.

Photo credit: Scala/Art Resource, NY.

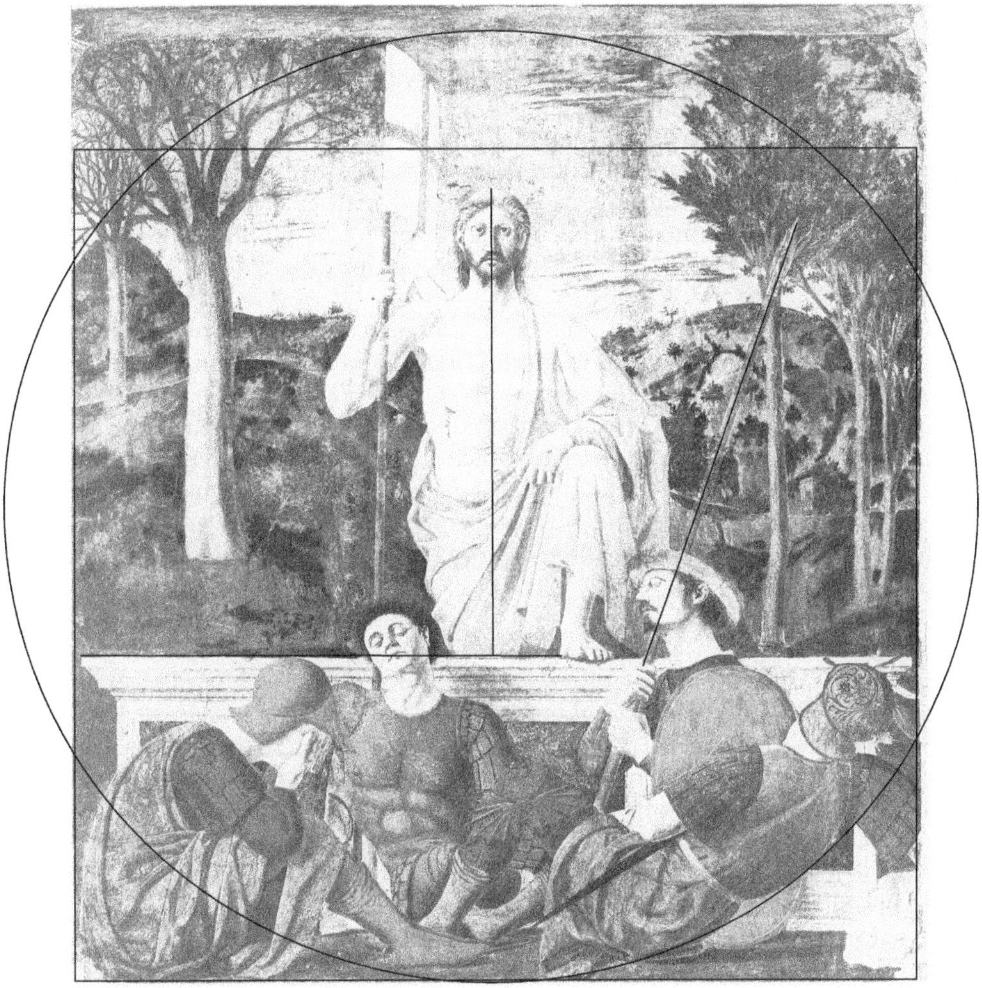

Figure 6.6 Piero della Francesca. *Resurrection*. Museo Civico, Sansepolcro. Conceptual diagram to be consulted in conjunction with the Appendix. The square based on the width between the columns closely approximates in area the circle whose height determines the height of space seen beyond the architectural enframement. The visible height of Christ's body, the length marked by his foot on the top of the tomb, and the top part of the lance measure the radius of this circle. Author's analysis based on Millard Meiss, *The Great Age of Fresco* (New York: George Braziller, 1970), 141.

Photo credit: Scala/Art Resource, NY.

square, implicating the idea of intersection.) The two geometric figures of square/circle initiate a semiotic resonance of associations: horizontal/vertical, earthly/celestial, temporal/eternal, mortal/immortal, human/divine. . . . Measurements of the fresco yielded values of 112–113 cm for the circular module (r) and 200.2 cm for the square-related width between the columns (x), values which correspond to the well-known formula for "squaring of the circle" ($x^2 = \pi r^2$, where x is the side of the square and r is the radius of the circle).[38] Nowhere are the implied circle, square, and pentagon seen *in toto,* however, as a design trussed upon the surface. Instead, Piero evokes the presence of the dimensions through a principle of *mathematical synecdoche* (the part for the whole).

Piero's proportional structure—highlighting Christ's victorious tread and bringing earth and heaven into proportion through the common presence of φ in both the square and circular dimensions—distills and mathematically fixes the essentials of St. Paul's fundamental text on the Resurrection (conflated with the Last Judgment):

> But now is Christ risen from the dead, and become the firstfruits of them that slept. For since by man came death, by man came also the resurrection of the dead. For as in Adam all die, even so in Christ shall all be made alive. But every man in his own order: Christ the firstfruits; afterward they that are Christ's at his coming. Then cometh the end, when he shall have delivered up the kingdom to God, even the Father; when he shall have put down all rule and all authority and power. For he must reign, till he hath put all enemies under his feet. The last enemy that shall be destroyed is death. For he hath put all things under his feet.
>
> (King James Vers., I Cor. XV.20–27)[39]

Using the dimensions of the earthly square, Piero proportions a box of mortality that traps the two farthest guards, waking in poses fashioned after captive barbarians and fallen soldiers in ancient Roman art.[40] Christ—whom St. Paul called *primitiae dormientium,* the firstfruits of the sleeping, and who marks the axis of revivification for the background trees—offers a dimension for "circling the square"—transforming mortality into immortality—in his resurrection, and the two sleeping guards nearest him rise above the tomb. Sleep, in St. Paul's text and in the painting, symbolizes death, but also, we may infer, baptism, the ritual enactment of death and rebirth. These figures' state of *vacatio* withdraws them from the mundane and orients them toward the supernatural, since, as Charles de Tolnay perspicaciously observed, these two figures "semblent dormir et rêver de la Résurrection."[41] With a paradox comparable to the title of one of Nicolaus Cusanus' tracts—*De Docta Ignorantia*—Piero reverses the traditional iconographical pattern of the Resurrection, in which a waking soldier serves as witness to the event; for it is the two sleeping soldiers, one senses, who are truly perceptive.

The figure to Christ's right, traditionally regarded as a self-portrait and axially pinned in the proportional structure, enters the space above the tomb at the level of his eyes and mind, a confident self-image for a devout painter/mathematician.[42] The lance-bearer, rising above the tomb at the line of his neck, is to be identified as Longinus, the converted centurion who—miraculously cured of blindness by a drop of blood from the wound he inflicted on Christ's side at the Crucifixion—appears in Byzantine legend as one of the guardians of the tomb and later received the martyr's palm through decapitation.[43] (Longinus was in the news around 1459 due to his veneration by Pope Pius II in Mantua, the site of an important church council.)[44] Punningly guillotined by the line of the tomb-top and haloed with his circular helmet, the saintly and metrical intercessor offers the picture's key dimensions through the Holy Lance, which crosses, in another visual *double-entendre*, the leafy background tree, a traditional metaphor for Christ.[45] At the level of the tomb, he offers a unit—partial and incomplete, as befits the earthly realm—for constructing the square of mortality but offers the radius of the celestial circle in fullness above the tomb. One moves from square to circle, whether rising vertically within the picture or entering the picture from without to within: Piero proportioned the height of the fictive entablature above the base on the viewer's side as equal to three times the height that the sarcophagus rises above the base (Figure 6.5); he proportioned the inner face of the entablature on Christ's side according to the diameter of the circle (Figure 6.6). Of course, the earthly equivalencies of square and circle are only approximations, since, as Cusanus wrote, "This equality, then, which they have presupposed, they have not seen with carnal eyes, but with mental ones."[46] Behind the plotting of the fresco's structure there lies an analogy: "As π links square with circle, so Christ links man with God." Founded on a meditation on types of vision, both the revelation of proportion and the manipulation of perspective literally translate words of St. Paul: "For now we see through a glass, darkly; but then face to face: now I know in part; but then shall I know even as also I am known" (King James Version, I Corinthians 13:12).[47] As Kenneth Clark was the first fully to articulate about the painting's perspective, Piero "deliberately made use of two points of vision. In the lower part the vanishing point is somewhere in the sarcophagus, so that we see the heads of the sleeping soldiers from below. But in the upper two thirds, our point of vision rises till it is level with the head of Christ. His figure is therefore able to maintain its frontal majesty, as if he were a Pantocrator from the apse of a Byzantine cathedral."[48] With mental eyes we see *facie ad faciem*.

I am willing to accept the possible role of the *Resurrection* as somewhat anomalous, the work by Piero most impacted with mathematical meaning. Like some other scholars, I accept a date of 1460 or later, after Piero's likely engagement in 1459 (and perhaps earlier) with the mathematical culture of papal Rome.[49] Yet, without shoehorning the variety of Piero's paintings into a formula, is it possible to identify some of the tendencies and mechanisms at work in the *Resurrection* in the artist's other works? These mechanisms include the enactment of measurement by beckoning characters in a painting, revelation through synecdoche, and liminal control of sacred space. Are these principles to be found in other of his works? Our concluding paragraphs will offer some *aperçus*.

In the *Nativity* (National Gallery, London), probably his latest surviving work, the two shepherds enter carrying staffs that meet at an angle of just under 52°, the same angle at which the straight line of Christ's upper left arm in the *Resurrection* would meet the horizontal of the tomb-top and the angle between the hypotenuse and base of a right triangle, where the ratio of hypotenuse to base equals φ:1 and where the other side of the triangle equals $\sqrt{\varphi}$ (Figure 6.1, Diagram IV).[50] This synecdochic detail, in a seemingly more relaxed composition than that of the *Resurrection*, may relate to some overall proportional organization, but I am willing to accept the shepherd's message as a more generic one: his one hand reveals divine measure, just as his other hand points to divine light.

In Federigo da Montefeltro's votive Urbino altarpiece of the *Madonna and Child with Saints and Angels* (Pinacoteca di Brera, Milan), St. John the Baptist points to three potential targets: the sleeping Christ Child, Battista Sforza's absence, and the ratio measured on his staff.[51] Measured from photographs, this seems to fall between 1.43 and 1.44, which encompasses two numbers of possible fascination, $\sqrt[3]{3}$ (approximately 1.44225) and $\sqrt{(\varphi\sqrt{\varphi})}$, the geometric mean between φ and $\sqrt{\varphi}$ (approximately 1.435). The code remains uncracked, yet some meaning seems to lurk in the ratio, since the upper part of the staff measures the image on the picture plane that the famous egg hangs from the half-dome's shell, and the lower part measures the length of the Christ Child's body.

An instance of sealing the limits of a picture in sacred measure occurs in a little-noticed detail of the *Annunciation* in the *True Cross Cycle* (San Francesco, Arezzo). All the other scenes in the chapel extend the full height between the fictive framing cornices that define the chapel's three tiers. In the *Annunciation*, however, the holiest scene of the chapel—indeed, an anomaly in its inclusion when compared to other cycles of the Legend of the True Cross—Piero adjusts the proportions by lifting the scene on a platform formed by a maroon stripe above a larger grayish-green one.[52] This separates the central column from the fictive plinth below and shortens the field so that the ratio of height to width is about 1.6, approaching φ. The irregular wavering edge of the chapel's corner at the left of the fresco frustrates precision and suspends this effort in the realm of intentionality. His approach here seems opposite to that of a realist like Andrea Mantegna: Piero distances the column from the architecture (and the scene from it narrative setting) and abstracts the image through proportion, whereas Mantegna, in the destroyed scene of the *Martyrdom of St. Christopher* (formerly Ovetari Chapel, Padua), rested a fictive central column directly on the plinth, as if to equate an image's sacredness with the intensity of its physical presence and proximity.

The fresco depicting Mary Magdalene (Duomo, Arezzo) offers the rare case of a nearly intact enframement in Piero's oeuvre, and the best opportunity to experience the pictorial mechanics of entry into a sacred space in its original setting (Figure 6.7). As we stand on the floor of the Arezzo Duomo staring upward, the projecting base on the lower forward side of the arched opening holds us at bay from the saint's space. Though missing a large patch of intonaco, the measurements of the lowest opening can be reconstructed from the intact side, since the perspective of the scene is symmetrical; this opening originally measured about 69.1 cm.[53] The final transition into the space is measured by the inner edges of the opening which rise into the haloing arch; this measures 77.5 cm. The area of circle with a diameter of 77.5 equals the area of a square with sides of 68.7. (The variance of 2 mm per side seems tolerable in the realm of the "fatter Minerva.") Moving from without to within, the square dimension is our gateway into the space of the picture and then dilates into the framing circle. Through π, the fictive architectural frame becomes a mediating diaphragm between tangent realms of reality, of earth and of heaven.

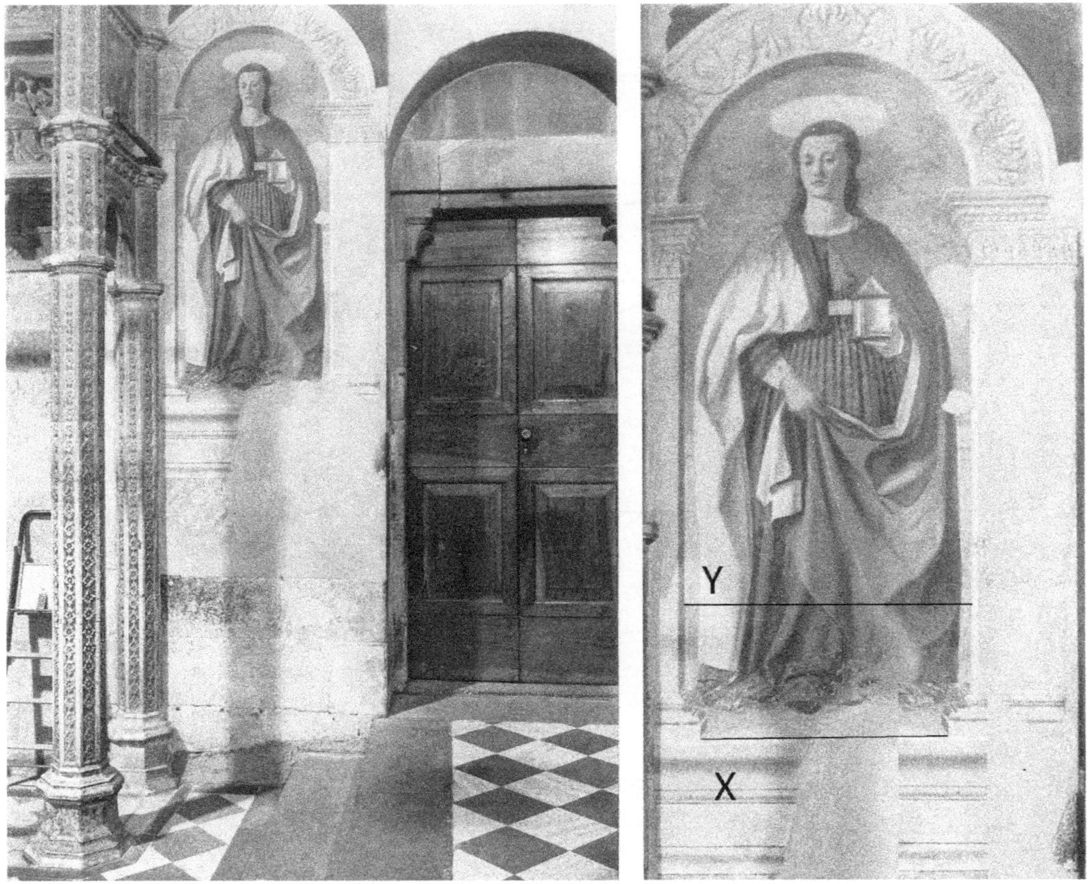

Figure 6.7 Piero della Francesca. *Mary Magdalene*. Duomo, Arezzo. The fresco *in situ* (left) and digitally "corrected" (right). A square with sides measuring "*X*" equals a circle with a diameter of "*Y*." See also Figure 6.2.

Photo credit: Perry Brooks.

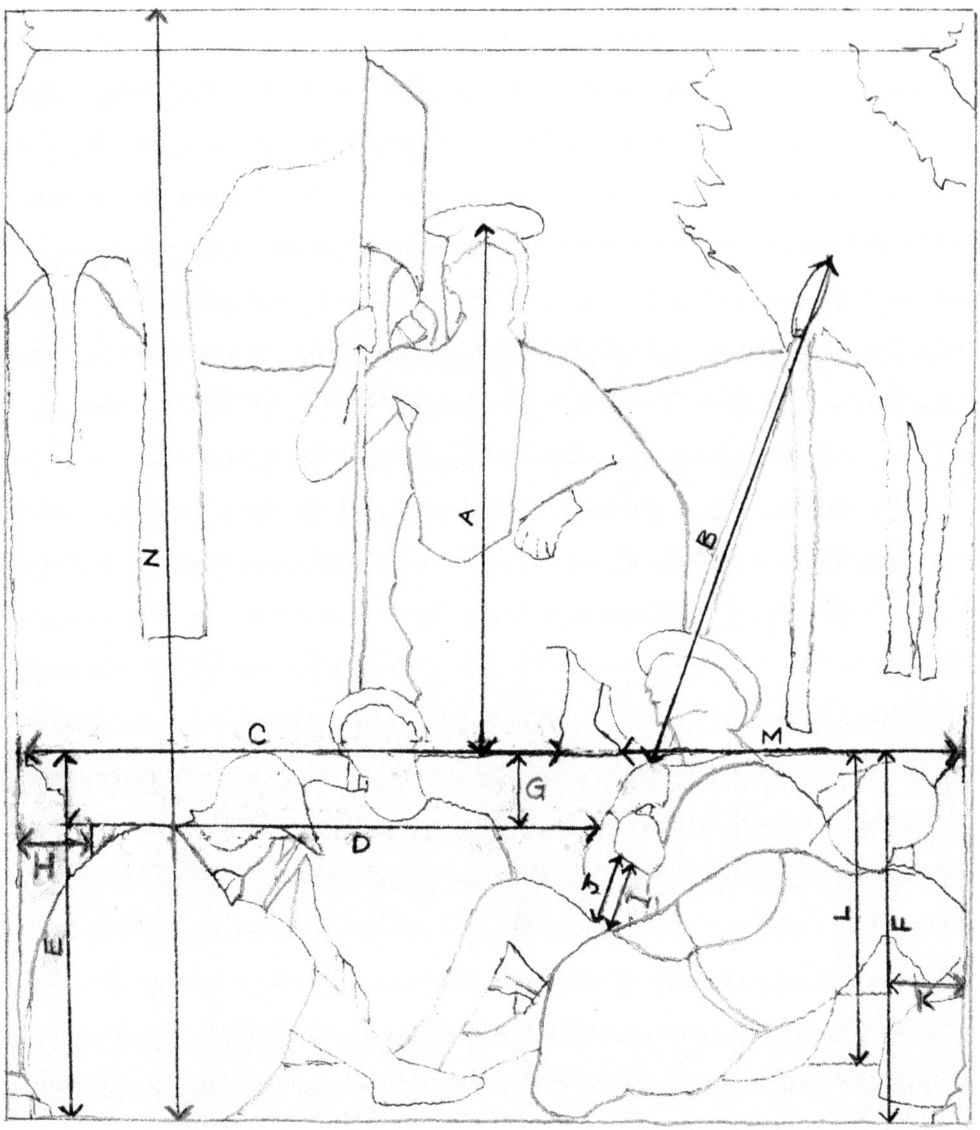

Figure 6.8 Piero della Francesca. *Resurrection*. Museo Civico, Sansepolcro. Schematic diagram with measured elements. See Appendix.

Appendix

Measurements of the *Resurrection* (See **Figure 6.8**)

NB: *"Left" and "Right" Refer to Viewer's Side*

A – Height of Christ above sarcophagus: 112 cm
B – Height of lance above hand: 113 cm
C – Length of sarcophagus top to left of Christ's foot: 113 cm
D – From left end of sarcophagus to lance-bearer's arm: 112 cm
E – Height of sarcophagus from base of picture, left end: 77.5 cm
F – Height of sarcophagus from base of picture, right end: 77.5 cm
G – Height of top moulding: 15.5 cm
H – From left column to right edge of moulding at left end of sarcophagus: 15.5 cm
I – Length of lance visible beneath hands, right side: 15.5 cm
J – Length of lance visible beneath hands, left side: 15.2 cm
K – From right column to left side of moulding at right end of sarcophagus: 16.3 cm
L – Height of sarcophagus, right end: 65.5 cm
M – Length of sarcophagus top to right of Christ's foot: 70.5 cm
N – Height of forward face of architrave above base (approx.): 230 cm
Width between columns at level of top of sarcophagus: 200.2 cm
Length of top of sarcophagus: 196.3 cm
Width of box of sarcophagus: 179 cm
Height of sarcophagus lid: 9.3 cm

Notes

1 This study derives from my Columbia University doctoral dissertation (directed by the lamented dedicatees), Perry Brooks, *Ut Pictura Mathesis: Studies in the Art of Piero della Francesca* (Ann Arbor: UMI, 1990), which in part goes back even further to my 1977 Columbia M.A. thesis, "Sacred Geometry: The Structure of Piero della Francesca's Fresco of the Resurrection." My dissertation, like the compact introductory monograph on the Arezzo Cycle partly based on it, *Piero della Francesca: The Arezzo Frescoes* (New York: Rizzoli, 1992), has hovered outside the scholarly bibliographic radar concerning Piero della Francesca. My research was facilitated by a Chester Dale Fellowship from the Center for Advanced Studies in the Visual Arts of the National Gallery of Art, Washington, DC, and a Whiting Fellowship from the Mrs. Giles Whiting Foundation, New York City. Jane Carter-Tresidder and Lucy Welliver Scanlon kindly allowed the reproduction of the drawings of B.A.R. Carter and Warman Welliver.

2 Influencing or attracting the attention of such notables as LeCorbusier and Paul Valéry, and treating the golden section in an ahistorical and occasionally hermetic manner, Matila Ghyka's works include: *Esthétique des proportions dans la nature et dans les arts* (Paris: Gallimard, 1927); *Le nombre d'or: rites et rythmes pythagoriciens dans le développement de la civilisation occidentale*, intro. by Paul Valéry, 2 vols. (Paris: Gallimard, 1931); and *The Geometry of Art and Life* (New York: Sheed & Ward, 1946), the latter also available in a Dover reprint.

3 Roger Herz-Fischler, *A Mathematical History of Division in Extreme and Mean Ratio* (Waterloo, Ontario: Wilfrid Laurier University Press, 1987), 111.

4 The pervasiveness of the golden section in the pentagon, and also in the star-shaped pentagon or pentagram, might be one reason for the use of the latter as a badge of recognition among the Pythagoreans and called by them "Health." Cf. Thomas L. Heath, *The Thirteen Books of Euclid's Elements*, 2nd ed., 3 vols. (1926; reprint, New York: Dover, 1956), 2: 99. In the Renaissance its status as an occult symbol was revived by Cornelius Agrippa's publication (though with no mention of "mean and extreme ratio") of the pentagram within a roundel inscribed in Greek "Health," as well as of a human figure inscribed within a circle and with his limbs conforming to a pentagram. See Henricus Cornelius Agrippa ab Nettesheym, *De Occulta philosophia sive De Magia Libri Tres* (1533; facsimile reprint with commentary by Karl Anton Nowotny, Graz: Akademische Druck- und Verlagsanstalt, 1967), clxiii, cclxxvi–cclxxvii (original pagination).

5 Herz-Fischler, *A Mathematical History*, 159–160.

6 See Ghyka, *Le nombre d'or*, plates XVIIff. and *The Geometry of Art and Life*, plates XXXVI–XXXVII and LXVI.

7 Rudolf Wittkower and B.A.R. Carter, "The Perspective of Piero della Francesca's 'Flagellation,'" *Journal of the Warburg and Courtauld Institutes* 16 (1953): 292–302.

8 Rudolf Wittkower, *Architectural Principles in the Age of Humanism* (London: Warburg Institute, 1949), with many later editions. See also his article, "The Changing Concept of Proportion," *Daedalus* 89 (Winter 1960): 199–215.

9 Herz-Fischler, *A Mathematical History*. This should be supplemented by Brooks, *Ut Pictura Mathesis*, 171–203, which is indebted to Herz-Fischler but expands it with a few more references and some corrections of Herz-Fischler's Latin translations. A more recent book, Mario Livio, *The Golden Ratio: The Story of Phi, the World's Most Astonishing Number* (New York: Broadway Books, 2002), follows the trend toward a more historically based popular account.

10 Luca Pacioli, *Divina proportione, Opera a tutti glingegni perspicaci e curiosi necessaria ove ciascun studioso di philosophia prospectiva pictura sculptura architectura musica e altre mathematice suavissima: sottile: e admirabile doctrina consequira: e delectarassi: con varie questione de secretissima scientia* (Venice: Paganius Paganinus, 1509). I have used the modern edition, *De divina proportione*, ed. Franco Riva, *Fontes Ambrosiani* 31 (Milan: Mediobanca, 1956) along with the two-volume edition that includes a facsimile of the Biblioteca Ambrosiana codex: *De divina proportione*, intro. by Augusto Marinoni, *Fontes ambrosiani* 72 (Milan: Silvana, 1982). Both the 1956 and 1982 editions include reproductions of the drawings of the regular bodies attributed to Leonardo da Vinci. Despite the 1509 date of

publication, the portion of the book treating directly the golden section is dated Milan, 14 December 1498.

11 J. V. Field, *Piero della Francesca: A Mathematician's Art* (New Haven and London: Yale University Press, 2005), assumes an agnostic attitude toward symbolic use of mathematics (including the "golden section") and explores the relevance of mathematics mostly in terms of intentions toward naturalism. James Banker, a scholar I greatly admire for revitalizing our knowledge of Piero's biography through his important documentary and manuscript discoveries of the last few decades, likewise downplays mathematics as a vehicle of symbolic statement and Platonic purpose in his recent monograph, *Piero della Francesca: Artist & Man* (Oxford: Oxford University Press, 2014), especially 218–219.

12 Cf. Pappus, *The Commentary of Pappus on Book X of Euclid's Elements*, trans. William Thomson, Harvard Semitic Series, VIII (1930; reprint, New York: Johnson Reprint Corp., 1968), 64 (I.2).

13 F. M. Cornford, *Plato's Cosmology* (1937; reprint, Indianapolis: Bobbs-Merrill, n.d.).

14 *Elements*, XIV: 10 (Campanus' numeration). My translation (here, as elsewhere, unless otherwise specified), from the dual Greek and Latin edition (which poses Campanus' older translation alongside the 1505 translation of Bartolomeo Zamberti) of Euclid, *Elementorum Geometricorum. Lib.XV. Cum Expositione Theonis in Priores XIII à Bartholomaeo Veneto Latinitate donata, Campani in omnes, & Hybsiclis Alexandrini in duos postremos* (Basel: J. Herwagen, 1537), 462:

> Mirabilis itaque est potentia lineae secundum proportionem habentem medium duoque extrema divisae. Cui cum plurima philosophantium admiratione digna conveniant, hoc principium vel praecipuum ex superiorum principiorum invariabili procedit natura, ut tam diversa solida tum magnitudine tum basium numero tum etiam figura, irrationali quadam symphonia rationabiliter conciliet.

Pacioli, *Divina proportione,* translated Campanus' passage without acknowledgment. See 1956 ed., 23: "gli è concesso maxime vedendo lei esser quella che tante diversità de solidi, sì de grandezza sì de moltitudine de basi, sì ancora de figure et forme con certa irrationale symphonia fra loro accordi."

The translation of Campanus given by Herz-Fischler, *A Mathematical History*, 171—who admits his lack of knowledge of Latin (xii)—mistakenly makes *simphonia* the subject of *conciliet*, rather than an ablative, and has *irrationali* modify *figura*, rather than *simphonia*. This is not how it was read by Pacioli.

15 Margaret Daly Davis, *Piero della Francesca's Mathematica Treatises: The "Trattato d'abaco" and "Libellus de quinque corporibus regularibus"* (Ravenna: Longo, 1977), 30–41, 46ff. Herz-Fischler, *A Mathematical History*, 144–149.

16 See Renzo Baldasso's and John Logan's chapter in this volume.

17 Pacioli, *Divina proportione*, 1509 ed., 3v.–4r. 1956 ed., 21–22.
Pacioli had briefly mentioned mean and extreme ratio and its singularity in his 1494 book, *Summa de arithmetica.* Since the text is omitted by Herz-Fischler, I present it here from the 1523 reprinting, *Summa de arithmetica geometria. Proportioni: et proportionalita: nouamente impressa . . .* (Tusculano: Paganino de Paganini, 1523), f.81v.:

> Rechiedarisse ancor qui dire de una certa, e sola specie separata de proportioni discrepante de le assignate, laqual da li philosophi se chiama proportione havente el mezzo, e doi extremi. De laquale Euclide in tertiodecimo e quartodecimo libro apieno tratta e virtualmente prima undecima del secondo sua forza dimostra, e in la vigesimanona del sexto.
>
> Que proportio habens medium duosque extrema in hoc maxime est differens ab aliis quia alie possunt ad minus in duobus terminis reperiri, ista autem nunquam in paucoribus quam tribus. Et in hoc similitudinem gerit proportionalitatis et cetera. Laquale de socto in questo nella mesura del corpo duodecim pentagonorum nel particulare Tractato de corpi regulari in fine a la domanda vigessima prima indurremo senza laquale la dimensione de ditti corpi non si po haver commo per Euclide se demostra etc.

18 Pacioli, *Divina proportione*, 1509 ed., 4r. 1956 ed., 21–22.

19 Pacioli, *Summa*, f.106r.-v:

> However, a method cannot be given for all operable cases, whether for lack of terms, unknown to us, or because the human intellect has not (yet) attained the end of its desire—as is said by all philosophers, especially Aristotle, of the squaring of the circle, that it is knowable, but not yet precisely found by anyone. However the practical operative method to its dimension with a certain approximation has been given by Archimedes of Syracuse. . . . But with time, if life goes on, man finds now one, now another rule, beyond those already found, just as the philosophers in ancient times, speculating on now one thing, now another, then had knowledge of many things. (Non pero a tutte cose operabili si po dare modo si per carentia de termini a noi ignoti, sianche perche lo perscrutare humano (finora) non a tinto el fine del suo desiderio, commo per tutti phylosophi, maxime Aristotele de la quadratura del cerchio se dici esser scibile, avenga che fin mo [*sic*] per nullo sia precisamente trovata. Quamtunche per Archimede siracusano el modo pratico operativo a sua dimensione con certa approximatione a noi sia dato. . . . Ma col tempo se la vita non scorta, lhomo ritrova, ora una, ora un altra regola, oltre le gia invente. Si commo antichamente fecero li phylosophi. Liquali ora una cosa, ora laltro speculando, di molte hebero poi noticia.)

20 Banker, *Piero della Francesca,* 188–193. See also Idem, "A Manuscript of the Works of Archimedes in the Hand of Piero della Francesca," *Burlington Magazine* 147 (2005): 165–169. As discussed by Banker, an important link between Piero and Roman mathematical culture was his distant cousin, the architect and papal scriptor Francesco del Borgo, whose role in architectural history was recuperated by Christoph Frommel, "Francesco del Borgo: Architekt Pius' II. und Pauls II.," *Römisches Jahrbuch für Kunstgeschichte* 20 (1983): 108–154, and 21 (1984): 129–138.

21 B. L. van der Waerden, *Science Awakening*, trans. Arnold Dresden (New York: Science Editions, John P. Wiley & Sons, 1963), 184–187, 204–206.

22 Nicolaus Cusanus, *De Docta Ignorantia* I.3, in *Opera*, 3 vols (Paris: Badius Ascensius, 1514), 1: 2r.:

> Intellectus igitur qui non est veritas, nunquam veritatem adeo praecise compraehendit quin per infinitum praecisius compraehendi possit habens se ad veritatem sicut polygonia ad circulum, quae quanto inscripta plurium angulorum fuerit tanto similior circulo, nunquam tamen efficitur aequalis etiam si angulos usque in infinitum multiplicaverit nisi in identitatem cum circulo se resolvat.

Useful introductions in English to the various facets of Cusanus' life and thought include Ernst Cassirer, *The Individual and the Cosmos in Renaissance Philosophy*, trans. Mario Domandi (Philadelphia: University of Pennsylvania Press, 1963); Joseph Hofmann, "Cusa, Nicholas," in reference to *Dictionary of Scientific Biography*, ed. Charles Coulston Gillespie, 18 vols. (New York: Scribner, 1970–90), 3: 512–516; Field, *Piero della Francesca: A Mathematician's Art*, 269–275; Bernard McGinn, *The Harvest of Mysticism in Medieval Germany (1300–1500)* (New York: Crossroad Publishing Company, 2005), 432–483; and Erich Meuthen, *Nicholas of Cusa: A Sketch for a Biography*, trans. David Crowner and Gerald Christianson (Washington, DC: Catholic University of America Press, 2010). A valuable online gateway to Cusanus' writings, in the original Latin as well as in English and German translations, is the "Cusanus Portal" sponsored by the Institut für Cusanus-Forschung, Trier (http://urts99.uni-trier.de/cusanus/index.php).

23 *De docta ignorantia*, III.4, in *Opera*, 1: 26v. He continues, 1: 26v.–27r.: "Si ipsa polygonia maxima esse debet qua maior esse non potest nequaquam in angulis finitis per se sibsisteret sed in circulari figura ita ut non haberet propriam subsistendi figuram etiam intellectualiter ab ipsa circulari & aeterna figura separabilem."

24 *Complementum Theologicum Figuratum in Complementis Mathematicis*, in *Opera*, 2: 94v.:

> Qui enim circuli quaesiverunt quadraturam; coincidentiam circuli & quadrati in aequalitate praesupposuerunt, quae certe in sensibilibus non est possibilis. Non enim dabile est quadratum, quod non sit inaequale omni dabili circulo in materia. Hanc igitur aequalitatem quam praesupposuerunt, non viderunt oculis carneis sed mentalibus.

25 Heath, *The Thirteen Books*, I: 153.

26 "Nos vero, quod sub aspectu rem positam esse volumus, pinguiore idcirco, ut aiunt, Minerva scribendo utemur." *De pictura*, I.1. See Leon Battista Alberti, *On Painting and On Sculpture: The Latin Texts of De Pictura and De Statua*, ed. and trans. Cecil Grayson (London: Phaidon, 1972), 36–37. Grayson translated the Latin as, "We, . . . who wish to talk of things that are visible, will express ourselves in cruder terms," while John Spencer rendered Alberti's Italian text as "we will use a more sensate wisdom," in Alberti, *On Painting*, trans. with intro. and notes by John Spencer (rev. ed., New Haven: Yale University Press, 1966), 43.

27 Piero della Francesca, *De Prospectiva Pingengi*, ed. Giusta Nicco Fasola, 2 vols. (Florence: Sansoni, 1942), 1: 65.

28 Nicholas of Cusa's name has been often invoked in relation to Piero della Francesca's, earlier in the sense of simultaneous manifestation of the *Zeitgeist*, more recently in terms of documentary evidence of the simultaneous presence in Rome in 1459 of Piero and Cusanus. Piero was paid on 12 April "per parte del suo lavor di certe dipinture fa nella camera della Santità di Nostro Signore Papa." See Eugenio Battisti, *Piero della Francesca*, 2 vols (Milan: Istituto Editoriale Italiano, 1971), 2: 224, doc. LVIII. On 11 January of that year Cusanus had been named Legate and Vicar General *in temporalibus* during the journey of Pope Pius II to Mantua; cf. Meuthen, *Nicholas of Cusa: A Sketch for a Biography*, 123. If we follow Vasari's biography, we may surmise that Piero returned to Borgo San Sepolcro after the death of his mother on 6 November 1459; cf. Battisti, *Piero della Francesca,* 2: 224, doc. LX.

On parallels or relations between Cusanus and Piero, see G. Nicco Fasola, "La prospettiva nell'estetica del rinascimento," introduction to *De Perspectiva Pingendi*, 17ff.; Wittkower and Carter, "The Perspective of Piero," 302; Battisti, *Piero della Francesca*, 1: 107, 476, n. 143; and Idem, "Una rilettura dopo 40 anni," introduction to reprint of Nicco Fasola ed. of *De Perspectiva Pingendi* (Florence: Le lettere, 1984), xx, n.41; and Daniel Arasse, "'Oltre le scienze di sopra': Piero della Francesca et la vision de l'histoire," in *Piero della Francesca and His Legacy*, ed. Marilyn Aronberg Lavin (Washington, DC: National Gallery of Art, 1995), 105–113.

Miklos Boskovits, "'Quello ch'e dipintori oggi dicono prospettiva': Contribution to Fifteenth Century Italian Art Theory," *Acta Historiae Artium Academiae Scientiarum Hungaricae* 9 (1963): 139–162, 147, 159, n. 78, judged Nicco Fasola's account as the most successful attempt to take account of the influence of Platonism in Piero's work, but "despite the numerous interesting ideas it moots not always convincing, in that it frequently links Piero's theoretical development to sources which only very much later influenced or could influence Italian art theory (e.g., Nicolaus Cusanus)." This judgment, of course, denies the most direct influence of personal contact.

29 See "Appendix I: Annotated Bibliographical Compendium of Mathematical Analyses of Piero della Francesca's Works Arranged by Location," in Brooks, *Ut Pictura Mathesis*, 259–275.

30 The line appears a second time (*Divina* Proportione, 1509 ed., 27r.) with vague indications that it can be used for proportioning the human body according to a ten-part Vitruvian system. For further discussion, see Brooks, *Ut Pictura Mathesis*, 83–88 (including a more problematic and unexamined use of the 1.85" module by Carlo Ginzburg).

31 Warman Welliver, "The Symbolic Architecture of Domenico Veneziano and Piero della Francesca," *Art Quarterly* 36 (1973): 1–30. The study also finds π relationships in Piero's *Montefeltro Altarpiece* (Brera, Milan).

32 For measurements see the Appendix. Francesco Comanducci of the Biblioteca Communale, Sansepolcro, kindly facilitated my measuring of the Resurrection in Summer 1980, with the assistance of Eloise Angiola (Professor Emerita of Art History, University of Alabama, Tuscaloosa) and Robert Beseda (Assistant Dean for Drama Emeritus, North Carolina School of the Arts). It should be noted that my analysis stays within the confines of the framing columns, which are mostly reconstructions and of which only the edges adjacent to the painted scene are original. Cf. Battisti, *Piero della Francesca*, 2: 33.

33 The painting in the Sansepolcro Cathedral served as high altar of the Badia of San Giovanni Evangelista. Cf. Banker, *Piero della Francesca*, 109–110 and Illus. 13.

34 The form of the sarcophagus seems to be determined by archeological considerations. Battisti, *Piero della Francesca*, 1: 285, 496–497, n. 311, noted the similarity of the sarcophagus, in architectural form and proportion, to the altar in the Chapel of the Holy Sepulcher built by Alberti for the Rucellai family in S. Pancrazio, Florence. The dimensions which Battisti measured—"molto empiricamente"—for the S. Pancrazio altar are 193 cm for the length of the top, 175 cm for the length of the base, and 77 cm for the height from base to top. The dimensions given by Battisti for the sarcophagus in the fresco are 196 cm for the length of the cover and 179 cm for the length of the box; these measurements conform exactly to measurements taken by the present author.

The measured height of the sarcophagus box is c. 65.5 cm. This measurement does not coincide with that of the Rucellai altar but closely agrees with the height given for the Holy Sepulcher in a medieval description: the Venerable Bede (*De locis sanctis*, 2: 2, in *Corpus Christianorum Series Latina CLXXV: Itineraria et alia geographica* [Turnhout: Brepols, 1965], 255) describes the sepulcher in Jerusalem as 7 feet long × 3 palmi high, which, if interpreted as Roman feet and palmi, equals c. 209 cm × 67 cm. The fresco also agrees with the medieval text in the color of the sarcophagus, which Bede describes as "albo et rubicundo permixtus."

Marco Dezzi Bardeschi, "Nuove ricerche sul S. Sepolcro nella Cappella Rucellai a Firenze," *Marmo* 2 (1963): 134–161, linked the Rucellai Chapel to a manuscript relating a 1457–58 mission to the Holy Land in which the Holy Sepulcher is described as 8 palmi long, 3½ palmi wide, and 4 palmi high, or in modern units as translated by Dezzi Bardeschi, 177 × 78 × 90 cm (using 22.3 cm as the equivalent of the Roman palmo, one arrives at 178 × 78 × 89 cm for the dimensions).

According to Lorenzo Coleschi, *Storia della città di Sansepolcro* (Città di Castello: S. Lapi, 1886), 192, in the oratory of the Church of S. Rocca, there was a stone Holy Sepulcher "conforme a quello che si venera in Gerusalemme, largo bracci 3 e lungo braccia 5 e ⅚ ed alto da pavimento alla volta braccia 4½ ed eseguito (per ciò che scrive il Lancis . . .) sul disegno di Leon-Battista Alberti fiorentino."

The occurrence of the dimensions of 175 cm and 179 cm in the Rucellai altar and the Resurrection sarcophagus, respectively, might have been determined by the Albertian concept that the ideal human figure was 3 Florentine braccia high, c. 175 cm. The conversion units used here are based upon those given by Angelo Martini, *Manuale di Metrologia* (Turin: E. Loescher, 1883): 1 Florentine braccio = 58.36 cm, 1 piede romano = 29.8 cm, 1 palmo romano = 22.3 cm.

On the fascinating relationship of local medieval measurements to the presumed length of Christ's body see Gustavo Uzielli, *Le misure lineari medioevali e l'effigie di Cristo* (Florence: Seeber, 1899).

35 Field, *Piero della Francesca: A Mathematician's Art*, 6, suggests that Christian artists would probably avoid the pentagon "because of its association with Islam," but I find this unconvincing for at least two reasons: firstly, the pentagon was not likely to be more prominent in the Islamic luxury goods available to European artists than other complex polygons, so this type of thinking, if valid, would seem to ban geometric decoration as a whole; secondly, Islamic associations did not prevent artists from decorating the Madonna's robes with pseudo-Kufic lettering and draping her heavenly settings with Turkish or Persian carpets. In the *Resurrection*, Piero makes reference to the pentagon without illustrating it.

36 Baldassare Boncompagni, ed., *Scritti di Leonardo Pisano, matematico del secolo decimoterzo*, 2 vols. (Rome: Tipografia delle scienze matematiche e fisiche, 1857–62), 1: 283–284. We do not know when the ratios of the successive Fibonacci numbers were first related to the golden section. Pacioli makes no overt reference to the series in his *Divina proportione*. The earliest written account whose author is known is a 1608 letter from Kepler to Joachim Tanckius. The earliest published account is Albert Girard's 1633 annotated translation of Diophantus' *Arithmetic*, Bks. V and VI, according to Edouard Lucas, *Théorie des nombres*, 2 vols. (Paris: Gauthier-Villars, 1891), 1: 4. According to L. E. Sigler, introduction to English trans. of Leonardo Pisano Fibonacci, *The Book of Squares* (New York: Academic Press, 1987), xviii, Lucas was the first to dub the series the "Fibonacci Series." For further discussion, see Herz-Fischler, *A Mathematical History*, 157–158, 160–162, and Brooks, *Ut Pictura Mathesis*, 194–199.

37 Leonard Curchin and Roger Fischler, "De quand date le premier rapprochement entre la suite de Fibonacci et la division en extrême et moyenne raison?" *Centaurus* 28 (1985): 129–138. For necessary corrections to their transcription, see Brooks, *Ut Pictura Mathesis*, 198, n. 40.

38 The measurement of 112–113 cm likely represents two Sansepolcro braccia. James Banker, the best authority on local practices, gives the measure for the Sansepolcro braccia as 56 cm (*Piero della Francesca: Artist & Man*, xvii), without however citing his source. In contrast, Battisti, *Piero della Francesca*, II: 216, suggested that the Florentine measure of 58.36 cm was in use. I conjectured (*Ut Pictura Mathesis*, 200–201) that perhaps the Sienese braccio, defined by Pacioli (*Summa de arithmetica*, 211r.) as in the ratio of 31:32 to the Florentine braccio, was also in use in Sansepolcro. This would make the Sienese braccio 56.536 cm and two braccia equal to 113.1 cm. This Sienese variant is absent in nineteenth-century tables such as Martini's *Manuale di metrologia*.

We may note, in keeping with our previous discussion of the "fatter Minerva," that discrepancies of 1–2 cm do not seem terribly disturbing, especially in view of the complex and contradictory fifteenth-century systems of measure that varied according to the object of measurement (e.g., cloth vs. land), the local political entity authorizing measurement, and whether that authority was secular or sacred. Sources for discrepancy include: 1) Inaccuracies on the part of the researcher: to protect the fresco surface, for some measurements it was necessary to use a flexible tape without millimeter calibration, so that centimeter fractions had to be estimated by eye. 2) Imprecisions inherent in the conversion between different standard units (e.g. braccia to meters). 3) Variations induced in the artist's tendency to express geometric quantities according to arithmetical/numerical values. 4) Impreciseness introduced in the process of transferring measured cartoons to the wall. 5) Ambiguities of edge basic to the painterly medium—where does one measure a brushed line: left edge, center, right edge?—as well as caused by the entasis of the columns framing the *Resurrection*. 6) Instability due to the physics of the materials in a 500-year period, especially when we note that there was a chimney behind the painting.

39 Vulgate, I Cor. XV.20–26:

> Nunc autem Christus resurrexit a mortuis primitiae dormientium, quoniam quidem per hominem mors, et per hominen resurrectio mortuorum. Et sicut in Adam omnes moriuntur, ita et in Christo omnes vivificabuntur. Unusquisque autem in suo ordine: Primitiae Christus; deinde ii, qui sunt Christi, qui in adventu ejus crediderunt. Deinde finis, cum tradiderit regnum Deo et Patri, cum evacuaverit omnem principatum, et potestatem et virtutem. Oportet autem illum regnare, donec ponat omnes inimicos sub pedibus ejus. Novissima autem inimica destruetur mors; omnia enim subjecit sub pedibus ejus.

40 Corneliius Vermeule, *European Art and the Classical Past* (Cambridge, MA: Harvard University Press, 1964), 41, pointed out that the soldier covering his face in the left foreground "is the barbarian found so often beneath trophies in Roman state relief, on sarcophagi, and on coins going back to the late Republic." Although Vermeule conjectured that the figure's antique pose is derived secondarily from Roman art by way of post-Giottesque painting, I believe the pose was actively sought out to transfer an ancient iconographical function into a new context. A figure similar to the leftmost soldier occurs on a sarcophagus in the Camposanto, Pisa; see Friedrich Matz, *Die antiken Sarkophagreliefs*, IV-4 (Berlin: G. Grote, 1975), no. 260. The rightmost soldier, leaning backward, also derives from ancient sarcophagi: he is a mirror image, reflected from behind, of a figure represented as being trampled underfoot in a battle sarcophagus in the Capitoline Museum, Rome; see Matz, *Die antiken Sarkophagreliefs*, IV-3 (Berlin: G. Grote, 1969), no. 238. Also cf. the similar figure of a fallen hunter in a sarcophagus with the *Wounding of Hippolytus* in Ince Blundell Hall, published by Richard Krautheimer, *Lorenzo Ghiberti* (Princeton: Princeton University Press, 1956), fig. 105.

41 Charles de Tolnay, "Conceptions religieuses dans la peinture de Piero della Francesca," *Arte antica e moderna* 6(23) (1963): 205–241, 230.

42 A. del Vita, "Il volto di Piero della Francesca," *Rassegna d'arte* 7 (1920): 109–112. Earlier mentions of the Resurrection self-portrait include G. F. Pichi, *La vita e le opere di Piero*

della Francesca (Sansepolcro: Becamorti & Boncompagni, 1892), 95; Evelyn Franceschi Marini, *Piero della Francesca* (Città di Castello: Lapi, 1912), 96; G. Mancini, ed., "L'opera 'De corporibus regularibus' di Pietro Franceschi detto Della Francesca usurpata da Fra Luca Pacioli," *Atti della Reale Accademia dei Lincei, anno CCCVI. Serie quinta. Memorie della Classe di Scienze morali, storiche e filologiche* 14 (1909–16): 479. Documents for comparison include Vasari's woodcut portrait and a late sixteenth-century portrait attributed to Santi di Tito from the collection of Piero della Francesca's descendants, who, until 1824, also owned a small self-portrait by Piero which was later lost. See W. Prinz, *Vasaris Sammlungen von Künstlerbildnissen* (Florence: Kunsthistorisches Institut, 1966), 79–81, and Creighton Gilbert, *Change in Piero della Francesca* (Locust Valley, NY: J.J. Augustin, 1968), 88ff., n. 41.

43 The figure in the fresco has been previously identified as Longinus by others, notably Gertrud Schiller, *Ikonographie der christlichen Kunst* (Gütersloh: Gütersloher Verlagshaus G. Mohn, 1968), 3: 81, without comment on the unusualness of Longinus' presence in the Western tradition Resurrection iconography. See G. Lucchesi, "Longino," in *Bibliotheca Sanctorum*, ed. Filippo Caraffa, 13 vols. (Rome: Istituto Giovanni XXIII nella Pontificia Università lateranense, 1961–706), 7: 90–95; and R. Peebles, *The Legend of Longinus in Ecclesiastical Tradition and in English Literature, and its Connection with the Grail* (Baltimore: J.H. Furst Co., 1911), 18, n. 40.

Two Greek accounts of the Longinus legend are published with Latin translation in J.-P. Migne, *Patrologia Graeca* (PG) (Paris: J.-P. Migne, 1857–1866): Hesychius Presbyter Hierosolymitanus, *Martyrium Sancti Longini Centurionis*, in PG 93: 1545–1560, an account from the first half of the seventh century; and Symeon Metaphrastes, *Martyrium Sancti et Gloriosi Martyris Longini Centurionis*, in PG 115: 31–44, an account from the tenth century based on the earlier hagiographical text. On Longinus' vigil at the tomb, cf. Hesychius, in PG 93: 1547: "Itaque cum Pilatu custodiam postulatus advigilaturam sepulcro unigeniti Filii Dei eamdem petentibus Judaeis concessit, dux eorum Longinus fuit."

Although the legend of Longinus' watch at the tomb was not common in Western iconography, there is a fresco of the Resurrection in Ferrara, Convento di S. Apollinare, published by Carlo Ragghianti, "Il Maestro di S. Apollinare, 2," *Critica d'arte* 20, fasc. 135 (May–June, 1974): 35–50, which Ragghianti dates to the early 1430s and sees as a precedent for Piero's version; a fragmentary inscription next to one of the sleeping soldiers reads: "14 . . . /mun . . . /LO. . . ." Could the bottom line have been "Longinus"? Both Pacioli and Vasari report that Piero worked in Ferrara; cf. Perry Brooks, "Observations on Piero della Francesca's Frescoes in Ferrara," in *Watching Art: Writings in Honor of James Beck. Studi di storia dell'arte in onore di James Beck*, ed. Lynn Catterson and Mark Zucker (Todi: Ediart, 2006), 95–102.

The relic of the Holy Lance now in St. Peter's, Rome, was transferred from Constantinople in 1492; cf. C. Rohault de Fleury, *Mémoire sur les instruments de la Passion* (Paris: L. Lesort, 1870), 272–275.

The proportional structure we have elicited in our discussion may have had a cramping effect in the rendering of the soldier and may be partly responsible for the figure's apparent "leglessness" discussed by Michael Baxandall, *Words into Pictures* (New Haven: Yale University Press, 2003), 117–161.

44 According to further elaboration of his legend, Longinus brought a vial of the Holy Blood to Mantua, where it was buried and later refound through a vision of St. Andrew's, and would become the chief relic of the Alberti-designed church of Sant'Andrea. In 1459 (while Piero frescoed walls in the Vatican Palace), Pope Pius II travelled to that city, where, at his instigation, a congress of European powers had convened to plan a new Crusade against the Turks. During the proceedings, the pope became ill and regained his health only after invoking the relic of the Holy Blood brought there by Longinus. Pope Pius then travelled to Gradaro, the place of martyrdom alleged by the local tradition, and had a marble column erected there; see Marita Horster, "Mantuae Sanguis Preciosus," *Wallraf-Richartz Jahrbuch* 25 (1963): 151–180, especially 162–163.

45 Cf. Cassiodorus, "In Psalterium Expositio," in J. P. Migne, *Patrologia Latina* (PL) (Paris: J.-P. Migne, 1841–55), 70, 30: "Bene ut arbitror, ligne fructifero comparatus est Dominus Christus, propter crucem quam pro hominum salute suscepit." This develops the metaphor

of Psalms 1.3: "And he shall be like a tree planted by the river of water that bringeth forth his fruit in good season: his leaf shall not wither."

46 *Complementum Theologicum Figuratum in Complementis Mathematicis*, in *Opera*, 2: 94v., quoted in n. 24 above.

47 Vulgate, I Cor. 13:12: "Videmus nunc per speculum in aenigmate: tunc autem facie ad faciem. Nunc cognosco ex parte: tunc autem cognoscam sicut et cognitus sum."

48 Kenneth Clark, *Piero della Francesca*, 2nd ed. (1951; London: Phaidon, 1969), 57.

49 Cf. Banker, *Piero della Francesca*, 82–83.

50 Cf. Brooks, *Ut Pictura Mathesis*, 171.

51 For further discussion, see Brooks, *Ut Pictura Mathesis*, 80–82, 265–267. This work has received a thorough φ-related proportional analysis with an elaborately trussed surface design (avoided in my examination of the *Resurrection*) in Elton M. Davies and Dean Snyder, "Piero della Francesca's Madonna of Urbino," *Gazette des Beaux-Arts* 75 (1970): 193–212.

52 For the use of φ in the *Annunciation* and the fictive architectural decoration of the Arezzo Chapel, see Brooks, *Ut Pictura Mathesis*, 164–170, 276–277 (with measurements taken *in situ*) and Brooks, *The Arezzo Frescoes*, col. 7. The measured width of the *Annunciation* at the bottom of the fresco is 196 cm. The published height of the fresco (including the red and gray/green stripes that shorten the pictorial field) is 329 cm. The stripes shorten the field by the measured amount of 15.2 cm. Thus the proportions of height:width are 313.8/196, or 1.60. Since the chapel's Gothic architecture is most irregular, the leftmost edge of the fresco wavers perceptibly, making φ (1.618 . . .) an ideal to be only approximated. For further speculation of an open-ended sort on the mathematics of the *Annunciation*, see Brooks, *Ut Pictura Mathesis*, 57–63.

53 See Brooks, *Ut Pictura Mathesis*, 223–226. The measurement of the inner edges (nearest the saint) of the opening is 77.5 cm. The lower molding projects 4.2 cm on the undamaged side, as seen in Figure 6.2. The opening nearest the viewer, and our line of entry, is thus 77.5–8.4, or 69.1 cm.

Part III

Euclid and Artistic Accomplishment

7 The Point and Its Line

An Early Modern History of Movement

Caroline O. Fowler

> Suppose that a book on the elements of geometry has always existed, each copy made from an earlier one . . . We can explain any given copy of the book in terms of the previous book from which it was copied; but this will never lead us to the complete explanation . . . For we can always ask: Why have there been such books? Why were these books written? Why were they written in the way they were?
>
> —G.W.F. Leibniz, "The Ultimate Origin of Things"

The point, an indivisible form defined by Euclid's *Elements* as "that which has no parts," was the foundation of not only fifteenth-century geometry but also artistic practice.[1] This initial mark materialized the physical world delineated in form, name and sensation. As artistic practices changed, however, from the fifteenth to the seventeenth century, artists stopped locating the foundation of their practice in the point and began focusing on the line as the limit of representation. While Euclid defined the point as "that which has no parts," the line was described as "length without breadth." In order to designate the delimitations of this "length without breadth," Euclid stated that points demarcate the termination of a line. Although Euclid designated the point as the line's terminus, throughout early modern translations of Euclid the point and the line increasingly were determined in an interrelationship of absence and movement. As opposed to defining the line as "length without breadth," the line became characterized as "a series of points" or as "the trace of the point's movement." The point and the line remained distinct entities although they existed in a necessary relationship, as fundamental as that of form to matter, space to geometry and numbers to arithmetic. This chapter will map out this trajectory from the point to the line, considering how and why the line surpassed the point as the foundation for artistic practice. While the interconnection between drawing and the line may seem obvious, it is historical. In the art theory of the fifteenth and sixteenth centuries, the point was the origin of the visual world of forms. The history of the point and the line tells a story about changing notions of magnitude/space and multitude/number that formatted a new structure of knowledge, in which the line became the foundation of artistic practice.

As will be seen, artists such as Leon Battista Alberti (1404–1472) and Albrecht Dürer (1471–1528) constructed theoretical systems founded on the point as the origin of the perceptible world. Moreover, this primary site was the means to measure the world, particularly the proportions of bodies, and the perspectival relationships in space between viewers and objects. The importance of the point as a tool for measurement continued into the seventeenth century, particularly in natural philosophy. As Michel Serres argues, the fixed point was decisive in the geometry, mechanics and cosmology of the seventeenth century.[2] In order to gauge the movement of planets, cannon balls or falling bodies, natural philosophers established an immobile point from which the trajectory of flight could be measured. Without a fixed point, there could be "no measure, no proportion, no order."[3] Yet as Serres elaborates, throughout the seventeenth century in the work of Girard Desargues (1591–1661), Blaise Pascal (1623–1662), and Gottfried Wilhelm Leibniz (1646–1716) the point changed from a motionless mark and became a designator of a multitude of optical relationships; the set point became the point of view (*Le point fixe est devenu point de vue*).[4] In artistic practice, the point began to function less a tool of measurement and more as the generator of the line, as a promise of presence constantly tracing its absence across space. The point, once the foundation of artistic practice, became obsolete. Artists shifted from discussing the point as a discernible instrument with which to measure bodies and the world and started writing about representing bodies in infinite varieties of animation, in which the line (as the gesture of the point) became the primary site for representation, form and artistic practice.

Geometry of Word and Image

In 1482, the Augsburg-printer Erhard Ratdolt (1442–1528) published in Venice the first printed edition of Euclid's *Elements*, presenting typographic mathematical diagrams and text on a single page (Figure 7.1).[5] The letter *P*—for *punctus*—commences the text. Just as all geometric forms unfold from the point, so the text begins with this primary point. Ratdolt placed the *P*-for-*punctus* within a historiated initial, juxtaposing the intricacy of the scrolling vines surrounding the P against the diagrammatic mathematical figure of the minute point. Ratdolt used an elaborate border indebted to the tradition of manuscript illumination to frame an innovation in print: the illustration of a mathematical text with diagrams. The chaotic growth of abstracted flora and fauna regulated into a pattern generates a visual play between the decorative presentation of the natural world and the geometric forms. The intricacy of the ornament stands out against the sparseness of the Euclidean figures, particularly the point as an almost imperceptible mark on the page. This opening page establishes geometry as a visual art both within the tradition of the illuminated manuscript and the representation of ideal mathematical figures. Yet the optical disconnection between the elaborate manuscript border and the mathematical diagrams also creates a visual disjunction on the page.[6] Later printed versions of the *Elements* do not present this same confluence of visual traditions. The intricacy of ornamental borders is omitted in favor of sparsely delineated mathematical figures. Nevertheless, the spare mathematical forms may be seen as the basis for the complexity of the ornament. Lines, circles and triangles become deformed and abstracted into acanthus scrolls and fleur-de-lis. When the abstract intellectual definitions of Euclid's *Elements* take form, particularly the point that has no parts, it leads not only to the line connecting two points but also to the infinite variety of the natural world and its styled representations.

The innovation for the "single geometry" of word and image as presented in the Ratdolt edition of Euclid (Figure 7.1) has been credited to the humanist polymath Alberti, who handwrote a treatise on painting between the years 1435 and 1436. This work contained no images and would not be printed until 1540.[7] Nevertheless, Alberti's *De pictura* commenced with Euclid's definitions of the point, the line and the surface. Alberti removed these definitions from the abstract realm of intellectual

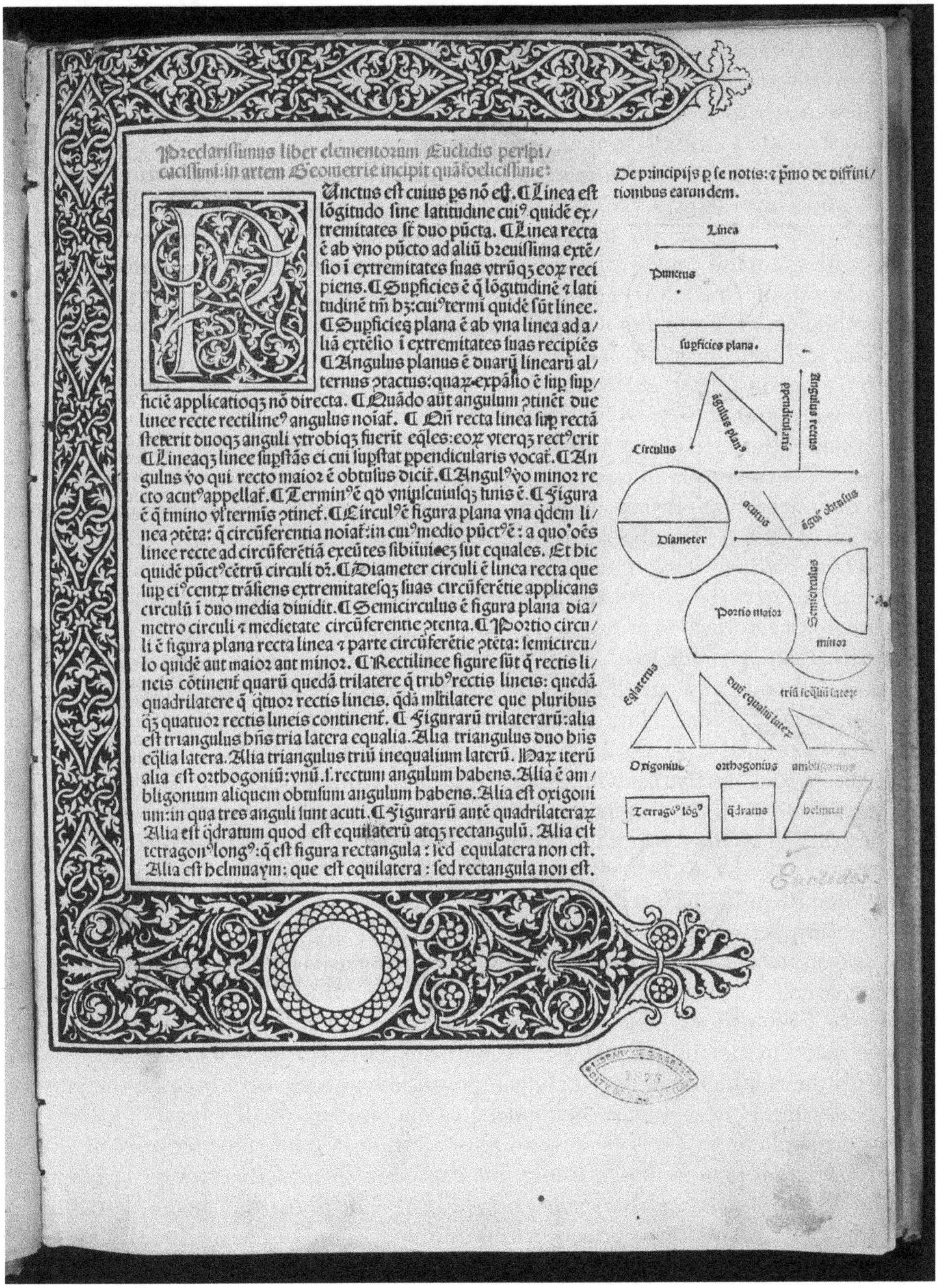

Figure 7.1 Erhard Ratdolt. *Preclarissimus liber elementorum*. Library of Congress, Washington, DC.

Photo credit: Library of Congress, Washington, DC.

figures and placed them to use as the rudiments of artistic practice, the fundamental tools necessary to picture the visible world. In the opening paragraphs of *De pictura*, Alberti stated that he will "take from mathematicians things that will seem to pertain to the subject," utilizing the point, the line and the surface.[8] As Alberti clarified, these mathematical forms are not abstract but grounded in physical existence. Alberti articulated himself "not as a mathematician but as a painter."[9] To clarify the difference, Alberti classified mathematicians as devoted to the intellectual world of numbers and ideal figures. In comparison, artists are bound to the finite limits of the material world: "Those [the mathematicians], in fact, measure figures and shapes of things with the mind only, without considering the materiality of the object."[10]

Contrary to the intellectual world of the mathematicians, Alberti considered the materiality of the object, beginning with the point. Just as Euclid's *Elements* commenced with the point, so Alberti began his text defining this intractable mark: "Before anything else, therefore, one must have understood that the point is a sign, so to speak, that in no way can be divided into parts."[11] Alberti then characterized the line: "The points will certainly make a line if they are joined without interruption, according to a sequence."[12] Alberti introduced the rudiments of painting with the Euclidean definitions of the point and the line. In this intellectual-yet-material geometrization of artistic practice, he separated it from a tradition of artisan's handbooks focused on technical procedures, such as Cennino Cennini's *Il libro dell'arte* (early fifteenth century). Instead Alberti situated painting within the intellectual realm of the *quadrivium* (geometry, arithmetic, music and astronomy).[13] Moreover, the point and the line provided the tools for Alberti to build his system of one-point perspective, "an open window" through which *historia* was viewed.[14] Alberti constructed this view with a quadrangle (the window) in the middle of which he placed a "point of sight" or "centric point." Through this point, all lines converged to create the simulation of three-dimensional space on a two-dimensional surface.[15] In Alberti, the point was both the necessary tool of the artist with which to establish the material world of representation and the means by which to order this world into a single-point perspective.

Alberti's *De pictura* travelled north to Nuremberg through the mathematician Regiomontanus (1436–76), who became acquainted with Alberti's work while in Italy. When Regiomontanus settled in Nuremberg with plans to start a printing house, his manuscripts from Italy were circulated among his intellectual circle, including the artist Dürer and the mathematician Thomas Venatorius (1448–1551), who wrote the opening epistle for the first printed edition of *De pictura* in 1540.[16] In this introduction to the 1540 printed *De pictura*, Venatorius stressed the relevance of Alberti's text to both mathematicians and painters. Venatorius asserted that Alberti began *De pictura* with the rudiments of the Euclidean point, line and so forth because the painter must understand "how man is the model and the measure of all things."[17]

This principle that man is "the measure of all things" guided the treatises of Dürer, both his *Underweysung der Messung mit dem Zirckel und Richtscheyt* (1525) and *Vier Bücher von menschlicher Proportion* (1528). The point and the line formed the basis for the construction of bodies, forms and images in Dürer's artistic theory, dedicated to defining man as "the measure of all things." For Dürer, the point marked the beginning and end of all lines, providing the tool to demarcate the finite definition of form. As Dürer explained in the opening paragraph to *Underweysung*, the point begins and ends not only all lines but also all living things that man would like to make or that man conceives with his mind.[18] Like Alberti, Dürer began his treatise with Euclid and his definitions of the point and the line. Like Alberti, Dürer struggled with the representation of that which has no size, width or thickness. Nevertheless, he acknowledged that the materialization of the point was necessary to the artist. Dürer conceded to make a point with a dab of the quill, presenting the corporeal point within the body of his introductory paragraph. In his illustration of this geometric figure, Dürer allowed a breadth of space between the words to offset the point, clarifying that this was the Euclidean point, which was the beginning of all sensory and bodily things (Figure 7.2). Dürer embedded the representation of the point within the text.[19] Although the point did not play as visual a role in Alberti's treatise as in Dürer's, the point was fundamental to both of their theoretical and perspectival systems.[20]

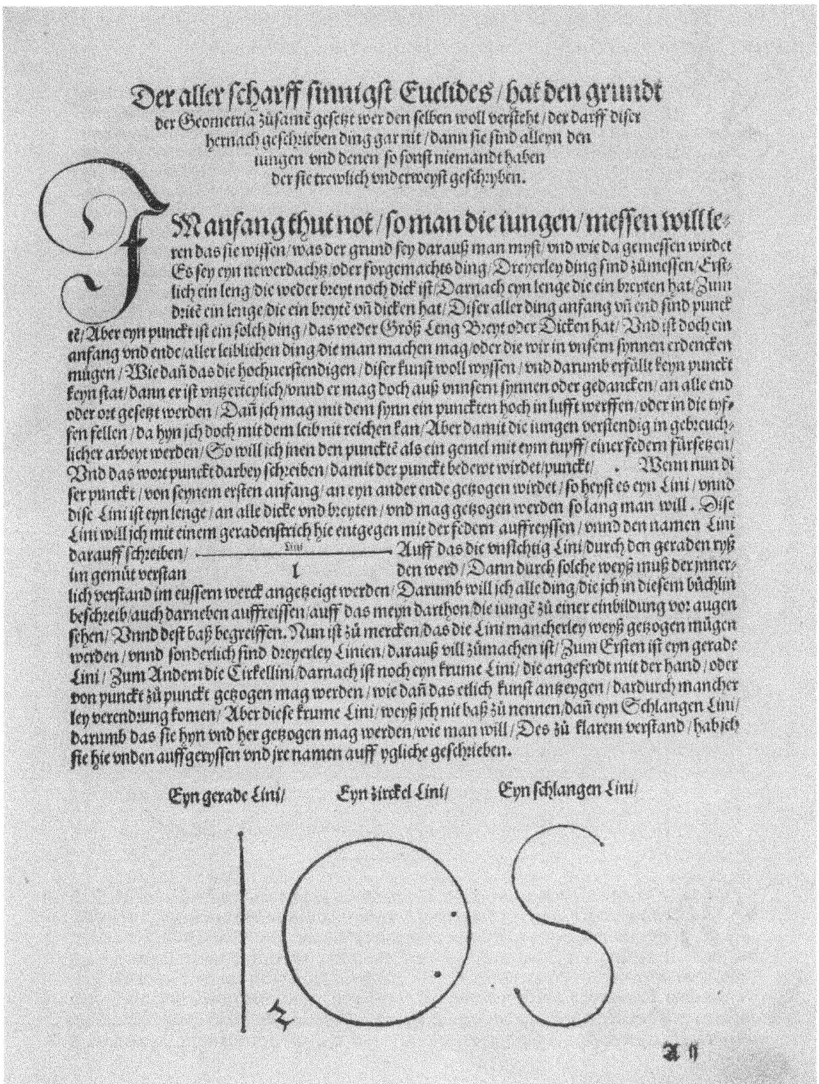
Der aller scharff sinnigst Euclides / hat den grundt
der Geometria zůsamē gesetzt wer den selben woll versteht / der darff diser
hernach geschriben ding gar nit / dann sie sind alleyn den
iungen vnd denen so sonst niemandt haben
der sie trewlich vnderweyst geschryben.

FManfang thut not / so man die iungen / messen will le-
ren das sie wissen / was der grund sey darauß man myst / vnd wie da gemessen wirdet
Es sey eyn newerdachts / oder forgemachts ding / Dreyerley ding sind zůmessen / Erst-
lich ein leng / die weder breyt noch dick ist / Darnach eyn lenge die ein breyten hat / Zum
dritē ein lenge / die ein breytē vñ dicken hat / Diser aller ding anfang vñ end sind punck-
te / Aber eyn punckt ist ein solch ding / das weder Größ Leng Breyt oder Dicken hat / Vnd ist doch ein
anfang vnd ende / aller leiblichen ding / die man machen mag / oder die wir in vnsern synnen erdencken
mugen / Wie dañ das die hochuerstendigen / diser kunst woll wyssen / vnd darumb erfüllt keyn punckt
keyn stat / dann er ist vntzerteylich / vnnd er mag doch auß vnnsern synnen oder gedancken / an alle end
oder ort gesetzt werden / Dañ ich mag mit dem synn ein punckten hoch in lufft werffen / oder in die tyf-
fen fellen / da hyn ich doch mit dem leib nit reichen kan / Aber damit die iungen verstendig in gebreuch-
licher arbeyt werden / So will ich inen den punckte als ein gemel mit eym tupff / einer federn fürsetzen /
Vnd das wort punckt darbey schreiben / damit der punckt bedewt wirdet / punckt / . Wenn nun di-
ser punckt / von seynem ersten anfang / an eyn ander ende getzogen wirdet / so heyst es eyn Lini / vnnd
dise Lini ist eyn lenge / an alle dicke vnd breyten / vnd mag getzogen werden so lang man will . Dise
Lini will ich mit einem geradenstrich hie entgegen mit der federn auffreyssen / vnnd den namen Lini
darauff schreiben / Lini Auff das die vnsichtig Lini / durch den geraden ryß
im gemůt verstan 1 den werd / Dann durch solche weyß muß der jnner-
lich verstand im eussern werck angetzeigt werden / Darumb will ich alle ding / die ich in diesem bůchlin
beschreib / auch darneben auffreissen / auff das meyn darthon / die iungē zů einer einbildung vor augen
sehen / Vnnd dest baß begreiffen. Nun ist zů mercken / das die Lini mancherley weyß getzogen mügen
werden / vnnd sonderlich sind dreyerley Linien / darauß vill zůmachen ist / Zum Ersten ist eyn gerade
Lini / Zum Andern die Cirkellini / darnach ist noch eyn krume Lini / die angeferdt mit der hand / oder
von punckt zů punckt getzogen mag werden / wie dañ das etlich kunst antzeygen / dardurch mancher-
ley verendrung komen / Aber diese krume Lini / weyß ich nit baß zů nennen / dañ eyn Schlangen Lini /
darumb das sie hyn vnd her getzogen mag werden / wie man will / Des zů klarem verstand / hab ich
sie hie vnden auffgerysen vnd jre namen auff ygliche geschrieben.

Eyn gerade Lini / Eyn zirckel Lini / Eyn schlangen Lini /

A ij

Figure 7.2 Albrecht Dürer. *Underweysung der Messung*. National Gallery of Art Library, Washington, DC.

Photo credit: National Gallery of Art Library, Washington, DC.

Accompanying his illustration to a lute drawn in proper perspective, Dürer outlined in the text how to picture this lute on a piece of paper. This process hinged on the artist's ability to utilize a thread to trace corresponding points between the object and its representation, a diligent procedure of matching a point on the physical lute with a point in space corresponding to the lute's image on the canvas. The artist and his apprentice measured the distance between the object and the canvas, "moving from point to point until the entire lute has been scanned and its points have been transferred to the tablet." Dürer then told the artist to "connect all the points on the tablet and you will see the result."[21] In this system of perspectival drawing, the point dominates the line as the image takes form through points as opposed to lines, which merely connect the pointed indentations (Figure 7.3).[22]

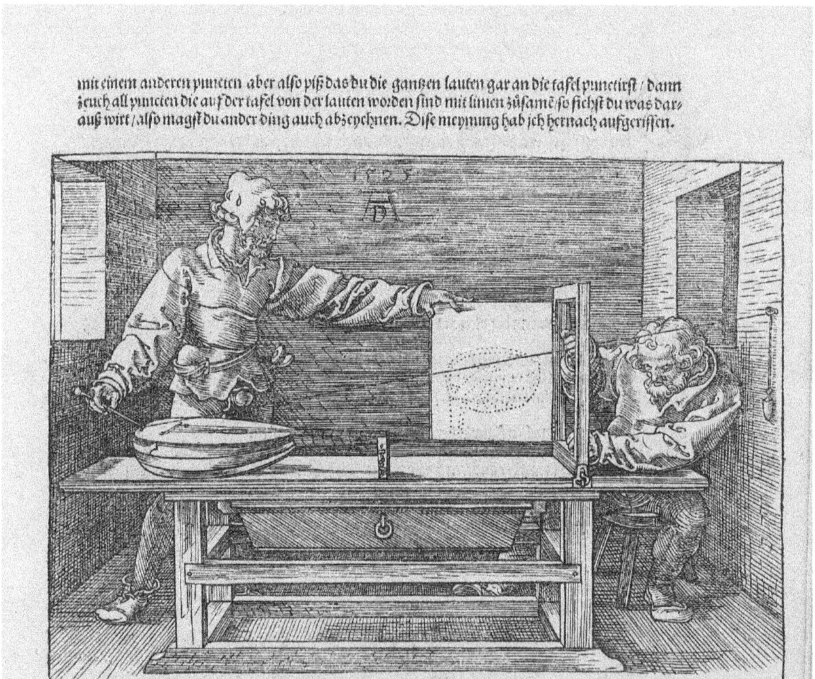
mit einem anderen puncten aber also piß das du die gantzen lauten gar an die tafel punctirst / dann zeuch all puncten die auf der tafel von der lauten worden sind mit linien zůsamẽ/so sichst du was darauß wirt / also magst du ander ding auch abzeychnen. Dise meynung hab jch hernach aufgerissen.

Vnd damit günstiger lieber Herr will jch meinem schreyben end geben / vnd so mir Got genad verleycht die bücher so jch von menschlicher proporcion vñ anderen darzů gehörend geschryben hab mit der zeyt in druck pringen/vnd darpey meniglich gewarnet haben/ob sich yemand vndersteen wurd mir diß außgangen büchlein wider nach zů drucken /das jch das selb auch wider drucken will / vñ auß lassen geen mit meren vnd grösserem zůsatz dañ ietz beschehen ist/darnach mag sich ein yetlicher richtẽ/Got dem Herren sey lob vnd eer ewigklich.

N iij

Gedruckt zů Nüremberg.
Jm. 1525. Jar.

Figure 7.3 Albrecht Dürer. *Men Drawing a Lute*. From "Unterweisung der Messung," Gedrückt zu Nuremberg: [s.n.], im 1525. Jar. "Institutiones geometricos." Spencer Collection.

Photo credit: The New York Public Library/Art Resource, NY.

Magnitudes and Multitudes

The conceptualization of the point as the foundation for artistic practice for both Alberti and Dürer was grounded in a systemization of knowledge in which the practices of geometry and arithmetic remained separate. Fundamental to Alberti's and Dürer's use of Euclidean geometry, especially the point and the line, were the concepts of magnitude and multitude. Euclid's *Elements* in the fifteenth century was part of a mathematical system in which geometry and arithmetic occupied separate spheres of the *quadrivium*. Geometry considered magnitudes. Arithmetic dealt with multitudes. Magnitudes (geometric figures such as the triangle) had the property of shrinking in size yet not losing their structure. Magnitudes were continuous. The point was not a magnitude, but it was the site from which all magnitudes unfolded. Bodies were magnitudes, and it was with the point and the line that artists measured bodies into proportions.

In comparison, numbers or multitudes were discontinuous entities composed of units. The foundational unit of multitudes (or number) was one. One was not a number. One was a unit from which numbers were composed. The first number was two. Just as the point was the beginning of all magnitudes but not a magnitude, so one was the beginning of all numbers but not a number. This relationship between magnitude and multitude shifted in the late sixteenth century, a modification that would eventually change the practice of representation. One of the most popular early modern editions of Euclid's *Elements*, published in English by the merchant and mayor of London, Henry Billingsley (d. 1606), with a lengthy preface by the Elizabethan occultist and mathematician John Dee (1527–1608/9), illustrates this epistemological distinction between magnitude and multitude, or continuous geometry of space and discontinuous numerical entities.

As Dee wrote, both *Number* and *Magnitude* "have a certaine Originall sede." The seed for multitude was the unit one. For magnitude, the seed was the point. Dee outlined the ways in which the unit and the point remained distinct. Although both were indivisible, a unit "is free, and can abyde no bondage." In contrast, the point must "have a certaine determined Situation: that is, that we may assigne, and prescribe a Point, to be here, there, yonder." Movement was not a quality of the unit because the unit was free. But as the point was tied to a "determined Situation," motion defined the point as it moved from here, there and yonder. Moreover, a point, "by his motion, produceth, Mathematically, a line."[23] While the point itself was not a magnitude, the motion of the point created the first magnitude: the line.

In this translation of the *Elements*, the point marked positions in space as they existed in the physical world. Moreover, this delineation of the relationship between the point and the line differed from Alberti's definition. For Alberti, a line was composed of a series of points: "the points will certainly make a line if they are joined without interruption."[24] For Billingsley and Dee, a line was composed of a point in motion. As Billingsley wrote: "*A lyne is the movying of a poynte*, as the motion or draught of a pinne or a penne to your sence maketh a lyne."[25] Billingsley's and Dee's definition of the line as the movement of the point was by no means original. Dürer's friend the mathematician and printer Regiomontanus wrote, "as the mathematicians duly assert, is a point by its imaginary flow or motion giving birth to lines."[26] The differences in definition, however, demonstrate that there are two competing classifications of the line: one as a series of points, the other as the movement of a point.

The definition of the line as series of points became problematized in the seventeenth century in the work of the Flemish mathematician and engineer Simon Stevin (1548–1620). In his *Arithmétique* (1585), Stevin argued that geometry and arithmetic, space and numbers were not separate spheres of knowledge. The separation of geometry and arithmetic in the *quadrivium* was no longer tenable as a means to describe the experience of space and the function of numbers in the world. As Jacob Klein argues in his study on early modern mathematics, Stevin's work as a bookkeeper allowed him to rethink mathematical language.[27] Stevin argued that "one" should be considered a number, and that both multitudes and magnitudes were continuous entities.[28] As stated previously, before Stevin numbers (multitudes) were discontinuous units and geometric shapes (magnitudes) were infinitely continuous.

To make his argument that one was a number, Stevin nuanced the definition of the point. Instead of likening the point to one, Stevin correlated the point to zero:

> Just as a point is an adjunct to a line and not in itself a line, so 0 is an adjunct to number, and not a number itself. Just as a point cannot be divided into parts, so 0 cannot be divided into parts. Just as many points, yea an infinity of them, do not make a line, so many 0's, even an infinity of them, do not form number.[29]

Just as a series of points do not make a line, so an infinite series of zeros do not make a number.[30] Whereas the point was once equivalent to one, it was now likened to zero. If the point was zero, the line could not be defined as a series of points "joined without interruption." The point has been likened to zero, and nothing makes nothing. Stevin's work demonstrates a crucial paradigm shift in the seventeenth century, signaling the breakdown of the *quadrivium*, in which geometry and arithmetic once occupied separate spheres of knowledge. Whereas both the point and one were the foundational parts from which magnitudes and multitudes unfolded (although neither was a magnitude nor a number), Stevin made one a number. In doing this, the point could no longer be likened to one (for it was still not a magnitude). The point became a signifier of nothing: zero. Instead of embodying the abstract concept of "that which has no parts" from which all the parts are formed, the point became a marker in space of something other than itself. The point became a place-holder for that which was once there or that which was yet to come. Rather than being the bare delimitation of that which had no parts, the point became likened to nothing. Nevertheless, Stevin was not the first mathematical theoretician to equate the point to zero. Nearly a century before Stevin's text, Leonardo wrote that the point was analogous to zero.

Noughts

Like artisans and craftsmen before him, Leonardo affirmed that art and geometry share the same starting place: the point. Unlike the artist theoreticians before him, however, Leonardo was less concerned with the point as a means of measurement and more interested in the relationship between the point and the line as a way to contour forms.[31] For Leonardo, "bodies have their origins in lines, the boundary of these surfaces."[32] Yet as Leonardo traced, these lines end in the point, for "we know that line has its conclusion in a point, and nothing can be smaller than that which is a point."[33]

Nevertheless, Leonardo differed in his writings from previous discussions of the point because he refused to acknowledge the coexistence of the intellectual mathematical point and the material point of the craftsman or painter. Leonardo clarified that the material point made by the painter was not a point:

> If you were to say that the contact made on a surface by the very tip of the point of a pen would create a point, this is not true. Rather, we would say that such contact as this would actually be a surface around a centre and that centre is the location of the point. And such a point is not of the material of the surface.[34]

A point centered the indentation made by the pen's ink, but this point was not part of the physical world. While the point was the starting site for all magnitudes, it was not a magnitude. Any physical realization of material form, no matter how small (such as the nib of the artist's pen), would nevertheless be a magnitude. Since a point cannot be a magnitude, this extension in space was not a point. Still the point's presence remained the site from which extension unfolded. Leonardo's understanding of the point as an impossible figure with no magnitude would lead him to argue (*contra* Alberti) that a line cannot be constructed by a series of points. As Stevin would write a century later, the point was not equal to one but to zero. For Leonardo: "If all the points that are potentially in the universe were to be united—should such a union be possible—neither they nor a single point would compose any part of a surface . . . This may be demonstrated by zero or nothing."[35] Leonardo's meditations on the point and its relationship to zero demonstrated a new concept of geometry and arithmetic that would not come together until Stevin's writings.

Yet the posthumous publication of Leonardo's writings in 1651 offers a lens with which to focus on the oncoming obsolescence of the point in favor of the line. While Leonardo's own theoretical framework likened the point to zero as a means to speak about the imperceptible world from which images unfold, this paradox became increasingly untenable as artists discussed the visual world not in terms of measurement but movement. The invisible form, which was once embodied in the point from which all magnitudes unfolded, lost its structural importance as artists began to consider visibility in relationship to the body in movement. Instead of considering the point from which all magnitudes extended, artists considered the impossibility of capturing the body in its infinite fluctuations of movement. The indiscernible form not was not the starting site of all magnitudes but rather the impossibility of the eye to grasp the flux of the human body as it moved from point to point. In this edited version of Leonardo's writings, the importance of the point in Leonardo's theory became obfuscated as the line takes precedence over the point.

The first published edition of Leonardo's writings was organized by later editors, including his student Francesco Melzi (1491–1568–70), Cardinal Francesco Barberini (1597–1679), Barberini's secretary Cassiana dal Pozzo (1588–1657), and the librarian and collector Raphaël Trichet du Fresne (1611–61).[36] While Leonardo declared the point's necessity to the painter, the first printed editions in Italian and French of Leonardo's writings only mention the geometric point briefly; for example, in an analysis of representing bodies from a distance. While the point played a pivotal role in Leonardo's artistic theory as the site from which the extended material world realized itself, in the *Trattato* the point appears in passing.

Although it is only in passing, the edited version of Leonardo's writings remained consistent with Leonardo's discussion of the point in describing the inability for the point to visibly materialize. In the *Trattato*, the geometric point occurs in an analysis on the boundaries between bodies:

> seeing that definitions of these bodies are nothing other than the definitions or terminations of their surfaces, and the definition of surfaces are lines . . . so it is not part of anything and is invisible, like the geometric point. This is why the painter should touch softly the contours of distant bodies.[37]

Leonardo used that which has no parts as a metaphor to extrapolate on the process of outlining figures, for both the point and the circumscribed line do not exist in the physical world. Yet the ability to softly contour these lines remains necessary for form to take shape, just as it is from the point's nonextended site that the visible world takes form. In this passage, Leonardo used the geometric point as a figure to describe the making of form with lines.[38] The line supersedes the point as the focus of artistic practice.

In the *Trattato*, the body and its infinitely changing proportions in movement become the object of attention. Before 1651, the point was fundamental to artistic practice. It was the cardinal tool for measuring the world and a means to conceptualize the materialization of form. Yet in the first published editions of Leonardo's writings, the interval between here and there (places marked by the point) could no longer be measured. In the face of this infinite incommensurability, artists concentrated on the point's movement: the line (the trace of a point's movement in space). In a diagram in the 1651 *Trattato*, a hand is placed within the contours of a triangle, whose side is formed by a line AB.[39] Yet as the passage tells the reader, the movement of the hand from point A to point B may never be measured:

> Now the Hand always changing its Figures and Aspect, as its Situation alters with regard to the Eye, it will be seen under as many different Aspects, as there are distinct Parts in the Motion; that is, the Aspects of the Hand are varied to infinity. And the Result wou'd be the same, if the Eye, instead of being lower'd from *A* to *B*, shou'd be rais'd from *B* to *A*; or if the Eye, were fixed, and the Hand, had its Motion.[40]

It is impossible to visualize the hand in its infinite variety of movements between points A and B. In its minute gesturing, the hand creates a multiplicity of viewpoints. Moreover, the eye also is mobile. There are two separate parts in motion: the hand and the eye. Representation mediates among many parts in transit both bodily and optical. At some point, the artist must choose a fixed view from which to picture the hand. While the point was fundamental to artistic practice in the fifteenth and sixteenth centuries, it lost its relevance in seventeenth-century artistic treatises, as seen in these editions of Leonardo's writings. The point became less necessary as artists let go of rigorous perspectival systems and began to celebrate the play of infinite points of view as a body moves through space, considering eyes and bodies that are both fixed and in motion.[41]

The oncoming obsolescence of the point revealed itself in other ways in the publication of Leonardo's manuscripts. In the Italian version of the *Trattato*, du Fresne included Alberti's *De pictura*, illustrated by Pier Francesco Alberti (1584–1638). While there were no accompanying illustrations in the first Latin printed edition of

De pictura (or Alberti's original manuscripts), this edition of *De pictura* included illustrations. The mathematical diagrams accompanying Alberti's treatise differ remarkably, however, from the illustrations for artists' treatises in the fifteenth and sixteenth century. As discussed, artists who included mathematical diagrams in their treatises followed the tradition of the *Elements*. They started with Euclid's definitions of the point, the line and the surface and accompanied these explications by pictorially realizing these rudimentary figures. In contrast, this version of Alberti's *De pictura* printed in 1651 does not begin with an illustration of the point. The first mathematical diagram presented is the circle (Figure 7.4). Alberti's text described the circle as: "a form of surface that a line encloses just like a crown. That is why, in the middle there will be a point." In this diagram accompanying Alberti's text, however, the circle's center remains empty. There is no point.[42]

In contradistinction to the earlier artistic treatises grounded in the point as a means to measure bodies and objects in space, most later seventeenth-century artistic treatises hardly discuss the "geometric point." Instead of measuring forms, theorists concentrated on an optical estimation. In Roger de Piles' *Les Premiers elemens de la peinture pratique* (1684), the third chapter on *Des Proportions* acknowledged: "the measurements of the body are the foundation of painting." Yet these proportions are not measured with the Euclidean point. Instead they are measured by the painter's eye, which grasps the moving proportions of a body by judging the interrelationships among a body's various members.[43]

Artists began to conceive of the point not as a fixed place from which to measure bodies but instead as a point-in-movement becoming the line. This signals a fundamental shift in early modern artistic practice from measuring the sensory world point by point to capturing with lines unbounded movement. Just as the diagram of the hand and the eye in Leonardo's *Trattato* demonstrated, the eye may remain fixed or in transit; either way the hand will realize an incalculable variety of forms from point A to point B. As the conjunction between the point and the line demonstrate, it was this indeterminable diversity of motility that came to dominate artistic practice in the seventeenth century. The history of the point and the line illustrates how the breakdown of the *quadrivium* and the strict separation between magnitudes and multitudes splintered in the seventeenth century. The point was once the basis of representation and extended form. Yet as the *quadrivium* fragmented in the seventeenth century, this irreconcilable point of that which has no parts became obsolete as both a tool of measurement and as a theoretical means to describe the making of images. Instead of focusing on the measurement of static bodies, artistic pedagogy and treatises dedicated themselves to the body in movement. The line, as the movement of the point, became the proper Euclidean form within which to ground a practice of representation devoted to bodies in movement.

No artist-theorist articulated this argument better than William Hogarth (1697–1764) and his *Analysis of Beauty* (1753), structured around the philosophy of the line. In a tradition going back to Alberti, Hogarth called on the diagrammatic world of the mathematician to clarify artistic practice. Yet Hogarth did not refer to the Euclidean point, line and surface but instead the line: "The constant use of lines by mathematicians, as well as painters, in describing things upon paper, hath establish'd a conception of them, as if actually existing on the real forms themselves." For Hogarth, the foundation of artistic practice was not the point, the line and the surface but the variety of lines: "the *straight line*, and the *circular line*, together with their different

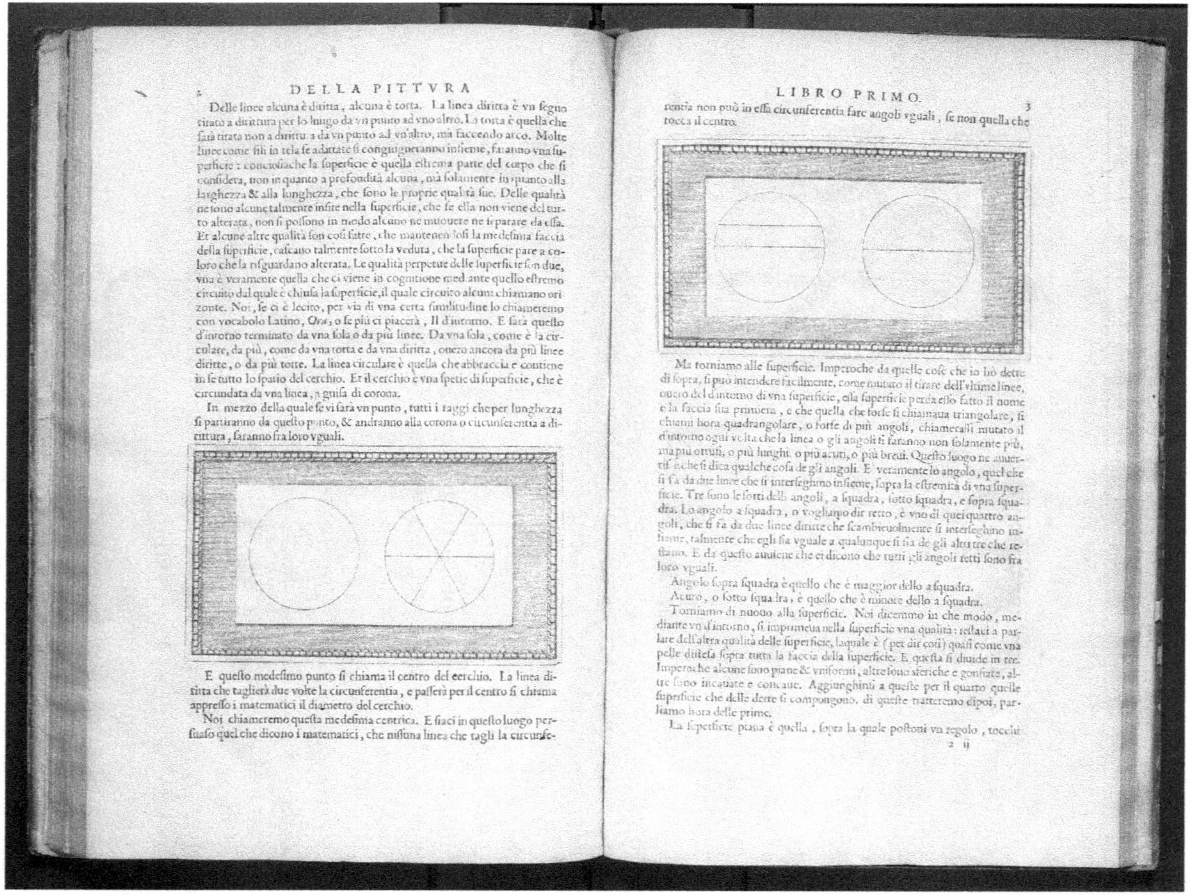

2 DELLA PITTVRA

Delle linee alcuna è diritta, alcuna è torta. La linea diritta è vn segno tirato a dirittura per lo lungo da vn punto ad vno altro. La torta è quella che sarà tirata non a dirittura da vn punto ad vn'altro, mà faccendo arco. Molte linee come fili in tela se adattate si congiugneranno insieme, faranno vna superficie: conciosiache la superficie è quella estrema parte del corpo che si considera, non in quanto a profondità alcuna, mà solamente in quanto alla larghezza & alla lunghezza, che sono le proprie qualità sue. Delle qualità ne sono alcune talmente insite nella superficie, che se ella non viene del tutto alterata, non si possono in modo alcuno ne muouere ne separare da essa. Et alcune altre qualità son cosi fatte, che mantenendosi la medesima faccia della superficie, cascano talmente sotto la veduta, che la superficie pare a coloro che la risguardano alterata. Le qualità perpetue delle superficie son due, vna è veramente quella che ci viene in cognitione mediante quello estremo circuito dal quale è chiusa la superficie, il quale circuito alcuni chiamano orizonte. Noi, se ci è lecito, per via di vna certa similitudine lo chiameremo con vocabolo Latino, *Ora*, o se più ci piacerà, Il d'intorno. E sarà questo d'intorno terminato da vna sola o da più linee. Da vna sola, come è la circulare, da più, come da vna torta e da vna diritta, ouero ancora da più linee diritte, o da più torte. La linea circulare è quella che abbraccia e contiene in se tutto lo spatio del cerchio. Et il cerchio è vna spetie di superficie, che è circundata da vna linea, a guisa di corona.

In mezzo della quale se vi sarà vn punto, tutti i raggi che per lunghezza si partiranno da questo punto, & andranno alla corona o circunferentia a dirittura, saranno fra loro vguali.

E questo medesimo punto si chiama il centro del cerchio. La linea diritta che taglierà due volte la circunferentia, e passerà per il centro si chiama appresso i matematici il diametro del cerchio.

Noi chiameremo questa medesima centrica. E siaci in questo luogo persuaso quel che dicono i matematici, che nissuna linea che tagli la circunfe-

LIBRO PRIMO. 3

rentia non può in essa circunferentia fare angoli vguali, se non quella che tocca il centro.

Ma torniamo alle superficie. Imperoche da quelle cose che io hò dette di sopra, si può intendere facilmente, come mutato il tirare dell'vltime linee, ouero del d'intorno di vna superficie, essa superficie perda esso fatto il nome e la faccia sua primiera, e che quella che forse si chiamaua triangolare, si chiami hora quadrangolare, o forse di più angoli, chiamerassi mutato il d'intorno ogni volta che la linea o gli angoli si faranno non solamente più, mà più ottusi, o più lunghi, o più acuti, o più breui. Questo luogo ne auuertisce che si dica qualche cosa de gli angoli. E veramente lo angolo, quel che si fa da due linee che si interseghino insieme, sopra la estremità di vna superficie. Tre sono le sorti delli angoli, a squadra, sotto squadra, e sopra squadra. Lo angolo a squadra, o vogliamo dir retto, è vno di quei quattro angoli, che si fa da due linee diritte che scambieuolmente si interseghino insieme, talmente che egli sia vguale a qualunque si sia de gli altri tre che restano. E da questo auuiene che ei dicono che tutti gli angoli retti sono fra loro vguali.

Angelo sopra squadra è quello che è maggior dello a squadra.

Acuto, o sotto squadra, è quello che è minore dello a squadra.

Torniamo di nuouo alla superficie. Noi dicemmo in che modo, mediante vn d'intorno, si imprimeua nella superficie vna qualità: restaci a parlare dell'altra qualità delle superficie, laquale è (per dir cosi) quasi come vna pelle distesa sopra tutta la faccia della superficie. E questa si diuide in tre. Imperoche alcune sono piane & vniformi, altre sono sferiche e gonfiate, altre sono incauate e concaue. Aggiunghinsi a queste per il quarto quelle superficie che delle dette si compongono. di queste tratteremo dipoi, parliamo hora delle prime.

La superficie piana è quella, sopra la quale postoui vn regolo, tocchi

a ij

Figure 7.4 Leon Battista Alberti and Pier Francesco Alberti, "Definition of a Circle," *De pictura*, National Gallery of Art, Washington, DC.

Photo credit: National Gallery of Art Library, Washington, DC.

combinations, and variations, &c. bound, and circumscribe all visible objects whatsoever, thereby producing such endless variety of forms."[44] The basis of Hogarth's fascination with the infinite variety of form found in the line developed from the inability to measure the human body: "the vast variety of intricately situated parts, belonging to the human form, will not admit of measuring the distances of one part by another, by lines or points, beyond a certain degree or number, without great perplexity in the operation itself, or confusion to the imagination."[45] Hogarth's text embodies a movement away from not only geometric points but also a concept of artistic practice as a means to measure. Hogarth's world is immeasurable, an incommensurability symbolized by the infinite movements of the human body embodied in the line.

Notes

1 In the West, the reception of Euclid's *Elements* in the twelfth and the thirteenth centuries occurred through the translation of Euclid's *Elements* from Arabic (translations of the original Greek) into Latin. Two of the most important translators were Adelard of Bath (1080–1152) and Campanus of Novarra (1220–1296). Novarra's version was the first published edition of the *Elements*, printed by Erhard Ratdolt in 1482. Although there was significant confusion about Euclid's historical identity in the early modern period, it generally was agreed that the *Elements'* author was Euclid of Megara, a philosopher contemporary to Plato c. 400 BCE. Leibniz's quote, which leads this chapter, expresses the consensus regarding Euclid's *Elements* that the original version had been corrupted through its many translations. Therefore, throughout the early modern period translators and editors of the *Elements* sought to right discrepancies in the text. One of the most important vernacular translators to do this was Niccolò Tartaglia, who corrected the definitions of ratio and proportion as defined in the earlier Latin versions of Euclid in his *Euclide megarense philosopho: solo introduttore delle scientie mathmatice* (Venice, 1543). For the reception of Euclid's *Elements* in the early modern period and a list of printed editions, see Antoni Malet, "Euclid's Swan Song: Euclid's *Elements* in Early Modern Europe," in *Greek Science in the Long Run: Essays on the Greek Scientific Tradition (4th c. BCE–17th c. CE)*, ed. Paula Olmos (Newcastle: Cambridge Scholars Publishing, 2012), 205–235. For the "medieval" Euclid, see: Menso Folkerts, *The Development of Mathematics in Medieval Europe: The Arabs, Euclid, Regiomontanus* (Aldershot and Burlington: Ashgate, 2006); John E. Murdoch, "The Medieval Euclid: Salient Aspects of the Translation of the Elements by Adelard of Bath and Campanus of Novara," *Revue de Synthèse* 49–52 (1968): 67–94; Marshall Clagett, "The Medieval Latin Translations from the Arabic of the *Elements* of Euclid, with Special Emphasis on the Versions of Adelard of Bath," *Isis* 44 (1953): 16–42.

2 Michel Serres, *Le système de Leibniz et ses modèles mathématiques*, vol. 2 (Paris: Presses Universitaires de France, 1968), 649–650.

3 Ibid., 2: 658.

4 Ibid., 656.

5 For an overview of typographic variations on the point, see: Michael J. Barany, "'That Small and Unsensible Shape': Visual Representations of the Euclidean Point in Sixteenth-Century Print," *Spontaneous Generations: A Journal for the History and Philosophy of Science* 6 (2012): 148–159.

6 Büttner, Damerow, Renn and Schemmel argue that Euclidean geometry was a foundation for early modern science particularly because it was a visualized mathematical tradition, inseparable from the construction of mathematical diagrams; see Jochen Büttner, Peter Damerow, Jürgen Renn, and Mattias Schemmel, "The Challenging Images of Artillery: Practical Knowledge at the Roots of the Scientific Revolution," in *The Power of Images in Early Modern Science*, ed. Wolfgang Lefèvre, Jürgen Renn and Urs Schoepflin (Basel: Birkhäuser Verlag, 2003), 3–27.

7 Friedrich Kittler, "Perspective and the Book," *Grey Room* 5 (2001): 38–53. Alberti wrote *De pictura* in both the Tuscan vernacular and Latin. There is debate among scholars about whether Alberti first wrote the treatise in Latin or the Tuscan vernacular. Although it has

been assumed that Alberti first wrote the Latin edition and then the vernacular edition for local craftsmen, Sinisgalli recently argued that Alberti wrote the vernacular first and then the Latin. For an overview of these arguments, see: Rocco Sinisgalli, *Leon Battista Alberti: On Painting*, ed. and trans. Rocco Sinisgalli (Cambridge: University of Cambridge Press, 2011), 7–14.

8 Leon Battista Alberti, *On Painting: A New Translation and Critical Edition*, ed. and trans. Rocco Sinisgalli (New York: Cambridge University Press, 2011), 22.

9 Ibid.

10 Ibid. There is also a Florentine commentary written on Euclid's *Optics* at the end of the fourteenth century that is careful to distinguish between the physical point and the imaginary mathematical point. The writer provides three definitions of the point: one geometric, one physical and one optical. On this treatise, see Jack M. Greenstein, "On Alberti's 'Sign': Vision and Composition in Quattrocento Painting," *The Art Bulletin* 79(4) (1997): 683; Graziella Federici Vescovini, *Studi sulla prospettiva medieval* (Turin: G. Giappichelli, 1965), 223–228.The words Sinisgalli translates into "materiality" from the 1540 printed version of *De pictura* include *materia, species* and *formas*.

11 Alberti, *On Painting*, 23.

12 Ibid.

13 For an important nuancing of this, see: Lon R. Shelby, "The Geometrical Knowledge of Medieval Master Masons," *Speculum* 47(3) (1972): 395–421. Shelby draws attention to the discussion of geometry by Villard de Honnecourt in his *Sketchbook* (c. 1230), in which Villard explicitly states the importance of geometry to drawing. Shelby also, however, contextualizes that Villard's "geometry" does not mean Euclidean geometry but "the ability to perceive design and building problems in terms of a few basic geometrical figures which could be manipulated through a series of carefully prescribed steps to produce the points, lines and curves needed for the solution of the problems" (420–421). While Alberti's text breaks new ground in its "intellectualizing" of the artist through grounding artistic practice in the language of mathematics, it should be recognized that the relationship between art and an idea of geometry extended to earlier craftsman traditions, especially the visualization of "points, lines and curves."

14 Alberti, *On Painting*, 39–40.

15 Another important discussion of the point in an early modern treatise is Piero della Francesca's in *De prospectiva pingendi*, written in the early 1480s. Like Alberti, Piero defines the point as "that which has no parts." Piero clarified that his treatise deals with the visual world of perspective, and therefore he must give another definition, by which the point is the smallest possible quantity that the eye may comprehend. In turn, the line is the extension from the point. Piero della Francesca, *De prospectiva pingendi*, ed. G. Nicco Fasola (Florence: Casa Editrice le Lettere, 1984), 65. Piero also had an exceptional relationship to mathematics, as he wrote two treatises devoted solely to mathematics: the *Trattat d'abaco* and *Libellus de quinque corporibus regularibus*. On Piero's mathematical works, see: Margaret Daly Davis, *Piero della Francesca's Mathematical Treatises: The "Trattato d'abaco" and "Libellus de quinque corporibus regularibus"* (Ravenna: Longo Editore, 1977); J. V. Field, "A Mathematician's Art," in *Piero della Francesca and His Legacy*, ed. Marilyn Aronberg Lavin (Hanover and London: University Press of New England, 1995), 177–197.

16 Venatorius dedicates the edition to the "teacher, doctor and very excellent mathematician" James Milichius. Venatorius stated that with his expertise in mathematics Milichius should be best equipped to write about painting; see Rocco Sinnisgalli, "La Dedica di Thomas Venatorius—The Dedication of Thomas Venatorius," *Il Nuova De pictura* di Leon Battista Alberti (Rome: Edizioni Kappa, 2006), 536–540. On Dürer and Alberti, see Idem., *Il Nuovo De pictura di Leon Battista Alberti / The New De pictura of Leon Battista Alberti* (Roma: Edizioni Kappa, 2006), 27–29; Peter Krüger, "Der Kompositionsbegriff und die Rhetorik: Alberti und Dürer," in *Dürers "Apokalypse,": Zur poetischen Struktur einer Bilderzählung der Renaissance* (Wiesbaden: Harrassowitz Verlag, 1996), 89–100; Hans Rupprich, "Die kunsttheoretischen Schriften L.B. Albertis und ihre Nachwirkung bei Dürer," *Schweizer Beiträge zur allgemeinen Geschichte* 18/19 (1960–61): 219–239.

17 Sinnisgalli, "La Dedica," 539.

18 Albrecht Dürer, *Underweysung der Messung mit dem Zirckel un[d] Richtscheyt in Linien, Ebnen und gantzen Corporen* (Nuremberg, 1525), n.p.

19 This double functionality of the point (as both the fundamental tool of measurement and as the key tool in perspective drawing) continues in less well-known works on artistic practice in the sixteenth century. In the tract Heinrich Lautensack, *Deß Circkels und Richtschents auch der Perspectiva und Proportion der Menschen* (Frankfurt, 1564), the goldsmith and painter Heinrich Lautensack (1522–1590) declared the point's fundamentality to both the introduction of the representable world and the basis for perspective. Like Alberti and Dürer, Lautensack began his treatise with the point: "To start with, I will speak about the point." As Lautensack told his reader, the point divides and measures: "In order to divide something, there must be the point because otherwise we would not know what is short or what is long." As Lautensack acknowledged, the point is fundamental not only to measurement but also to perspective, for the *Augpuncten* must be taken into consideration because without it nothing can be brought into perspective. Other *Kunstbüchlein* also begin with the point, such as Sebald Beham's *Das Kunst und Lere Büchlin Sebalden Behems* (Frankfurt, 1552).

20 For an overview of the "distance point," see: Lyle Massey, "Configuring Spatial Ambiguity: Picturing the Distance Point from Alberti to Anamorphosis," in *The Treatise on Perspective: Published and Unpublished*, ed. Lyle Massey (New Haven and London: Yale University Press, 2007), 161–176.

21 Albrecht Dürer, *The Painter's Manual: A Manual of Measurement of Lines, Areas and Solids by Means of Compass and Ruler*, trans. Walter L. Strauss (New York: Abaris Books, 1977), 391–393.

22 Carmen Bambach discusses a series of Italian drawings attributed to Luca Signorelli (1445–1523), which were formed through the proportional construction method outlined by Piero della Francesca in *De prospectiva pingendi* (a text/method that influenced Dürer's own). As Bambach argues, this is a concrete example of a drawing that took shape first through pricking the outline of the design with holes and then using lines to connecting these pricked indentations. Bambach describes this method as "a means of teaching spatial relationships by developing visual judgment. Thus, by mastering the exercise, painters and designers of decorative arts could learn the rudiments of perspective and foreshortening." Carmen Bambach Cappel, "On '*La Testa Proportionalmente Degradata*': Luca Signorelli, Leonardo da Vinci and Piero della Francesca's *De Prospectiva Pingendi*," in *Florentine Drawing at the Time of Lorenzo the Magnificent*, ed. Elizabeth Cropper (Bologna: Nuova Alfa Editoriale, 1994), 28.

23 John Dee, "John Dee His Mathematicall Praeface," in *The Elements of Geometrie of the Most Ancient Philosopher Euclide of Megara*, trans. Henry Billingsley (London: John Daye, 1570), n.p.

24 Alberti, *On Painting*, 23.

25 Billingsley, "Definition 2," *The Elements of Geometrie*, n.p.

26 Regiomontanus, *On Triangles. De triangulis omnimodis*, trans. Barnabas Hughes (Madison: University of Wisconsin Press, 1967), 37.

27 Jacob Klein, *Greek Mathematical Thought and the Origin of Algebra*, trans. Eva Braun (Cambridge, MA and London, UK: MIT Press, 1968), 186. See also: Katherine Neal, *From Discrete to Continuous: The Broadening of Number Concepts in Early Modern England* (Dordrecht, Boston and London: Kluwer Academic Publishers, 2002), 1–12, 33–36; Wolfgang Schäffner, "The Point: The Smallest Venue of Knowledge in the 17th Century (1585–1665)," in *Collection, Laboratory, Theater: Scenes of Knowledge in the 17th Century*, ed. Helmar Schram, Ludger Schwarte and Jan Lazardig (Berlin: Walter de Gruyter, 2005), 57–74; Malet, "Euclid's Swan Song: Euclid's *Elements* in Early Modern Europe," 205–235.

28 Malet points out that several early modern editions of Euclid's *Elements* laid the groundwork for Stevin's new definition of numbers and magnitudes. In particular, Malet looks at Clavius, *Commentaria in Euclidis Elementa Geometrica* (Mainz, 1611); Henry Billingsley, *Euclides: The Elements of Geometrie* (London, 1570); Tartaglia, *Euclide megarense philosopho . . .* Antoni Malet, "Renaissance Notions of Number and Magnitude," *Historia Mathematica* 33 (2006): 63–81. For an elucidation of early modern mathematical

terminology, see: Henk J. M. Bos, *Redefining Geometrical Exactness: Descartes' Transformation of the Early Modern Conception of Construction* (New York: Springer Verlag, 2001), 119–134.

29 Simon Stevin, *The Principal Works of Simon Stevin*, ed. D. J. Striuk, vol. IIB (Amsterdam: C.V. Sets & Zeitlinger, 1958), 498.

30 Klein, *Greek Mathematical Thought*, 192–193.

31 For Leonardo on the point, see: Augosto Marinoni, "L'Essere del Nulla," *Letture Vinciane* I–XII (1960–1972): 9–28; Frank Fehrenbach, "Leonardo da Vinci: Leonardo's Point," uploaded on February 16, 2012, The Warburg Institute, http://www.youtube.com/watch?v=XwwkvjCpauQ.

32 Leonardo da Vinci, *Leonardo on Painting*, trans. Martin Kemp and Margaret Walker (New Haven and London: Yale University Press, 1989), 13–14. On Leonardo and "continuous quantity" see: Pierre Duhem, "Leonard de Vinci et les deux infinis," *Études sur Léonard de Vinci* (Paris: A. Hermann, 1906–1913), 2: 3–53; Martin Kemp, "Leonardo e lo spazio dello scultore," *Lettura Vinciana* 27 (1988): 5–23.

33 da Vinci, *Leonardo on Painting*, 15.

34 Ibid.

35 Ibid. Leonardo does not argue (as does Stevin) that because the point is equivalent to zero *both* multitudes and magnitudes may be seen as continuous. Leonardo kept separate the traditions of discrete discontinuous units found in arithmetic and the continuous extended bodies of geometry.

36 The *Trattato della pittura* consists of excerpts compiled from Leonardo's writings by his student Francesco Melzi. This compilation, in turn, became abridged into a series of manuscripts that circulated throughout Italy. Finally, Cardinal Francesco Barberini and Cassiano dal Pozzo instigated the publication of Leonardo's *Trattato* from several of the abridged manuscripts. Famously, Poussin drafted a series of figural drawings to the publication and Pier Francesco Alberti made the diagrammatic images, both series engraved by Charles Errard. Raphael Trichet de Fresne edited the publication and included dedicatory letters, biographies of Leonardo and Alberti, and Alberti's treatises on painting and sculpture. On this publication history, see: Leonardo da Vinci, *Treatise on Painting: Codex Urbinas Latinus 1270 by Leonardo da Vinci*, trans. Philip McMahon (Princeton: Princeton University Press, 1956); Kate Traumann Steinitz, *Leonardo da Vinci's 'Trattato della pittura': A Bibliography of the Printed Editions 1651–1956* (Copenhagen: Munksgaard, 1958); Jean Paul Richter, "History of the 'Trattato della Pittura'," in *Leonardo's Writings and Theory of Art*, ed. Claire Farago (New York and London: Garland Publishing, 1999), 315–322; Claire Farago, "Introduction: The Historical Reception of Leonardo da Vinci's Abridged *Treatise on Painting*," in *Re-Reading Leonardo: The Treatise on Painting across Europe, 1550–1900*, ed. Claire Farago (Burlington: Ashgate, 2009), 407; Farago, "Who Abridged Leonardo da Vinci's *Treatise on Painting*?" *Re-Reading Leonardo*: The Treatise on Painting across Europe (Farnham and Burlington: Ashgate, 2009),77–106; Idem., "How Leonardo da Vinci's Editors Organized His Treatise on Painting and How Leonardo Would Have Done it Differently," *The Treatise on Perspective*, 21–52; Catherine M. Sousloff, "The *Vita* of Leonardo da Vinci in the Du Fresne Edition of 1651," *Re-Reading Leonardo*, 175–196; Donatella Livia Sparti, "Cassiano Dal Pozzo, Poussin and the Making and Publication of Leonardo's 'Trattato'," *Journal of the Warburg and Courtauld Institutes* 66 (2003): 143–188.

37 Leonardo da Vinci, *Trattato della pittura / Traité de la peinture* (Paris: Langlois, 1651), chapter 306.

38 In later editions, however, this analogy loses its resonance, for later translators changed the point to the "Mathematical line." In this same passage, the "Geometrical Point" becomes the "Mathematical Line" as the limit of the visible world: "Now the termination or bounding of one thing upon another, is in reality no more than a Mathematical Line; not having the Properties of a Physical one." The variations in language between the 1651 editions in Italian and French and a later 1721 edition in English demonstrate the coming obsolescence of the point. Leonardo da Vinci, *A Treatise of Painting by Leonardo da Vinci: Translated from the Original Italian* (London: Printed for J. Senex . . . and W. Taylor . . ., 1721), 174–175.

39 While there has been considerable attention to Poussin's illustrations of human figures for the *Trattato*, there has been little consideration of the geometric illustrations for both the *Trattato* and Alberti's *De pictura*. For articles on the changing role of word and image in the *Trattato*, see Martin Kemp and Juliana Barone, "What Might Leonardo's Own *Trattato* Have Looked Like? And What Did it Actually Look Like Up to the Time of the *Editio Princeps*," *Re-Reading Leonardo*, 39–60; Michael Cole, "On the Movement of Figures in Some Early Apographs of the Abridged *Trattato*," *Re-Reading Leonardo*, 108–125.
40 da Vinci, *Trattato*, chapter 271.
41 This passage illustrates Serres' argument that the "fixed point" became crucial in the seventeenth century only to transform into the "point of view." Similar to Serres' argument about the point in natural history, in which the fixed point was central for establishing limits and laws, in their study of "Baroque Science" Ofer Gal and Rez Chen-Morris also argue that the system of mathematical perspective that relied on a "fixed point" (such as Alberti's window) regulated this infinite movement of form precisely "to provide an anchor of stability in a world of shifting images." Ofer Gal and Raz Chen-Morris, *Baroque Science* (Chicago and London: University of Chicago Press, 2013), 134–145. Gal and Chen-Morris specifically connect this "fixed" point to Albertian concepts of beauty. This becomes more complicated when looking at an artist such as Albrecht Dürer, who also relied on the "fixed point" for perspectival instruction, yet Dürer also understood beauty in a way distinctly separate from Albertian conceptions. See: Erwin Panofsky, "Dürer as a Theorist of Art," in *Life and Art of Albrecht Dürer* (Princeton: Princeton University Press, 1955).
42 Alberti, *On Painting*, 24.
43 Roger de Piles, *Les premiers elemens de la peinture pratique* (Paris: Chez Nicolas Langlois, 1684), 5.
44 William Hogarth, *The Analysis of Beauty*, ed. Ronald Paulson (New Haven and London: Yale University Press, 1997), 41.
45 Ibid., 64.

8 Between the Golden Ratio and a Semiperfect Solid

Fra Luca Pacioli and the Portrayal of Mathematical Humanism

Renzo Baldasso and John Logan

The presence of the *Portrait of Luca Pacioli and Gentleman* (Museo Nazionale di Capodimonte, Naples; hereafter referred to as *Portrait*) at the important exhibition *Dürer—German Master* confirmed the popularity of this painting among curators and the general public.[1] Even though not a great masterpiece, the *Portrait* enjoys a sort of iconic status. Thanks primarily to the many mathematical references it contains, it appears "modern" and familiar to the general public, who may also consider it an epitome of the achievements of the Renaissance, as linear perspective, naturalism, printed books, and the rebirth of the scientific knowledge of the Ancients are all themes present in the *Portrait.* In spite of the broad appeal, its contents remain poorly understood. Recent research has detailed the connections between its iconography and mathematical humanism and brought into clearer focus the mathematics displayed in the panel. Building on these results and hoping to advance our understanding of the *Portrait*, this chapter addresses two outstanding puzzles: the numbers and line segments on the bottom left of the tablet and the addition problem on the bottom right, and the two remarkable solids appearing respectively in the lower right and upper left. The numbers and segments on the tablet will be shown to relate to different powers of the golden ratio, while the solids will be analyzed on their own terms and their presence explained contextually from the perspective of the history of complex geometric solids in the Renaissance. The present study confirms that the iconography of the *Portrait* (Figure 8.1) was carefully composed by Pacioli, at a moment—retrospectively—that marked a turning point in his life: after 1495, the friar will transform himself from itinerant teacher of introductory and classical mathematical knowledge into courtier and creator of new and exciting mathematical insights, which eventually bore fruit in the pages of the *Divina proportione*.

Because of the status the *Portrait* has acquired—starting from its celebration on the cover of the catalog of the blockbuster exhibition *Circa 1492: Art in the Age of Exploration*—and its frequent inclusion in international exhibitions, in recent years the painting has received considerable scholarly attention.[2] In an article that appeared in the *Art Bulletin* in 2010, Renzo Baldasso pursued an integral interpretation of the *Portrait*, which also brought into focus the subject of geometric figures and the visual reasoning they demand and the differences between the painted diagrams and those printed in the two editions of the *Elements* by Euclid available in 1495.[3] In another 2010 publication, Enrico Gamba revisited some of the painting's outstanding issues including the authorship, the identity of the second figure, and the mathematics displayed, while also considering various hypotheses explaining the half-drawn line to which the friar points.[4] Notably, Baldasso and Gamba agree on two important

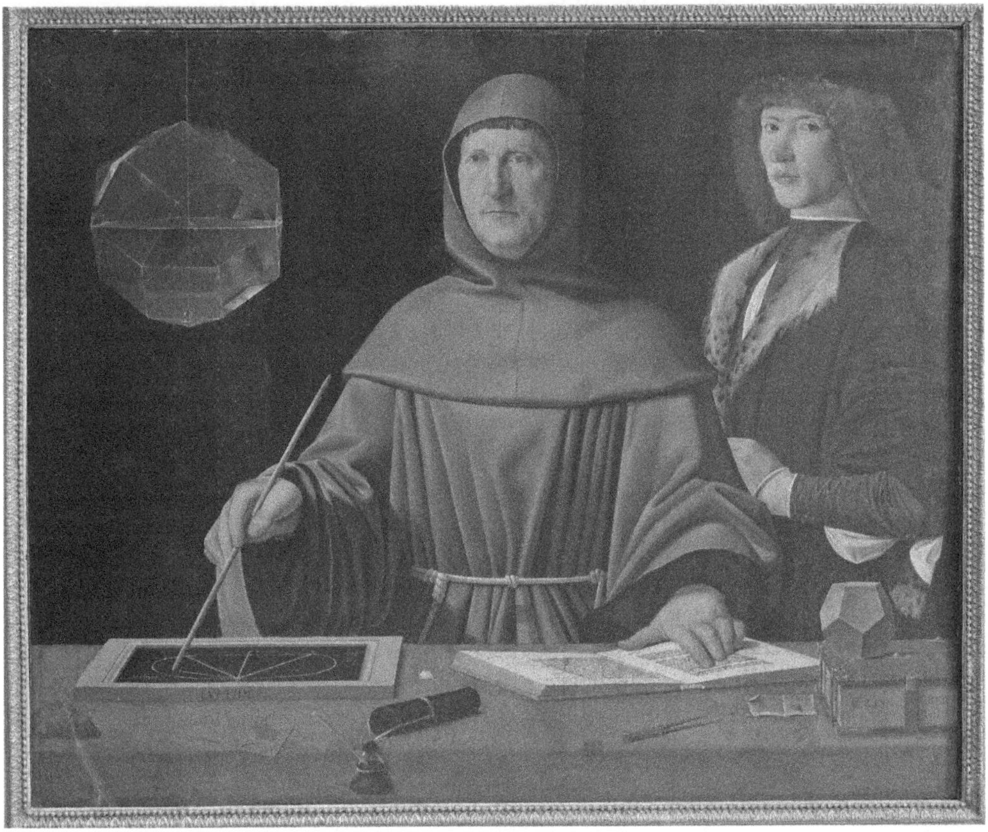

Figure 8.1 Jacopo de' Barbari (?). *Portrait of Luca Pacioli and Gentleman*, 1495, 98 × 108 cm. Museo e Real Bosco Capodimonte, Naples.

Photo credit: Renzo Baldasso. Reproduced by courtesy of the Ministero dei Beni e delle Attività Culturali e del Turismo.

points: both attribute the complex iconographic program to the friar and recognize as an important aspect of the painting the fact that it challenges the viewer to learn about and work through the figures that Pacioli is considering. In a 2011 article, Argante Ciocci, an expert on Pacioli, reviewed the painting with particular attention to the attribution challenge and, integrating the results from the just-mentioned publications, proposed a reading of the *Portrait* as a metaphor of the various aspects of Pacioli's mathematical interests and activities.[5]

Notwithstanding various unresolved questions, the following have become firm points in the scholarship of the *Portrait*. The attribution to Jacopo de' Barbari is generally accepted, including by the curators of the Museo di Capodimonte, which had maintained the safer "Jaco.Bar" as author in the last printed catalog of the collections but now list on its website the Venetian as the painter.[6] Concerning the iconography of the painting, the consensus is that de' Barbari followed Pacioli's directions: it is the friar and not the painter that devised the iconographic program for the *Portrait*—the results later in the chapter offer further support for this claim. The painting presents the friar as an authority on Euclid and Euclidean geometry, and particularly on the geometric figures of the *Elements*.[7] Standing taller and partly behind the friar, wearing a fur robe, the second figure does not appear to be a student; rather, his illuminated hand and enticing stare at the viewer pose him as a patron of Pacioli and his research and, more generally, of the education and mentality offered by mathematical humanism. The mathematician and his knowledge are pictured in a space that for size and illumination resembles a *studiolo*, a setting assumed to be Urbino's—a milieu also suggested by the painting's provenance. In turn, this hypothesis implies that the nobleman is Duke Guidobaldo, who was first a pupil and later a patron of Pacioli. In addition to the friar's knowledge and researches, the painting displays recent and remarkable mathematical objects: a printed copy of the *Elements*, Pacioli's *Summa*, and a perfect and a semiperfect solid. Finally, the figure drawn on the tablet, together with those on the pages of the opened book, may also be read as novel and fashionable because they bring to the attention of the beholder the problem of the correctness of geometric figures while underscoring the importance of the skill of visual reasoning according to the Euclidean grammar.

A noteworthy addition to the history of the *Portrait* is its entry in the 1582 inventory of the *guardaroba* of the Urbino palace. The current scholarly literature takes the first document related to our painting to be Bernardino Baldi's puzzling description from the late 1580s. Baldi's words are problematic because he attributes the panel to Piero della Francesca (a claim that is incompatible with the painting style and the *cartellino* of the Capodimonte panel) and inexplicably ignores the second figure.[8] Beyond predating Baldi's reference, the 1582 *guardaroba* entry is important because of its precision: "A painting of Fra Luca dal Borgo with the Duke Guido, *felicis memoriae*, painted on panel, large and high about two braccia, with a frame."[9] The accuracy of this description is validated by the size mentioned, basically that of a square panel, which corresponds to the dimensions of the *Portrait* in Capodimonte (and strengthens the hypothesis that the original panel was cut down slightly, as two sides have a perfect edge). Above all, this formal and precise entry helps lay to rest the question of the identification of the second figure, as it is very unlikely that an official of the Urbino court would have misidentified Duke Guidobaldo.[10]

In addition to the articles mentioned earlier, new research on the remarkable solids painted in the *Portrait* has appeared since 2010. A model of the crystal rhombicuboctahedron, a solid that continues to fascinate historians and mathematicians alike, was built by the artist Claude Boehringer; its geometric accuracy was called into question first obliquely in an article in *Scientific American* (March 29, 2011) and more recently and more directly by Carlo Séquin and Raymond Shiau.[11] These two scientists proved that in spite of its beauty, the rhombicuboctahedron of the *Portrait* fails to correctly render both the reflection patterns and the faces submerged by the fluid; they also noted that its perspective is inconsistent with that of the rest of the painting.[12] Even though some of the perspective problems likely arose because de' Barbari drew an independent preliminary study of this solid and then copied it on the panel—the piecemeal nature of this process compromised the exactness of the integrated perspective of the scene—it is useful to pursue the question of what knowledge and texts the evaluation of this complex solid and of its perspective would have entailed for a beholder in 1495.[13] Moreover, the assessment of the geometric and optical accuracy of the crystal may help to further probe the meaning and function of the two solids in our panel. The dodecahedron and the rhombicuboctahedron solids in the painting are not presented as direct, open challenges, as in the case of the line and the figures to which Pacioli points, but the questions of whether they are correct from a mathematical standpoint and whether the patters of reflections and refractions in the crystal are accurate should not be dismissed as simply "modern" and irrelevant for historians of art and science. Surely, the questions of whether the rhombicuboctahedron portrays a real crystal (perhaps owned by Duke Guidobaldo) and what the reflections painted on its faces actually picture would have been asked by original Renaissance viewers, two questions that naturally pair with others about the presence and meaning of such solids in this double portrait.

Before delving into these questions, it is important to stress that the two geometric solids in the *Portrait* serve several purposes, qualifying the artist, the viewer, and the cultural space supported by Pacioli and his patron. First, they determine the ideal position as well as the intellectual range of the beholder, while their geometric complexity and the perspective challenge implicit in their representation confirm the artist as a master of perspective.[14] Specifically, the level of the liquid in the crystal clarifies that the eyes of the ideal viewer meet those of the nobleman straight on (and that the ideal viewer also stands higher/is taller than the friar). Second, the audience is also challenged to identify these two solids—both naming them properly and recognizing the wooden one as a perfect or Platonic solid and the crystal as a semiperfect or Archimedean solid—and to assess their perspectival correctness. Notably, for viewers in 1495, the identification and assessment of the perspective of the two solids invoked knowledge offered by Pacioli's recently published *Summa*, the first printed book that illustrates and discusses perfect solids; it may not be casual that this tome appears in the *Portrait* and supports the wooden dodecahedron.[15] Third, the reflections on the faces of the crystal of the light coming into the room through a window allow beholders to glimpse outside and discern the wing of a princely palace. Finally, the presence of the two solids, one crystalline and hanging in a seemingly independent space, the other wooden, tangible and familiar for the sitters as suggested by its casual presence serving almost as a paperweight, objectifies a facet of the knowledge and education supported by the cultural space pictured and projected by the painting. Some of this knowledge reflects back onto the painter, whom the audience would have readily deemed a master of perspective.

At first blush, the Renaissance viewer of the *Portrait* would probably have explained the inclusion and the prominence of the dodecahedron and rombicuboctahedron by fitting them in the tradition of complex geometrical solids in the visual arts. In this regard, the *mazzocchio*—the structure supporting fashionable and extravagant turban-like headpieces—exemplifies what contemporary viewers may have recognized as demonstrations of an artist's mastery of perspective.[16] Though present also in the tarsias of Urbino *studiolo*, the most famous instances of *mazzocchi* occur in Paolo Uccello's *Battle of San Romano* and *Flood and Retreat of the Waters* (the first is a large painting from circa 1438–40 preserved in separate sections at the National Gallery in London, Uffizi, and the Louvre, and the second a fresco dating to circa 1447 in the Chiostro Verde of Santa Maria Novella in Florence).[17] Easily identified against the confusion that surrounds them thanks to their perfect shapes, these fancy fashionable headgears help validate the space and the historical events Uccello depicted by attesting to his knowledge of perspective—and implicitly of geometry—which is the grammar of the *costruzione legittima*.[18]

As Uccello's complex solids explicitly connect fifteenth-century artists and mathematical humanists, then the *mazzocchio* in the Urbino *studiolo* establishes the special association between complex solids and learned patrons in that court.[19] Implicitly, both associations characterize the *Portrait*.[20] Naming some of the prominent intellectuals active at the interface of art and science and present in Urbino during the second half of the fifteenth century clarifies the relevance of these connections for our painting. Among the artists and mathematicians that interacted at this court we find Paolo of Middleburg, who maintained that the challenges of the solids would fill a book.[21] He was a close friend of Piero della Francesca, the author of the *Libellus de quinque corporibus regolaribus*, which is dedicated to the Duke of Urbino.[22] Piero instilled interest in the complex solids on Pacioli, who incorporated the artist's results in the *Summa* and also gave a set of wooden solids to Guidobaldo when he was his pupil.[23] Finally, a recurrent summer guest of Duke Federico da Montefeltro was Leon Battista Alberti, the dean of Quattrocento art and architectural theory and a believer in mathematics as foundational for these arts. Likely, Alberti was also Pacioli's first teacher in matters of art and mathematical sciences, having hosted the friar as one of his *familiares* in Rome during the early 1470s.[24] These intellectuals and their writings are the basis for the claim that Urbino was a center of mathematical humanism supporting the integration of art and mathematics.[25] The roots of this modern claim are intertwined with the *Portrait*: the dedicatory letter of the *Summa* (published November 10, 1494) presents the Urbino court as encouraging the ongoing scholarly discourse in all disciplines, from the arts to mathematics and the sciences, and tangibly through a rich library of classical as well as contemporary works. Most likely, Pacioli composed this dedication while he was "inventing" the iconography of the *Portrait*.[26] In sum, in 1495, a knowledgeable viewer in Urbino would interpret the mathematical objects painted in the *Portrait* as references to both the larger tradition in the Renaissance visual arts and the special cultural environment supported by the court of the Montefeltros.

These contextual references, from Piero's *Libellus* to the wooden set gifted by Pacioli, explain the presence of the two solids in the *Portrait*. Unfortunately, the later history of complex solids has obfuscated the novelty of these two, and particularly of the crystal one, while also diluting their direct connection to Urbino. Because the identity of this court as a center for mathematics was solidified by the intellectuals that flourished there during the sixteenth century—including Federico Commandino, Guidobaldo del Monte, and Bernardino Baldi, each a leading figure in the field in his own way—we consider these two solids through the lens of the sixteenth-century history of mathematics. Early in the Cinquecento, the domain of complex solids expanded thanks to Leonardo's famous drawings completed in Milan (1496–98) and the Venetian publication of Pacioli, *Divina proportione* (1509) and the second edition of the *Summa* (1523). Their newly gained fame found visual expression in the examples included in various intarsias (in Santa Maria in Organo (Verona), Lodi cathedral, and abbey of Monte Oliveto Maggiore (Asciano, Siena)), as well as in famous prints such as Durer's *Melancholia I* (1514) and Jacopo Caraglio's *Diogenes* (1524–27). Later in the century, the allure of complex geometric solids was popularized by the appearance of several manuals by influential writers, including Dürer (whose *Institutiones geometricae* was translated into Italian by Cosimo Bartoli in 1537), Daniele Barbaro, and Wemzel Jamnitzer, which demystified their construction, eroding their potential to display an artist's mastery of perspective and knowledge of mathematics.[27]

Similarly, it is worth noting that complex solids also expanded their reach at court during the sixteenth century: for instance, we find them incorporated in the frescoes of Francesco I de' Medici and among the holdings of the cabinet of curiosities at the Innsbruck court.[28] In sum, the expansive history of complex geometrical solids during the sixteenth century has made the two solids in the *Portrait* nonspecific, hampering the answer to the question of why Pacioli placed these two solids in the painting.

Sitting on the *Summa* and figuratively supported by the knowledge and the learning offered by Pacioli's *opus magnus*, the dodecahedron (Figure 8.2) references the wooden set of Platonic solids that the friar had given to Guidobaldo when the Duke was his pupil, thus becoming a record of the long-standing relationship between the two sitters and of their love for mathematics.[29] Its specific appearances may further qualify this relationship: the painter has characterized this solid through its material dimension by carefully making visible the veining of its wood. The same wooden solidity characterizes the depiction of the pages of the *Summa* that supports the dodecahedron; notably, the book was also a gift from the friar to Guidobaldo, presented publicly through the dedication but perhaps also privately through the copy depicted in the *Portrait*.[30] Their wooden quality is shared by the painting: as the guardaroba entry reminds us, it is a "tavola," a panel painting, which suggests that, like the other wooden objects, the *Portrait* too may be a gift of the friar to the Duke.

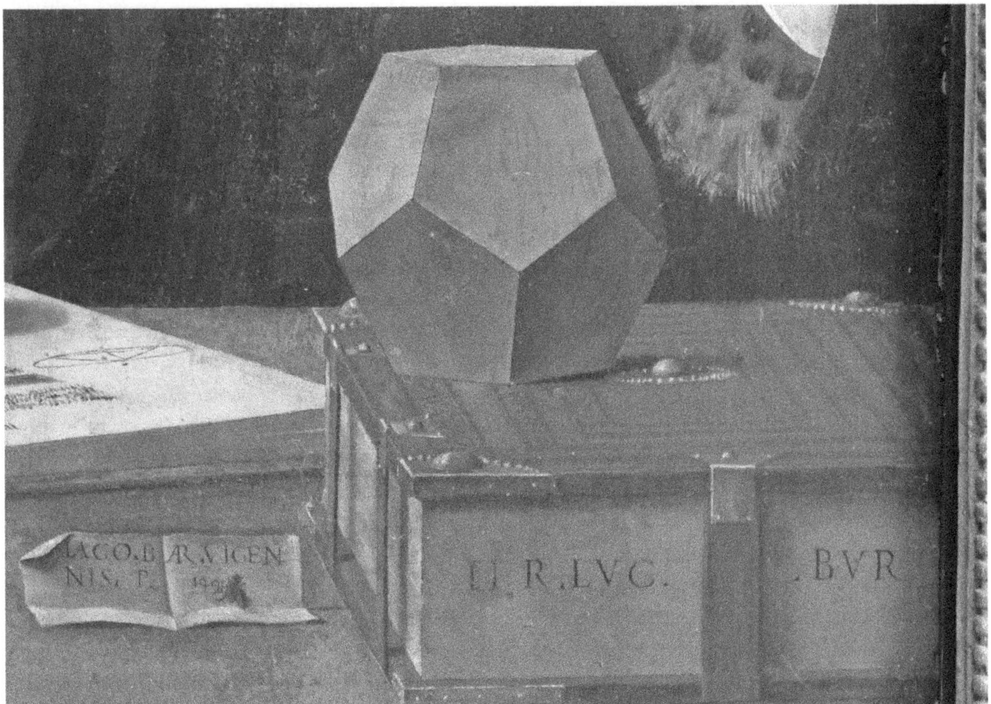

Figure 8.2 Dodecahedron, Summa, and Cartiglio. Detail of Jacopo de' Barbari. *Portrait of Luca Pacioli and Gentleman*, 1495. Museo e Real Bosco Capodimonte, Naples.

Photo credit: Renzo Baldasso. Reproduced by courtesy of the Ministero dei Beni e delle Attività Culturali e del Turismo.

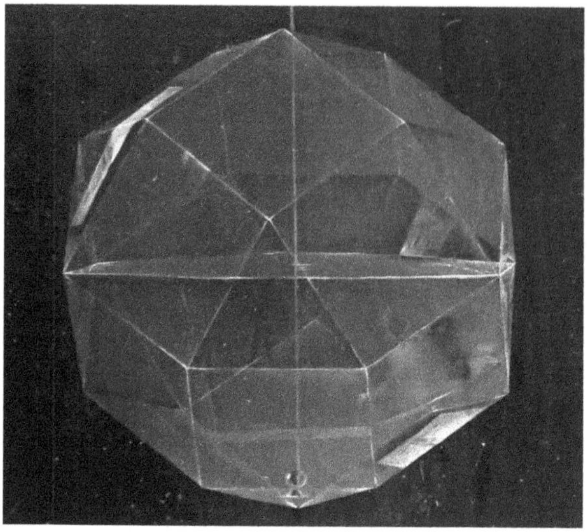

Figure 8.3 Rhombicuboctahedron. Detail of Jacopo de' Barbari. *Portrait of Luca Pacioli and Gentleman*, 1495. Museo e Real Bosco Capodimonte, Naples.

Photo credit: Renzo Baldasso. Reproduced by courtesy of the Ministero dei Beni e delle Attività Culturali e del Turismo.

The hanging rhombicuboctahedron (Figure 8.3) is a remarkable object and a mathematical novelty of the same caliber as the other mathematical novelties displayed in the painting. Even though *prima facie* it appears an addition introduced to secure a visual balance on the upper register of the painting, which is dominated by the heads of the two sitters, its value goes well beyond the incidental. Considering that the *Portrait* was likely "invented" by the friar during his stay in Venice while overseeing the printing of the *Summa*, this crystal would have been an appropriate present for the Duke: Murano was quite possibly the only place where one may have such an object made; it was there that the Baroviers had invented and perfected crystal glassmaking in the preceding decades.[31] Could this be another gift from Pacioli to his pupil and patron? Given the technical difficulty of making such an extraordinary crystal, the beholder wonders whether it is fictitious or a real object.[32] In turn, this invites careful scrutiny of the Archimedean solid. The reflections on its faces, together with its size, which is somewhat awkward because it is larger than the heads it counterbalances, suggest that the painter depicted this marvelous crystal copying a real object. However, a second look challenges this supposition. If the liquid it contains is water, the considerable weight of the half-filled crystal would likely stress the cable that supports it beyond its limit. In fact, further examination reveals that the line supporting it is attached to an improbable hook floating at the bottom, which is not glued to a corner or a face of the crystal. The ease with which this problem could have been resolved by the painter introduces another puzzle. Even before thinking through the effects of the liquid on the reflection patterns—for instance, the submerged section of the solid would appear different—one comes to realize that the painter rendered them selectively. Three faces bear the reflection of the view from the room's window, but other faces of the crystal reflect elements from the room's interior. On the front and lower right square faces we discover first the green tablecloth, then the friar's gray frock and the dark clothing of the second figure, and, finally, Pacioli's hand and stick. The only other reflection from the interior of the room is the sketch of the dark figure in the triangular face of the crystal parallel to the picture plane, which may well be a self-portrait of the painter with his easel: his head remains just about the level of the water, setting his eyes precisely in line with those of the ideal viewer.[33] These details indicate that the representation of this crystal is not true to reality; rather, it pursues instead a "selective naturalism" intended to bring specific information to the attention of the beholder—the palace outside, the "ad vivum" portrayal of the two sitters by the painter, and Pacioli's act of pointing.

A systematic examination of the solid reveals faults in the reflection patterns, which should be altered by the water, and in the hanging line, which should pass through a different point on the top face. These mistakes mirror others in the painting, including the impossible folds of the *cartellino*. These imperfections might be meaningful, but they may also be shortcoming—understandable for a young artist—introduced by the painting process, based on independent preliminary studies. More importantly, scrutiny of the reflections on the rhombicuboctahedron clarifies that because the room's interior appears on the crystal's front faces, the solid is behind the two sitters, a significant choice that goes beyond confirming that Pacioli is thinking abstractly rather than staring at the crystal.[34] Overall, by being situated behind the two sitters and being both higher than them and larger than their heads, the solid becomes a sort of dean presiding over the scene. Beyond functioning as a mirror to document and verify the scene, the crystal brings emphasis to Pacioli's action of pointing with the stick and to the palace outside, which, it is important to recall, in the dedicatory letter of the *Summa* is described as "l'admiranda fabrica del degno palazzo del V.D.S."[35] Moreover, in the *Divina proportione*, Pacioli presents the rhombicuboctahedron as important for architecture (Figure 8.4): "Un

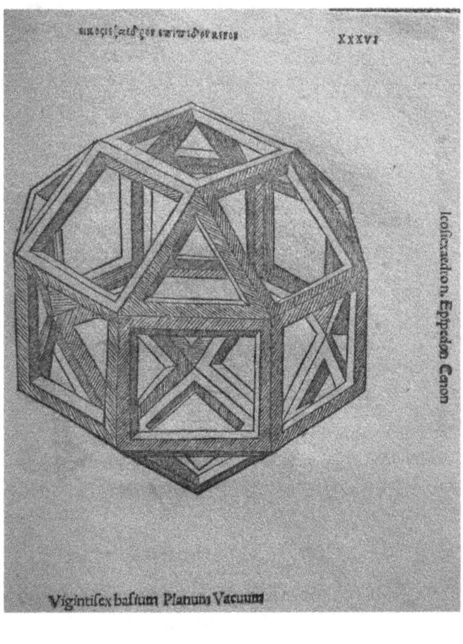

Figure 8.4 Rhombicuboctahedron from Luca Pacioli *Divina proportione* (Venice: Paganino Paganini, 1509), Part III, Figure 36. Houghton Library, Cambridge MA, shelfmark Typ 525.09.669.

Photo credit: Renzo Baldasso.

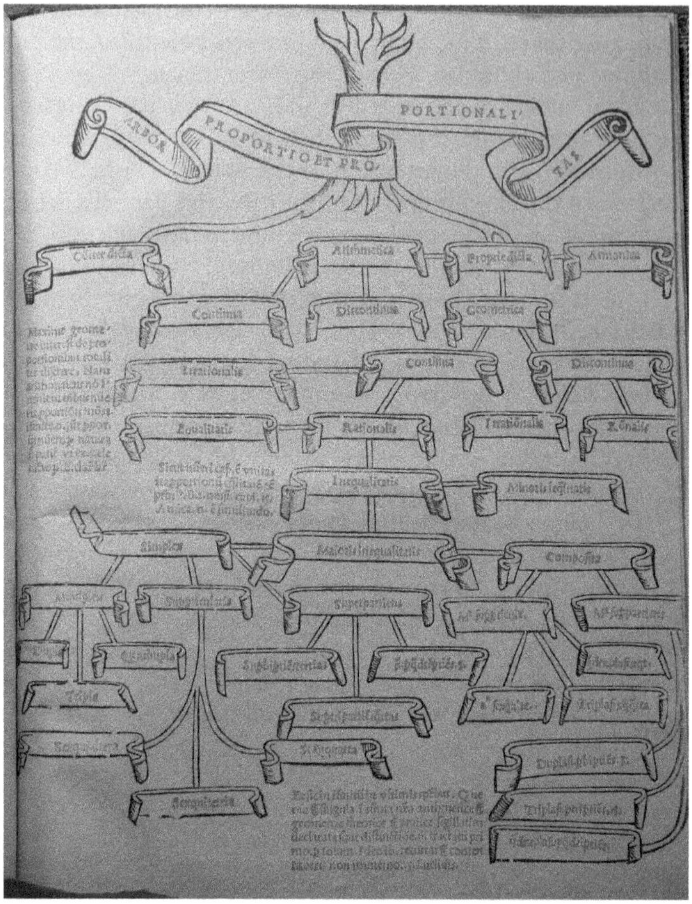

Figure 8.5 Arbor proportio et proportionalitas from Luca Pacioli *Divina proportione* (Venice: Paganino Paganini, 1509), Part III, Figure 62. Houghton Library, Cambridge MA, shelfmark Typ 525.09.669.

Photo credit: Renzo Baldasso.

altro corpo . . . de 26 basi . . . la sua scientia in molte considerationi utilissima a chi ben la sa accomodare maxime in architectura."[36]

The principles of knowledge embedded in this solid are not immediately clear, but proportion, the "mother and queen" of all of the arts, is very likely its cornerstone.[37] A special kind of proportion, the *divina proportione*, is what Pacioli points at with his stick.[38]

The numbers on the slate board (Figures 8.6 and 8.7 [detail]) remind us that this portrait is not a realistic rendition of a double portrait set in an intimate teaching scene but a programmed display of mathematical humanism.[39] These numbers are clearly readable, and the care with which they were painted implies that they are not random numbers. The committed spectator finds ready confirmation of this: despite having to read the numbers upside down, it is easy to compute their sum and prove

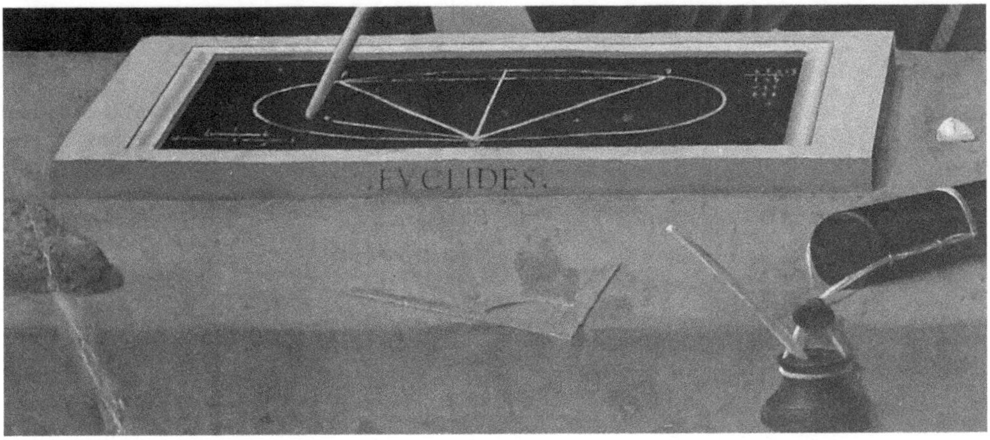

Figure 8.6 Slate tablet. Detail of Jacopo de' Barbari. *Portrait of Luca Pacioli and Gentleman*, 1495. Museo e Real Bosco Capodimonte, Naples.

Photo credit: Renzo Baldasso. Reproduced by courtesy of the Ministero dei Beni e delle Attività Culturali e del Turismo.

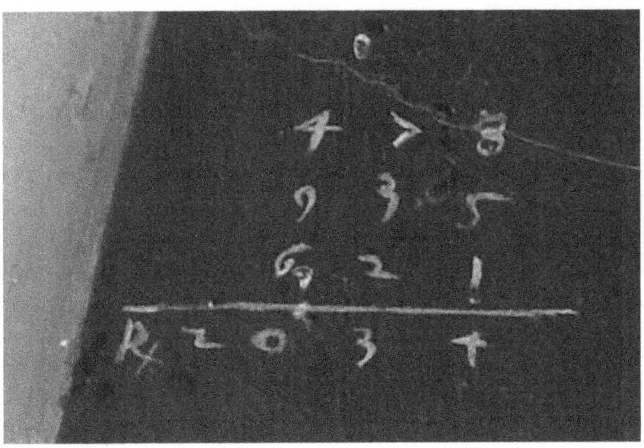

Figure 8.7 Sum. Detail of Jacopo de' Barbari. *Portrait of Luca Pacioli and Gentleman*, 1495. Museo e Real Bosco Capodimonte, Naples.

Photo credit: Renzo Baldasso. Reproduced by courtesy of the Ministero dei Beni e delle Attività Culturali e del Turismo.

its correctness: 478 plus 935 plus 621 total, indeed, 2,034. While the computational accuracy of this sum validates these numbers as an ensemble, it also questions their presence: the painting's iconographic program as a whole is erudite, and such a simple calculation would be out of place if it did not have a second, more profound meaning. As we shall see, this secondary meaning is neither simple to see nor easy to corroborate without paper and pen, patience, and an eye for proportionality.[40]

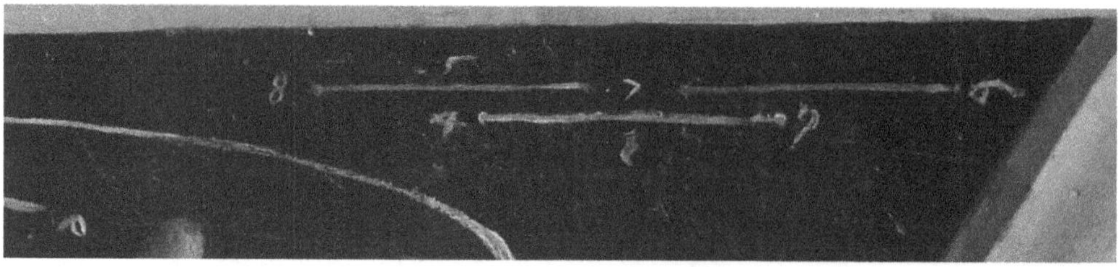

Figure 8.8 Segments with numbers. Detail of Jacopo de' Barbari. *Portrait of Luca Pacioli and Gentleman*, 1495. Museo e Real Bosco Capodimonte, Naples.

However, one larger issue becomes apparent to the wider audience once the intrinsic numeric significance of these numbers is questioned: their presence next to the figure drawn on the slate tablet presses the spectator to ponder how these numbers relate to it—-and, similarly, how the array of line segments to its left (Figure 8.8) relates to the figure.

In inspecting the numbers and computing the sum, one comes to realize that they are special in one way: the three addends contain nine different digits, each used only once. The peculiarity that no digit is repeated might immediately lead to the hypothesis that the numbers to be added would form a magic square.[41] As it may be recalled from the famous example of a 4 × 4 magic square that appears in the top right of Dürer's engraving *Melencolia I* (1514), a magic square has the property that adding each column or row gives the same sum.[42] Yet, in this case, even though the top horizontal and first vertical columns do add to 19, a complete check falsifies the hypothesis. While this may be discouraging, a second look at the "graphic" context of these numbers confirms that a magic square would be out of place: on the tablet labeled "Euclides," together with the computations, appears the figure of an inscribed triangle with a half-way drawn line and several proportional segments. Several clues encourage the informed beholder to seek connections in the contents of the tablet through the relationship between the perfect solids and the Fibonacci sequence of numbers that approximate the golden ratio.[43] First, the reference to perfect solids, already an established subject in the painting thanks to the wooden dodecahedron, is made clear by the label "Euclides" inscribed in the tablet and by the copy of the *Elements* laid next to it, which is open to Book XIII, the book where Euclid discusses perfect solids.[44] Second, Book XIII addresses also the connection between the golden ratio and the dodecahedron.[45] Thirdly, the property of incommensurability of the two segments giving rise to the golden ratio may be visually hinted at by the line drawn in the figure on the slate, which, despite being labeled with a letter, does not connect, intersect, or relate to the others in the figure, and therefore results in a very strange line in a geometric figure.

Those familiar with texts such as the *Liber abaci* by Leonardo Pisano (Fibonacci) and, closer to the Urbino context, the *Trattato d'abaco* by Piero della Francesca, both well known to Pacioli and very likely to his mentee and patron Guidobaldo da Montefeltro, might recall the connection between Fibonacci numbers—solutions to the famous problem of the Rabbit Puzzle in Book XII of the *Liber abaci*—and the golden ratio.[46] Specifically, the ratio of Fibonacci numbers progressively approximates the golden ratio, a number that Pacioli will term "*divina proportione*" and which in the *Elements* is known as "division in extreme and mean ratio."[47] This property of the Fibonacci numbers was certainly known by the early decades of the sixteenth century, as suggested by a manuscript note on a copy of Pacioli's edition of the *Elements* (Venice: Paganino Paganini, 1509; BnF V.104) at the passage that first refers—implicitly—to the division in extreme and mean ratios or DEMR.[48] Given where it appeared, quite possibly, this result was known during Pacioli's time.

Searching for phi in the addends painted on the tablet proves rewarding. Specifically, if we read the addends in reverse, i.e. written right to left as in Leonardo's manner, and take their ratios, something amazing is found: 874/539 is 1.62152 . . .; 539/126 is 4.2778; and 874/126 is 6.93651 . . . Notably, 1.62152 . . . is close—within less than 1%—to the value of phi, which is 1.61803 . . .; 4.2778 . . . is close to the value of phi to the power of 3, which is 4.23604 . . .; and 6.93651 . . . is close to the value of phi to the power of 4, which is 6.85403 . . . Needless to say, these are not fortuitous ratios. Rather, they display a result from Pacioli's active research agenda.[49] Beyond the wonder effect, these ratios could have been seen as useful findings, while the fact that they are not exact would not have been a problem for the Renaissance audience, as approximations through ratios of integers for numbers such as π were universally accepted in calculations.[50]

In addition to the discovery that the nine integers can be arranged in sets of three so that their ratios are close approximations of the powers of phi, the slate presents a second result of Pacioli's researches on the *divina proportione*. Three segments of equal length (Figure 8.8) are found to the left of the figure; they are qualified by numbers at their ends and middle points in the following way: the two in sequence are 8___5___7 and 7 ______ 6, while the single one is marked by 4___x___9 . The x-digit may be a reversed 3 or an incomplete 8; it is impossible to decipher. Again, taking Euclid's *Elements* as reference, the graphic format of its many theorems connected with constant proportionality resembles the slate's segments.[51] In turn, this suggests that proportion, i.e. ratios, between segments is the interpretative key and that Pacioli may be seeking to connect it with the golden ratio.[52] Much like the sum of the numbers on the right of the figure, here the constant equal length and positioning of the lines serves to hide a higher truth. To partly anticipate, the result that Pacioli presents is based on the following challenge—construct a line ABCDEF such that AB is to BC as AC is to AB as AE is to AC as BC is to CD as BD is to BC as BE is to BD—and on how the segments of such a ABCDEF line relate to the *divina proportione*.

Let's consider the three equal segments superimposed on one line that is to be divided into the mean and extreme ratio. The single segment is superimposed on the first two in such a way that it creates a gap between these two equal to the original length of each of the two aligned segments less twice their minor segments when divided into the extreme and mean ratio. The lower line illustrates the length of that gap, overlapping each of the upper lines by an amount equal to the shorter segment defined by the golden ratio. To clarify: recreating the same diagram and superimposing the bottom line on the top two and lettering the critical points:

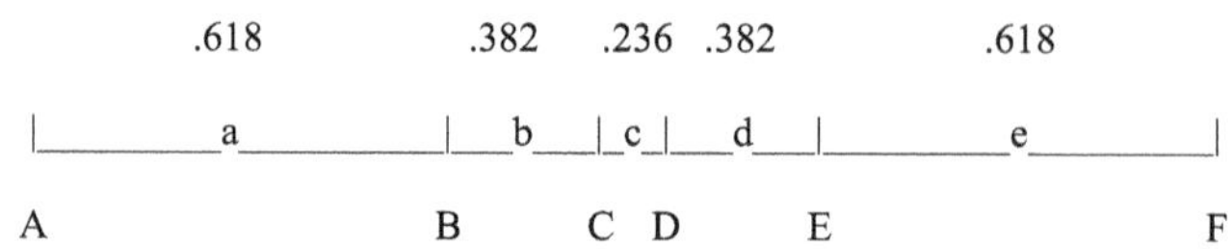

The results: the ratios between the segments composing such a line are more easily understood if one identifies the line segments, as shown here above the line, by lower case letters. For further convenience, we assume a length of one unit for each of the three original lines and assign the approximate values: .618 to a and to e since each is the major segment of its original line, and .382 to b and d, their minors. Therefore c, the "gap," is .236, since that is what is left of one unit once two minors are subtracted (i.e. $1 - (.328 \times 2)$). These approximations do not affect the validity of this demonstration but make it much easier for us to follow.

Now we calculate the ratios between the lengths of various pairs of line segments. Using the given approximate values, the skeptical reader can easily replicate these calculations on a calculator.

$$a/b = (a + b)/a = (a + b + c + d)/(a + b) = b/c = (b + c)/b = (b + c + d)/(b + c) = \text{phi} = 1.618\ldots$$

The ratios required to satisfy our Euclid-inspired constraints hold. A bonus result is that if we are willing to accept noncontiguous numerators and denominators:

$(a + b + c)/(b + d)$, $(a + b + d)/(a + c)$, $(a + b + d + e)/(a + b + c)$, and $(a + b + c + d + e)/(a + b + d)$ also yield phi.

The same ratios hold if we read from right to left: i.e. instead of a/b we can begin at the other end, e/d, repeating them in the reverse direction. In the ten ratios above, eight different denominators are involved.

According to this reading, the proportional segments in the bottom left of the slate become a visual demonstration of a relationship that Pacioli probably discovered while pursuing his studies of Euclid's *Elements*. Moreover, the segments relate to the addition problem on the left side: a to the third power yields c; (a+b+c) to the fourth power yields (a+b+c+d+e); and a squared gives b, as with any line divided into major and minor segments according to the *divina proportione*. Overall, through these mathematical "games" illustrated on the slate, Pacioli displays profound insights hidden behind trivialities, as both contain encoded examples that illustrate relationships among the powers of phi.

While adding to the painting's celebration of mathematics and mathematical humanism and their relevance at court at the closing of the fifteenth century, retrospectively the presence of phi in the *Portrait* links it also to the works embodying the *divina proportione* by Piero della Francesca, Leonardo, and Raphael, showing the status of this mathematical notion in Renaissance intellectual culture.[53] More importantly, these results offer new insights into the contents of the *Portrait* that both further limit the value of labeling it simply a double portrait and press the question about its original function. Its display of mathematical novelties, from the printed book to the Archimedean crystal, to Pacioli's research on the *divina proportione*, brings to the fore the why-question, i.e. Why the *Portrait*? In other words, what was the intention behind the making of this painting and its iconography? Answers may follow by asking how did it function in its original context. As implied by our analysis, its interactive purpose, confirmed also by the pen offered to the beholder at the edge of the table, supported a wide range of responses: while learned Renaissance beholders would have recognized the solids and the Euclidean reference to the connection between perfect solids and the golden mean, the number "game" is not one that was going be solved readily. Only those in the know had the solution. Acting as such a person and as the master of the beholder interaction through his firm stare, the second figure confirms his role as patron with an illuminated, enlightened hand. This suggests that the *Portrait* was created to serve as a gift in the patronage economy—a gift that served as a reminder of previous gifts while offering new ones—and that the friar presents the second figure as his patron, implicitly proposing that this beneficial arrangement secured status and fame to both. The fineness of this proposition is confirmed by the careful posing of the second figure who stands taller and partly behind—and thus visually backing—Pacioli, and whose stillness transforms him into another of the mathematical objects displayed. Notably, their careful placement emphasized another relational position and hierarchy: both the friar and his patron are reflected in the crystal, a marvelous object that projects the beauty and timelessness of mathematical truths and principles, while its semiperfect nature—one seemingly incommensurate to phi—is an apt reminder that the *Portrait*'s scene reflects the imperfect, corruptible reality of the figures and objects that live in it.

Notes

1 The exhibition was held at the Städel Museum in Frankfurt am Main from October 23, 2013, to February 2, 2014. For the *Portrait*'s entry in the catalogue, see Jochen Sander, ed., *Albrecht Dürer: His Art in Context* (Munich: Prestel, 2013), 190–191. The *Portrait* is at the time of this writing part of the show *Aldo Manuzio. Il rinascimento di Venezia* on view March 19 to June 19, 2016, at the Gallerie dell'Accademia in Venice. Its earlier appearance in the Marco Palmezzano exhibition of 2005–6 offered an occasion to compare portrait painting styles, rendering of folds in *cartellini*, and examples of complex solids in contemporary paintings. See Antonio Paolucci, Luciana Prati and Stefano Tumidei, eds., *Marco Palmezzano: il Rinascimento nelle Romagne* (Milan: Silvana Editoriale, 2005), and specifically the *Portrait* at 178–179, the *Enthroned Madonna with Saints* (circa 1497, Faenza, Pinacoteca Nazionale), includes stellated polyhedrons at 230–235, the *Portrait of a Man* (circa 1495, Vienna, Gemaldegalerie der Akademia bildenden Künste) at 248–249, and *cartellini* with folds at 90, 201, 229, 237, and 295.

2 See Jay A. Levenson, ed., *Circa 1492: Art in the Age of Exploration* (Washington, DC: National Gallery of Art, 1991); Martin Kemp discusses the *Portrait* at 244–246.

3 See Renzo Baldasso, "*Luca Pacioli and Disciple*: A New, Mathematical Look," *The Art Bulletin* 102 (2010): 83–102. The *editio princeps* of the *Elements* was published in 1482 by Erhard Ratdolt in Venice; the second edition was issued in 1491 by Leonardus Achates of Basilea and Guilielmus of Pavia in Vicenza.

4 See Enrico Gamba, "Proviamo a rileggere il 'Doppio ritratto' di Luca Pacioli," in *Le tre facce del poliedrico Luca Pacioli*, ed. Francesca Maria Cesaroni, Massimo Ciambotti, Enrico Gamba and Vico Montebelli (Urbino: Quaderni del Centro Internazionale di Studi Urbino e la Prospettiva, 2010), 81–97. Previously, Gamba had considered the painting in an exhibition catalogue essay, "Pittura e storia della scienza," in *La ragione e il metodo:*

Immagini della scienza nell'arte italiana dal XVI al XIX secolo, ed. Marco Bona Castellotti, Enrico Gamba and Fernando Mazzocca (Milan: Electa, 1999), 43–53; in this beautifully illustrated essay, Gamba identified the Capodimonte panel as a marker of the awareness of Renaissance intellectuals to be the makers of the "new science" and, thus, agents of the Scientific Revolution. Most recently, Gamba has reviewed the painting in a recent exhibition catalogue essay, "L'Umanesimo matematico a Urbino," in *La città ideale: l'utopia del Rinascimento a Urbino tra Piero della Francesca e Raffaello*, ed. Alessandro Marchi and Maria Rosaria Valazzi (Milan: Electa, 2012), 233–247, with the entry for the *Portrait* at 238–240.

5 See Argante Ciocci, "Il Doppio Ritratto del Poliedrico Luca Pacioli," *De computis. Revista Española de Historia de la Contabilidad* 15 (2011): 107–130. More generally on Pacioli and the cultural context in which he operated, see Ciocci, *Luca Pacioli e la matematizzazione del sapere nel Rinascimento* (Bari: Cacucci 2003), and *Luca Pacioli tra Piero della Francesca e Leonardo* (Sansepolcro: Aboca Museum, 2009).

6 Two dissenting voices on the attribution are Ciocci, who speaks of the attribution being in a "historiographic fog" (see "Il Doppio Ritratto," 114), and Alessandro Angelini, who argues for the inclusion of the *Portrait* in the oeuvre of Jacometto Veneziano; see "Jacometto Veneziano e gli umanisti. Proposta per il 'Ritratto di Luca Pacioli e di Guidobaldo da Montefeltro' del Museo di Capodimonte," *Prospettiva* 147/148 (2012): 126–149. In the only modern monograph on the artist, Simone Ferrari accepts the *Portrait* as de' Barbari's first known work; see Ferrari, *Jacopo de' Barbari: Un protagonista del Rinascimento tra Venezia e Dürer* (Milan: Mondadori, 2006), 23–27, 80–82. Ferrari has reaffirmed this attribution in "Dürer e Jacopo de' Barbari: persistenza di un rapporto," in *Dürer, l'Italia e l'Europa*, ed. Sybille Ebert-Schifferer and Kristina Hermann Fiore (Milano: Silvana Editoriale, 2011), 39–46.

7 In this respect, the *Portrait* signals a different emphasis on Euclidean studies than the one that had been presented in the *studiolo*, where the dedication at the base of the Euclid portrait defines his main achievements, the measurement of the Earth and the determination of its center, which are more appropriate for a geographer than a mathematician. For the dedication, see Gamba, *La ragione e il metodo*, 45.

8 See Baldasso, "*Luca Pacioli and Disciple*," 83, n. 5, and Gamba, "Proviamo a rileggere," 60–62.

9 See Fert Sangiorgi, *Documenti Urbinati: inventari del Palazzo Ducale (1582–1631)* (Urbino: Accademia Raffaello, 1976), 40–41. Specifically, this inventory refers to the panel as follows: "Un quadro di frà Luca dal Borgo con il duca Guido f(elicis) m(emoriae), dipinto in tavola, largo et longo braccia doi in circa, con le sue cornige attorno."

10 The issue of the different features between this and another portrait of Duke Guidobaldo, which has been brought forth in the past to challenge such identification, may be explained by recalling that the *Portrait* most likely was not painted in Urbino but by Jacopo de' Barbari in Venice, where he did not have access to the Duke and may have relied on a description given by Pacioli.

11 See Dirk Huylebrouck, "Lost in Triangulation: Leonardo da Vinci's Mathematical Slip-Up," *Scientific American* 29 (March, 2011). Relevant earlier online work by mathematicians includes that of Joost Rekveld, "The Ghost of Luca Pacioli," http://www.joostrekveld.net/?p=615 uploaded April 18, 2016; Herman Serras, "Mathematics on the 'Ritratto di Frà Luca Pacioli'," http://cage.ugent.be/~hs/pacioli/pacioli.html, uploaded April 18, 2016; and George Hart, http://www.georgehart.com/virtual-polyhedra/pacioli.html, uploaded April 18, 2016.

12 See Carlo Séquin and Raymond Shiau, "Physically Correct Rendering of Pacioli's Rhombicuboctahedron," (2014), http://www.cs.berkeley.edu/~sequin/BIO/talksSorted.html. Another publication on the *Portrait* just appeared, but unfortunately it is too late to include its contribution here; see the article-discussion by Mara-Lisa Kinne, Roland Krischel and Stefan Neuner, "Annäherung an ein Rätsel: Das *Bildnis des Luca Pacioli* von Jacopo de' Barbari," *Wallraf-Richartz-Jahrbuch* 76 (2015): 79–115.

13 The same perspective problem affects the open book on the table between the figures, as noted by Baldasso, *Portrait of Luca Pacioli*, 83, 98, n. 10. Like the crystal, the open book

with its detailed illustrations was also very likely first studied through an independent preliminary drawing and then copied on the panel.

14 See Baldasso, *Portrait of Luca Pacioli*, 95, and Gamba, "Proviamo a rileggere," 68.

15 Pacioli's *Summa* was the only published text that included descriptions and illustrations of perfect solids, but in the Urbino context, an alternative source—in fact also the source of the *Summa*'s chapters on the solids—was Piero's *Liber*, a copy of which was in the Duke's library.

16 See Margaret Daly Davis, "Carpaccio and the Perspective of Regular Bodies," in *La prospettiva rinascimentale: codificazioni e trasgressioni*, ed. Marisa Dalai Emiliani, vol. 1 (Florence: Centro Di, 1980), 183–200, especially 184–186.

17 Three are present in the Uffizi section of the *Battle* (the "Berndino della Ciarda Unhorsed") and two in the Louvre's ("The Counter Attack on Micheletto da Cotignola"); two more appear in the Santa Maria Novella fresco. Various complex perspectival drawings as well as the mosaic on the floor of the church of St. Mark in Venice, all traditionally attributed to Uccello, have been listed instead as "uncertain attributions" in the most recent monographic study on the artist; see Hugh Hudson, *Paolo Uccello: Artist of the Florentine Renaissance Republic* (Saarbrücken: VDM Verlag, 2008), 313–323; Hugh discusses Uccello's battle paintings in Chapter 3, 158–175.

18 The importance of the mathematical connections of Uccello's generation is celebrated by the painting known as "Five outstanding men" in the Louvre, representing (according to Giorgio Vasari) Giotto, Brunelleschi, Donatello, Uccello, and Giovanni Manetti. Notably, Manetti, the only nonpainter in the line-up, was qualified by Vasari as "the most excellent mathematician of his time," and a "friend with whom [Uccello] often conferred over Euclid's theories." See Vasari, *Le Vite de' più eccellenti pittori, scultori, ed archittettori*, ed. Gaetano Milanesi, vol. 6 (Florence: Sansoni, 1878–85), 273. The importance of this subject and this painting is confirmed by the change in interpretation between the first and the second edition of the *Vite*: Vasari revised the attribution from Masaccio to Paolo Uccello. On the Louvre painting, see Filippo Camerota, *La prospettiva nel Rinascimento: Arte, architettura, scienza* (Milan: Electa, 2006), 49–50. The Uccello-Manetti relationship is not unique, and it had a famous precedent: after his return to Florence in 1425, Pier Paolo Toscanelli became both mentor and the best friend of Filippo Brunelleschi, as indicated by Antonio di Tuccio Manetti in his *Life of Brunelleschi*; in it, Toscanelli qualifies his interaction with the artist as the "greatest association of my life"; see *The Life of Brunelleschi, by Antonio di Tuccio Manetti*, introd., notes, and critical text ed. Howard Saalman, trans. Catherine Enggass (University Park: Pennsylvania State University Press, 1970), 54–57.

19 In this context, it is worth mentioning that Pacioli was directly connected to Lorenzo di Lendinara: the famous *instarsiatore*, representative of a class of craftsmen that became associated with the representation of complex perspectives and polyhedral, was his godfather.

20 On the Urbino (and Gubbio) *studiolo*, see Luciano Cheles, *The Studiolo of Urbino: An Iconographic Investigation* (University Park: Pennsylvania State University Press, 1986), which suggests the connection between the symbolic meaning of the *mazzocchio* and the portrait of Euclid that stood above it, 90. It should be noted that the complex figures appearing in the tarsias in Verona and Lodi were created after the appearance of Pacioli's *Summa* and of the *Portrait*; the case of those in Pisa's cathedral is less clear, but they are assigned to Guido da Seravallino and dated to 1510–13 in Adriano Peroni, ed., *Il Duomo di Pisa/The Cathedral of Pisa* (Modena: Panini, 1995), 2: 922–923; their entry in the catalog edited by Filippo Camerota, *Nel segno di Masaccio: L'invenzione della prospettiva* (Florence: Giunti, 2001), 110, suggests that they were drawn independently from Leonardo's drawings and Pacioli's figures, noting also that payment to this tarsia master were issued since 1490.

21 See Gamba, *La ragione e il metodo*, 46. On Paolo of Middleburg and the connections between mathematics and astrology as well as on these connections between Urbino and Rome, see Patrizia Castelli, "Matematici e astrologi tedeschi alla 'corte' dei Montefeltro," in *Die Kunst und das Studium der Natur vom 14. zum 16. Jahrhundert*, ed. Wolfram Prinz and Andreas Beyer (Weinheim: Acta Humaniora—VCH, 1987), 237–251.

22 On Piero della Francesca and his *Libellus*, see James R. Banker, *Piero della Francesca: Artist & Man* (New York: Oxford University Press, 2014), 198–205.

23 As Pacioli notes in the *Summa*, he gave the set of wooden polyhedra to Guidobaldo in 1489 when the Duke went to Rome to visit the new Pope Innocent VIII; see, *Summa*, 2nd part, 68v.

> Questi son quelli Magnanimo Duca di quali le forme materiali con asai adornezze nelle proprie mani di V.D.S. nel sublime palazzo del Reverendissimo cardinale nostro protectore Monsegnor de San Pietro in vincula quando quella venne ala visitatione del Summo pontefice Innocentio 8°, negli anni della salute nostra 1489, del mese de aprile, che già sonno 5 anni elapsi. E insiemi con quelli vi foron molti altri da ditti regulari dependenti. Quali fabricai per lo Reverendo Monsegnor meser Pietro de Valetarij de Genoa, dignissimo vescovo de Carpentras, al cui obsequio alora foi deputato in casa de la felicissima memoria del R.mo Cardinale de Fois, nel palazzo ursino in campo de fiore.

See also Bernardino Baldi, *Le vite de' matematici*, ed. Elio Nenci (Milan: FrancoAngeli, 1998), 333–334. Another set of such solids was commissioned to Pacioli by the Florentine Republic in the early years of the sixteenth century in connection to Leonardo's *Battle of Anghiari* commission; see Carlo Pedretti, *Leonardo: A Study in Chronology and Style* (Berkeley: University of California Press, 1973), 83.

24 In addition to the dedication of *De re aedificatoria* to Federico da Montefeltro (as claimed by Bernardino Baldi), Alberti's connections to Urbino, including even attributions of the design of palace and painting of the Ideal City panel, have been invoked by various modern scholars. Inter alias, see Liane Lefaivre, *Leon Battista Alberti's Hypnerotomachia Poliphili: Re-Cognizing the Architectural Body in the Early Italian Renaissance* (Cambridge: MIT Press, 1997), 149, and Franco Borsi, *Leon Battista Alberti* (New York: Harper & Row, 1977), 193–199. In his biographic study of Duke Federico, Baldi recounts also the cultural reach of his court; see *Vita e fatti di Federigo di Montefeltro Duca di Urbino* (Rome: Salvioni, 1824), 3: 239–242.

25 The connection between *l'humanisme mathématique* and Urbino was first proposed by André Chastel, who also remarked that the *Portrait* is "l'image perfaite de l'humanisme mathématique que l'on pratiquait à Urbin et de ses prestiges," see *Renaissance Méridionale: Italie 1460–1500* (Paris: Gallimard, 1965), 41, 50; for other references to mathematics and applied mathematics in Urbino, see Gianni Volpe, "Scienza, arte e architettura a Urbino. Proposta per un itinerario," in *L'arte della matematica nella prospettiva*, ed. Rocco Sinisgalli (Perugia: Cartei & Bianchi, 2009), 101–113. Moreover, it is worth recalling that Piero della Francesca, who completed several paintings for the Duke of Urbino between 1468 and 1486, was termed by Vasari "the best geometer of his times." Concerning architectural theory at this court, beyond Alberti's, it is worth recalling Laurana's commitment to mathematics as the foundation of architecture.

26 Pacioli, *Summa*, 2r.

27 Bartoli's translation of Dürer's *Institutiones geometricae* (from the French edition of 1532) is preserved in an autograph manuscript in the library of the Academy of Sciences in St. Peterburg, shelfmark Sobr. Muzeja Prijenisej Skogo Kraja 69.

28 For the Medici frescoes, see Valentina Conticelli, *«Guardaroba di cose rare et preziose»: lo studiolo di Francesco I de' Medici: arte, storia e significati* (La Spezia: Agorà, 2007), 37–39; she includes an illustration of Giovanni Stradano's frescoes, Tav. 10, which confirms the poor state of conservation of the work. On the wooden models at the court of Archduke Ferdinand II, see Bret Rothstein, "Making Trouble: Strange Wooden Objects and the Pursuit of Difficulty ca. 1596," *The Journal for Early Modern Cultural Studies* 13 (2013): 96–129.

29 In the dedication of the *Summa*, Duke Guidobaldo is presented as "grecis latinisque litteris ornatissumum et mathematice discipline cultorem ferventissimum" *Summa*, 3r. It should be noted that his father, Duke Federico, was also learned in geometry and arithmetic, according to his biographer Vespesiano da Bisticci; see his "Commentario de la vita del Signore Federico d'Urbino," in *Le Vite*, ed. Aulo Greco (Florence: Istituto nazionale di studi sul Rinascimento, 1970–76), I: 383: "Di geometria et d'arismetrica n'aveva buona peritia, et aveva in casa sua uno maetro Pagolo, tedesco, grandissimo filosofo et astrolago. Et di geometria et d'arismetrica aveva bonissima notitia. Et non molto tempo inanzi che si morissi, si fece legere da maestro Pagoo opera di geometria ed d'arismetrica." A recent translation is

available in *The Vespasiano Memoirs: Lives of Illustrious Men of the XVth Century*, trans. William George and Emily Waters (1926; reprint, Toronto: University of Toronto Press, 1997).

30 See the discussion of in Baldasso, *Portrait of Luca Pacioli*, 94–95.

31 Angelo Barovier, the "inventor" of *cristallo*, was more than a craftsman, as suggested by the fact that he attended the lectures of Paolo della Pergola and by the mention he received in Filarete's *Trattato di architettura* (where he is presented on the same footing of Piero della Francesca, Fra Lippo Lippi, and Mantegna); see Liliana Grassi and Anna Maria Finoli, eds., *Antonio Averlino, Trattato di Architettura* (Milan: Il Polifilo, 1972), 257–258, 667–668. The foremost authority on the history of Venetian glass has deemed our crystal, which is interpreted as the true protagonist of the *Portrait*, an unicum, unexplained by anything that preceded it or anything else in the Renaissance; see Rosa Barovier Mentasti, ed., *Trasparenze e riflessi: il vetro italiano nella pittura* (Verona: Banca Popolare di Verona e Novara, 2006), 6–7, 35–40.

32 It is worth recalling that Baldi, probably following hearsay in Urbino rather than from direct knowledge of the Portrait, mentions that it includes "corpi regolari *finti* di cristallo appesi in alto." See Bernardino Baldi, *Le vite de' matematici: Edizione annotata e commentata della parte medievale e rinascimentale*, ed. Elio Nenci (Milan: FrancoAngeli, 1998), 330–45, at 344 [emphasis added].

33 Several scholars saw a knight with a shield in this figure; Maria Grazia Ciardi Dupré Dal Poggetto suggested that it represents Duke Guidobaldo, see her "Il Ritratto di Luca Pacioli e di Guidobaldo da Montefeltro," in *Piero e Urbino, Piero e le Corti rinascimentali*, ed. Paolo Dal Poggetto (Venice: Marsilio, 1992), 197–200, especially 197.

34 The claim that Pacioli looks at the crystal has been recently maintained by Elena Filippi, *Umanesimo e misura viva: Dürer tra Cusano e Alberti* (Venice: Arsenale, 2011), 32. While this may not be correct, Filippi's conclusion that the solid is a Pythagorean reference to the basic components of the world, the point, the straight line, the triangle, the square, and the fifth element might have been discussed by the original audience of the painting. See idem., 37. Conversely, Pacioli's meditation has been argued convincingly by Hannah Baader, who noted that Franciscans wore their hood in this fashion when meditating; see her "Das fünfte Element oder Malerei als achte Kunst das Porträt des Mathematikers Fra Luca Pacioli," in *Der stumme Diskurs der Bilder: Reflexionsformen des Ästhetischen in der Kunst der Frühen Neuzeit*, ed. Valeska von Rosen, Klaus Krüger and Rudolf Preimesberger (Berlin: Deutscher Kunstverlag, 2003), 182–183. Precisely this point clarifies also that the sitters are not engaged in conversation.

35 See Pacioli, *Summa*, 2r.

36 See Chapter LIII in Luca Pacioli, *Divina proportione* (Venice: Paganino Paganini, 1509), 15v. "Un altro corpo . . . de 26 basi. Da principio e origine ligiadrissimo . . . e l'origine da questo fia da lo exacedron uniforme secondo ogni suoi parti tagliato commo similmente a l'ochio la sua material forma ci demostra. E fia la sua scientia in molte considerationi utilissima a chi ben la [sa] accomodare maxime in architectura e questo a notitia del suo solido piano e vacuo." More generally on Pacioli's interest in envisioning complex solids as architectural principles, see Byrna Rackusin, "The Architectural Theory of Luca Pacioli: *De Divina Proportione*, Chapter 54," *Bibliothèque d'Humanisme et Renaissance* 7 (1977): 479–502.

37 See Pacioli, *Summa*, 68v, as well as his edition of Euclid's *Elements* (Venice: Paganino Paganini,1509), 30v. The importance of proportion of architecture and allied arts and sciences is the subject of the second section in the *Divina proportione*, 23v–33v, while the foundational role of proportion in Pacioli's thought is epitomized by the image appearing the last page of this publication and representing the "Arbor Proportio et Proportionalitas" (Figure 8.5). Moreover, it is also worth recalling that Piero della Francesca employed the principle of extreme and mean ratio (the golden mean) to create some of the imperfect solids by truncating Platonic ones, as he indicates in the *Libellus*. See on this Banker, *Piero della Francesca*, 204. It is possible that then existed the hypothesis that the rhombicubocatahedron was produced by the same principle of the golden mean.

38 On the general importance of the *divina proportione* for this cultural setting, see Patrizia Castelli, "La proporzione 'divina'. Alcuni aspetti della cultura scientifica alla corte di

Federico da Montefeltro," in *Presenze filosofiche in Umbria II. Dal Medioevo all'età contemporanea*, ed. Antonio Pieretti (Milan: Mimesis, 2012), 127–146. In the section entitled *La matematica come archetipo*, Pieretti suggests that the friar was conversant with the Platonic tradition, in particular the tradition of the *Timaeus*, and he will articulate the "power" of the *divina* in 1509, which may have been shared *in nuce* in 1495 because in the *Summa* Pacioli is primarily committed to mathematics as the undergirdment of all knowledge; for larger philosophical connections, see Pieretti, "Platonismo e Aristotelismo tra Umanesimo e Rinascimento," in *Presenze filosofiche in Umbria I. Dal Medioevo all'età contemporanea*, ed. Antonio Pieretti (Milan: Mimesis, 2010), 111–134, eespecially 120.

39 The interpretation of the *Portrait* as a teaching scene is often repeated; see for instance Annalisa Perissa Torrini, "Leonardo da Vinci: L'armonica proporzionalità, la quale è composta de divine proporzioni," in *Ut Pictura Poësis: Per una storia delle arti visive/For a history of the visual arts*, ed. Rocco Sinisgalli (Poggio a Caiano: CB Edizioni, 2012), 187–200, especially 194.

40 The eye for approximating and guesstimating quantities and their ratios was rightly assimilated into the "period eye" by Michael Baxandall; see his *Painting and Experience in Fifteenth Century Italy* (Oxford: Clarendon Press, 1972), especially 86–93.

41 For an overview of the history of magic squares, see Menso Folkerts, "Zur Frühgeschichte der magischen Quadrate in Westeuropa," *Sudhoffs Archiv Kiel* 65 (1981): 313–338; the article includes references to several fourteenth- and fifteenth-century manuscripts that give examples of magic squares. Pacioli himself included the subject among the mathematical games in the manuscript (composed between 1496 and 1508) titled *De viribus quantitatis* (Biblioteca Universitaria, Bologna Ms 250 and available in facsimile reproduction by Aboca (Sansepolcro, 2009)); see on this text Amedeo Agostini, "Il De viribus quantitatis di Luca Pacioli," *Periodico di Matematiche* 4(1924): 165–192.

42 The bibliography on Durer's *Melencolia I* is vast. An important early source is Raymond Klibansky, Erwin Panofsky and Fritz Saxl, *Saturn and Melancholy: Studies in the History of Natural Philosophy, Religion and Art* (London: Thomas Nelson & Sons, 1964), 284–402. For the recent bibliography, see the catalogue entry by Karoline Feulner in Jochen Sander, ed., *Albrecht Dürer: His Art in Context* (Munich: Prestel, 2013), 262–263. For the larger context, see Laurinda Dixon, *The Dark Side of Genius: The Melancholic Persona in Art, ca. 1500–1700* (University Park: Pennsylvania State University Press, 2013).

43 Euclid's definition (*Elements*, Book VI, Definition 3) of the golden ratio, or *divina proportione* to use Pacioli's label, is that of division in extreme and mean ratio: "A straight line is said to have been cut in extreme and mean ratio when, as the whole line is to the greater segment, so is the greater to the lesser." (In algebraic terms: AC / CB = AB / AC, and the whole line is AB).

44 In Book XIII Euclid covers the five Platonic solids, and this book is part of the last section of the work, which is dedicated to stereometry; specifically, Book XI covers the basics of solid geometry through parallelepipeds and Book XII concentrates on cones, pyramids, and cylinders. Notably, in Greek intellectual culture the equivalent of modern stereometry had distinguished standing because in the *Republic* (528a-e) Plato notes that mathematics, geometry, and the science of the cubes and figures with depth are disciplines the leaders of the state should master.

45 The dodecahedron relates to the *divina proportione*, phi, in two ways: its surface measures 15phi / $\sqrt{3}$–phi, and its volume 5phi^3/(6 – 2phi).

46 In fact, Baldi mentions that a manuscript of geometrical contents by Fibonacci was available in the library of Federico da Montefeltro, which, as he notes, was used by Pacioli; see Bernardino Baldi, *Cronica dei matematici* (Urbino: Monticelli, 1707), 107. In both Fibonacci's *Liber abaci* and Piero della Francesca's *Trattato d'abaco*, the golden ratio is present in the solution of various problems. Pacioli will draw on their works (usually giving credits) in the *Summa* when covering problems on the dodecahedron. See Mario Livio, *The Golden Ratio: The Story of Phi, the World's Most Astonishing Number* (New York: Broadway Books, 2002), 96–109, 127–128.

47 The division in extreme and mean ratio is abbreviated as DEMR in mathematical writings. In algebraic terms, Fibonacci's rabbit problem asks for the solution to the following equation: $U_{n+1} = U_n + U_{n-1}$. The ratio of successive U, i.e. U_{n+1} / U_n, progressively approximates phi as U becomes larger, for instance 453/280 = 1.6178, 733/453 = 1.6181.

48 See Leonard Curchin and Roger Herz-Fischler, "De quand date le premier rapprochement entre la suite de Fibonacci et la division en extrême et moyenne raison?" *Centaurus* 28 (1985): 129–138; they based part of their argument on an annotation at b6v (page 14v) on the copy of Pacioli's edition of the *Elements* of Euclid (Venice: Paganino Paganini, 1509) preserved at the Bibliotheque National de France, shelf number Res. V 104. This volume has been copiously annotated throughout in an easily legible handwriting. The author of the notes is unlikely to be connected with Pacioli because nothing is noted either on the pages of the title and dedication or on those describing the 1508 lecture of Pacioli in Venice (respectively a1-a3 and d6-d7). The notes themselves might have been copied rather than written directly on this volume: this is suggested by the neatness and precision of the position of the blocks of notes, in most cases self-contained. Notes by two other hands are included in the form of separate leaves: the first is titled on the squaring of the circle, and it was probably also copied, as suggested by the catch-word at the bottom of the first page; the second is present in several small leaves added in the text: the contrast of writing styles suggests that those on the pages proper were copied systematically and do not record the learning process done in the book by its author.

49 It is worth recalling that the manuscript version of the *Divina proportione* (which survives in two copies, one at the Bibliothèque de Genève and the other at the Biblioteca Ambrosiana in Milan) was completed in 1498, and Pacioli dedicated it to Guidobaldo da Montefeltro.

50 The scientific study of approximation, today a separate field of study in mathematics known as Approximation Theory, has its roots in the Enlightenment, in the work of Euler, Laplace, and Fourier. However, approximation was a problem faced already in antiquity; see for instance David Fowle and Eleanor Robson, "Square Root Approximations in Old Babylonian Mathematics: YBC 7289 in Context," *Historia Mathematica* 25 (1998): 366–378. While a complete survey of its history for the premodern period remains a desideratum, several good pages on the subjects are found in Enrico Gamba, "La matematica dei tecnici tra oralità e scrittura, tra esattezza e approssimazione," in *L'arte della matematica nella prospettiva*, ed. Rocco Sinisgalli (Florence: C.B. Cartei & Bianchi, 2009), 229–246, passim and especially 238–246.

51 Euclid first considers constant proportionality in Book V, Definition 5, and dedicates the rest of that book to it.

52 A simple example of proportionality that Pacioli includes in the *Divina proportione*, Chapter VII, at 5v, is 9 is to 6 as 6 is to 4. Each line is divided in half respectively by the 5, the x/8/3, while the 7 divides the top two lines taken as a whole.

53 The relevance of the *divina proportione* in Renaissance art has been underscored by recent research. Rocco Sinisgalli has convincingly argued that Leonardo's drawing of the Uomo Vitruviano is based on the golden mean; see "La sezione aurea nell'*Uomo vitruviano* di Leonardo," in *I disegni di Leonardo da Vinci e della sua cerchia nel Gabinetto dei Disegni e Stampe delle Gallerie dell'Accademia di Venezia*, ed. Carlo Pedretti (Florence: Giunti, 2003), 179–181, and Enrico Gamba found it in Raphael's *School of Athens* fresco, see his "Scuola di Atene: Note sulla lavagnetta di Euclide-Bramante," published online by http://urbinoelaprospettiva.uniurb.it/

9 Mathematical Imagination in Raphael's *School of Athens*

Ingrid Alexander-Skipnes

In *Le Vite*, Giorgio Vasari makes one of the earliest comments on the praiseworthy mathematical aspects of Raphael's *School of Athens* when he writes, "It is not possible to describe the beauty of those astrologers and geometricians drawing many figures and characters on tablets with their compasses."[1] He goes on to give a description of Euclid: "a figure bends toward the ground with a pair of compasses in his hand and turning them on a tablet."[2] In these short descriptions, Vasari draws our attention to the lively and elegant mathematical themes and activities in the fresco's foreground.

When Raphael was called to the court of Pope Julius II (r. 1503–13) in 1508, it was to undertake what would become one of his most ambitious projects, the decoration of the Stanza della Segnatura. As a monumental *tour de force* of elegant gestures, Raphael's *School of Athens*, on the east wall of the room, demonstrates not only the artist's skill at creating a sophisticated narrative design but also his mathematical knowledge and ability to construct a spacious concourse where a mastery of perspective underpins the architectural setting (Figure 9.1). Vasari called attention to the aesthetic powers of the fresco's perspective when he wrote that the artist had "adorned this work with a perspective and many figures, so delicately and finely finished."[3] The majestic architectural background with its three barrel vaults is arranged to accommodate what is essentially two bands of figures rhythmically placed within the space. The figures nearest to Plato and Aristotle stand close to one another in a row on each side of the two Greek philosophers and at the same time contribute to a visual perspectival recession. These tightly conceived groups in the background gradually unfold as our eye moves away from the two central philosophers and towards the foreground. Like Plato's Academy, individuals of different ages mingle and interact, read and write, and have animated discussions.

The most loosely arranged groups of figures appear at the edge of the fresco and in the foreground mathematicians. Significantly, from the perspective of geometry, the group around Euclid opens up so the viewer is able to see the figures more clearly. Another notable feature of the fresco is how occasionally Raphael manipulates the mathematical premises of perspective in order to achieve certain visual effects. Plato and Aristotle stand majestically on a square inscribed in a polygon. To enhance an illusion of depth, the white marble square is foreshortened and appears rectangular. The marble square inside the polygon creates flanking triangles (on each side of the square) and can remind the viewer that Plato wrote on the five regular bodies in the *Timaeus*, the book he holds in his left hand. This promenade of polygons on which the two philosophers stand changes to a pattern of red and white squares in the fresco's foreground.

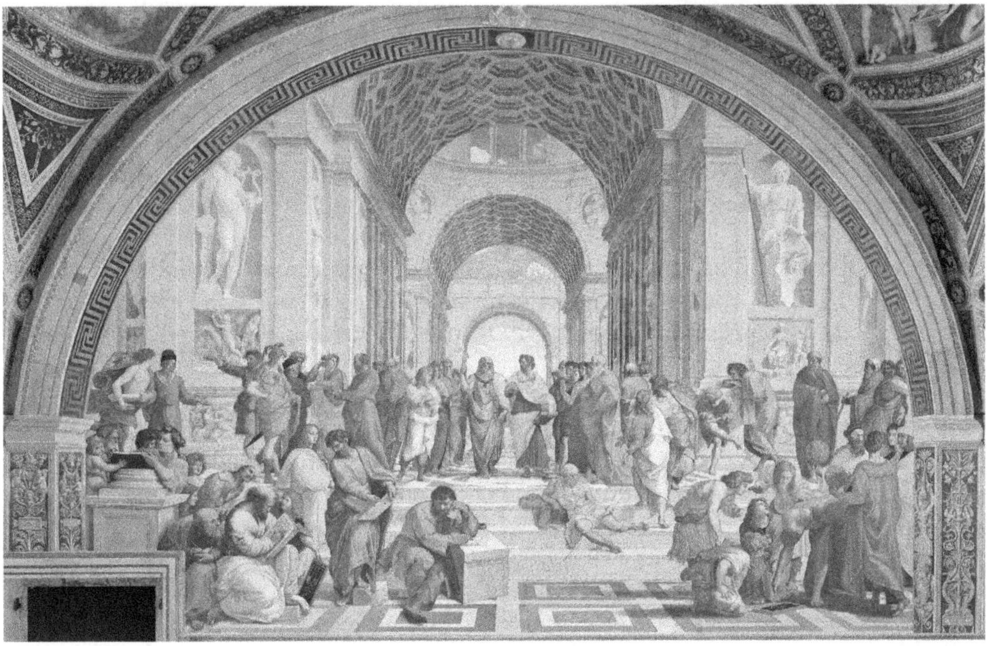

Figure 9.1 Raphael. *School of Athens*, c. 1509–10. Stanza della Segnatura, Vatican Palace.

Photo credit: Vatican Museums.

Of the 58 figures in the *School of Athens*, mathematicians figure prominently, and they are readily recognizable by their gestures, the activities they are engaged in and the objects they hold. They are among the most identifiable figures. Raphael's ability to shape and direct the viewer's attention towards the mathematical activities in the fresco is remarkable, where every detail contributes to the overall harmony in the painting. The richly illustrated mathematical activity in the fresco appears quite close to the picture plane, positioned directly at the height of the viewer. The two foreground groups of figures display studiousness, curiosity and discovery, a perfect ambiance in fact when we consider that the erudite setting for the fresco was the papal library, where eventually books, including rare mathematical texts, would be housed.

Two Greek mathematicians of antiquity, Pythagoras on the left and Euclid on the right, anchor the fresco's composition and frame the foreground for the *due Principi della Filosofia* (two princes of philosophy),[4] Plato and Aristotle appear to stroll comfortably toward the lower band of figures, where the mathematicians and other scientists are located. Interestingly, the only two objects where mathematical diagrams are clearly drawn and unmistakably visible at eye level for the spectator are the tablets in the foreground. A great deal has been written about the slate in front of Pythagoras, but there is little agreement on the meaning of the drawing on Euclid's slate. Raphael has placed a particular emphasis on this group of scientists, which includes not only Euclid but also Ptolemy and Zoroaster, in the painting's lower right area, and especially the figure of Euclid.

When scholars have studied the mathematics in the *School of Athens*, their discussions have often centered on Raphael's use of linear perspective. While numerous studies have discussed the drawing on Euclid's tablet, surprisingly, many of the interpretations have been dominated by modern mathematical thinking and little attention given to how mathematics was viewed or understood in the Renaissance humanist environment.[5] Even though several authors have commented on the figures of Plato and Aristotle, they have not explored the mathematical content in the *Timaeus* and in the *Ethics* and therefore have not related the mathematical discourses in these two books to the fresco.

This chapter will show that there was a significant increase in the number of mathematical texts in the Vatican Library in the period. With a focus on the mathematicians in the right foreground, this chapter will also examine the mathematical themes in the fresco, the significance of the geometrical diagram on Euclid's slate and Raphael's presence among a group of mathematicians. Furthermore, this chapter will shed light on the power of mathematics to engage the viewer both visually and intellectually.

The Stanza della Segnatura was part of a suite of rooms in the papal apartments whose decoration began under Pope Julius II and continued under Pope Leo X (r. 1513–21). In a somewhat ambiguous phrase, according to Raphael's earliest biographer, Paolo Giovio, the frescoes were painted "*ad praescriptum Iulii pontificis.*" This phrase does not clearly inform us as to what extent Pope Julius II directed the commission or what role he played in the overall decorative scheme of the room.[6]

Building on earlier hypotheses, Franz Wickhoff argued in a seminal article that the room was the private library of Pope Julius II.[7] And John Shearman presented a convincing idea that in contrast to the other Stanze, in the Stanza della Segnatura, the lower edge of the position of the fresco corresponded to the top of the bookshelves.[8]

The library of the Stanza della Segnatura was arranged according to a system based on humanist libraries at the time. The system was devised by Tommaso Parentucelli, who later became Pope Nicholas V (r. 1447–55), where books of sacred literature, the largest portion of the collection, would be placed next to works of philosophy. These were followed by mathematics, music and the *studia humanitatis*, history, grammar, rhetoric, poetry and moral philosophy. Parentucelli's bibliographic canon was adopted for the well-known library of Federico da Montefeltro in Urbino, a library that the young Raphael may have known. Regarding the humanist interest in mathematics, Leon Battista Alberti played an important role in nurturing this interest at the papal court of Nicholas V and brought together the interest in mathematics in Florence and Rome.

Pope Nicholas V began in earnest to increase the number of mathematical texts in the Vatican Library through his efforts to have many important Greek mathematical texts translated into Latin. In the inventory of 1455, 12 manuscripts of a mathematical content are among the 414 Greek manuscripts brought together by Nicholas V.[9] Pope Sixtus IV (r. 1471–1484), added to Nicholas' collection, as can be seen in the inventories compiled in 1475 and 1481 (Vat. lat. 3954 and Vat. lat. 3947, respectively). A third inventory for Pope Innocent VIII in 1484 (Vat. lat. 3949) showed that the total number of Greek mathematical texts had risen considerably. It is clear, indeed, that the interest in the mathematical sciences lies with the Greek manuscripts. The three inventories counted several codices of Euclid's *Elements* and Ptolemy's *Almagest* that covered mathematical astronomy and summarized his scientific theories.[10] A volume of various mathematical texts that was bound in green and consisting of a *Euclidis*

Geometria, et Ledomana (*sic*), *et prospectiva* (Vat. gr. 192?) was later rebound in black during the pontificate of Julius II.[11] During the Julian pontificate, a *Ptolomei Almagestus* (Vat. gr. 180) and another *Almagestus* (*Syntaxis mathematica*) *cum Theone* (Vat. gr. 184) were rebound as well as a *Euclidis Geometria completa* (Vat. gr 193), which also comprised a work entitled *Theodosii sphaericorum* and *Claudii Ptolemaei geographiae*.[12] In the 1475 inventory, mathematical manuscripts are listed under *Astrologi Greci*, while in the 1481 inventory, mathematical texts are to be found under both *Bibliothecae Grecae in Philosophia* and *Mathematici et Astrologi*.[13] Noteworthy in the 1475 inventory are Latin translations of Ptolemy's *Almagest* along with works by Euclid.[14]

In 1510, Tommaso Inghirami (Fedra) took charge of the Vatican Library of Pope Julius II. An inventory of the Greek manuscripts was drawn up by Fabio Vigili in which several manuscripts of Ptolemy's *Cosmographia* and *Almagest* are listed, as well as a *Strabonis Geographiae* (Vat. gr. 174) and a *Euclidis Catoptrica* together with a *Elementis*.[15] Of the many texts by Ptolemy, one *syntaxis mathematica* was noted in particular as especially beautiful: "*liber est pulcherrimus*" (Vat. gr. 180).[16] The ninth-century Euclid's *Elements* (Vat. gr. 190), a treasure among the Euclid texts, seems to have been overlooked during the Renaissance.[17] It is noteworthy that at the time geometrical, geographical and astronomical instruments decorated the library shelves "pour le plaisir des yeux autant que pour l'utilité pratique."[18] In the register of loans compiled by Maria Bertòla, codices by Ptolemy on astronomy, mathematics and geography were often loaned.[19] A Greek codex described as *Cosmografie et axtronomie Phtolomei in grecho* are among the mathematical codices lent by the Vatican Library during the pontificate of Julius II.[20] Euclid's *Elements* was also loaned, and a work by Galen with the insignia of Julius II was borrowed in 1521.[21] The increase in mathematical texts in the Vatican Library in the period demonstrates interest in Greek science in the humanist environment of the papal court. Most importantly, manuscripts of the works of Euclid and Ptolemy seem to have been particularly prized, given the number of these codices among the collection. As reflected in the Vatican Library, Euclid and Ptolemy have also found a prominent position in representing the mathematical sciences in the fresco.

Let us turn our attention now to the private library of Pope Julius II. Léon Dorez carried out an early study of Julius' library.[22] As Paul Taylor has shown for the *Disputa*, there was a close correlation between the books depicted there and the papal library.[23] Although there were no Greek manuscripts among the circa 220 codices in Pope Julius' private library, there were important texts such as a Latin translation by Argyropoulos of Aristotle's *Ethics*.[24] When it comes to the mathematical and astronomical manuscripts, there were fine texts of Ptolemy's *Cosmographia* and *Tabule* and Strabo's *Geographica*. The largest number of books in the papal library was those devoted to theology and philosophy, and presumably they were placed beneath their respective frescoes.[25] Pope Julius II had manuscripts of Gregory the Great and Albertus Magnus, authors who wrote on theology and philosophy. Pope Gregory's *Moralia*, for example, might be placed either among the philosophy books or among the books on theology.

The often-repeated quote in Vasari's life of Michelangelo recounts that when Julius was asked about which attribute should be applied to a statue of him, Julius declared that he would rather hold a sword and not a book, as he is no scholar.[26] While Julius seemingly preferred to not be depicted with a book, his library belies an interest in intellectual pursuits. Moreover, the military pursuits of any leader at the time should not be seen as a reason to exclude his interest in learning, as the library of the Duke of Urbino can certainly attest. Federico da Montefeltro was painted wearing armor while reading a book. Even as a mercenary captain, Federico showed an interest in mathematics, and the environment at the court has been described as a center of mathematical humanism.[27] Vespasiano da Bisticci recounts that Federico enjoyed discourses on mathematics. Even Aristotle, who was a teacher of Alexander the Great, wrote on the importance of mathematics in warfare. Furthermore, Pliny the Elder, most known for his *Natural History*, was also a commander in the army of the early Roman Empire; his first work developed the treatment of the javelin as a cavalry weapon. Pliny also had broad intellectual interests.

The four walls of the Stanza della Segnatura are decorated with frescoes depicting four main areas of learning or faculties: philosophy, theology, poetry and jurisprudence. Female figures that personify these branches of learning are painted in roundels on the ceiling just above the fields of learning that they represent. The two largest frescoes in the room, the *School of Athens* on the east wall and the *Disputa* on the west wall, reflect the large number of books on theology and philosophy, which included the mathematical sciences, among the papal holdings. Furthermore, a visitor to the room would see the *Disputa* first. This also reflected the hierarchic placement of books in Parentucelli's bibliographic canon, as mentioned above, where the largest portion of books in humanist libraries would be devoted to sacred literature, followed by works of philosophy. Thus the mathematical texts together with philosophical codices would be placed in front of the *School of Athens*.

The figures in the *School of Athens* suggest that Raphael's concept for the fresco is based on the tradition of the *uomini illustri* (Illustrious Men) and the Seven Liberal Arts. The tradition of the *uomini illustri* had its origin in antique biographical writings in which orators, statesmen and generals were praised for their contributions. Inspired by Petrarch's *De viris illustribus*, visual representations of famous men began in the Middle Ages, and in fact Petrarch was instrumental in the revival of humanist interest in the subject during the Renaissance. Of interest for this study is the fact that Petrarch's *De viris illustribus* was also in Julius II's library.[28] The earliest cycle of *uomini illustri* was painted by Giotto during the 1340s in a period where the popularity of this tradition flourished. This is clearly demonstrated by the several cycles of heroes depicted in Padua, Florence, Naples and Rome.[29]

Traditionally, mathematicians and other scientists were not included in the writings or visual representations of *uomini illustri*. In his biography of Marcellus, the Roman general who conquered Syracuse, Petrarch, however, included comments on Archimedes. Petrarch described Archimedes as a *mechanicus summus*, stressing his practical knowledge and echoing the praise of earlier authors.[30] Archimedes was primarily known for his mechanical achievements, which did not have quite the same appeal as Euclid in the cultural *milieu*. Euclid's name was synonymous with geometry, and the theoretical foundations of his work, as well as its intellectual *gravitas*, were held in high esteem. Mathematics was prized in learned circles because of its method of dialectic, where logical deductions derive from initial premises in order to arrive at conclusions. The conclusions are true if the premises are true. The reliability of mathematics gave it a certain *cachet*.

For our study, Petrarch's remarks on Archimedes are interesting. They delineate a distinction between the applied and the theoretical aspects of mathematics, which would become more important during the early modern period. Mathematics would be appreciated as an intellectual activity for its own sake. Interestingly, the period's intellectual interests in mathematics have a parallel in the changing view of the arts from a craft tradition to one in which the intellectual and inspired skill of the artist were praised.

Two notable representations of *uomini illustri* that Raphael is likely to have known are the 28 portraits painted by Joos van Ghent and Pedro Berruguete in the *studiolo* of Federico da Montefeltro in the ducal palace in Urbino from c. 1476 and the frescoes by Perugino in the Collegio del Cambio in Perugia dated 1500. Mathematicians among *uomini illustri* seem to have been illustrated for the first time in Urbino. The *studiolo*, with its rich literary sources, was decorated with famous men not only from the past but also with contemporary figures. In the *studiolo* a portrait of Euclid of Megara, frequently confused with the Greek mathematician Euclid of Alexandria, appears together with those of Ptolemy and Boethius, a philosopher whose writings included texts on geometry, astronomy and arithmetic. Euclid of Alexandria is clearly the figure intended in the Urbino portrait, as the inscription confirms: *Euclidi Megaren*[*si*], *ob comprehensa terrae spacia lineis centroq*[*ue*] *Fred*[*ericus*] *dedit, invento exactissimo* (To Euclid of Megara, because he grasped the spaces of the earth by means of lines and the compass. Federico gave this for a very precise invention). As mentioned, the theoretical basis of Euclid's works found favor among humanists. Thus it is not surprising that Euclid's portrait is in the *studiolo* while Archimedes's is not.

Among the *studiolo* portraits, Federico included his teacher, Vittorino da Feltre, and Cardinal Bessarion, friend and promoter of philohellenism in Urbino. Similarly, Raphael painted an assembly where ancients and contemporaries mingle. And as some scholars have proposed, Raphael's fellow artist, Sodoma, and Julius II's nephew, the duke of Urbino, Francesco Maria della Rovere, are present in the *School of Athens*.[31] Although Giovanni Santi had closer ties with the court of Urbino than his son did, Raphael painted portraits and other paintings for members of the court including Duke Guidobaldo, Duchess Elisabetta Gonzaga and also for Giovanna della Rovere and her son Francesco Maria della Rovere.[32] Thus Raphael had several occasions to return to Urbino not only to attend to family affairs but also to pay visits to the court, where he would have probably more than one occasion to see the *studiolo*.

The flourishing interest in Greek science at the court of Urbino may account for the inclusion of mathematicians among the portraits of the famous men. While Federico was never fluent in Greek, his library reflected his intellectual aspirations and a certain pride in exploring the literature beyond the Italian peninsula.[33] Euclid's *Elements* and Ptolemy's *Cosmographia* and other texts that relied on mathematics were among the mathematical codices in the Urbino library, including an elegant translation of Euclid's *Optics*. The cycle of portraits of *uomini illustri* in the *studiolo* reveals a close correspondence between these figures and the books in the library. The famous *studiolo* provided refuge for contemplation and study, but it also was a showcase where distinguished visitors were treated to a display of imagined and real objects. Undoubtedly, the Stanza della Segnatura had a similar role for Pope Julius II.

There is also a close correlation between Perugino's frescoes in the Collegio del Cambio in Perugia and Raphael's concept for the *School of Athens*. In Perugino's decoration of the space, female personifications of the virtues sit on banks of clouds and are accompanied by winged putti carrying inscribed tablets.[34] Heroes, who exemplify the virtues above, stand below. Perugino's Heroes are identified with scrolls and inscriptions. Whereas Perugino's Heroes, based on traditional depictions of Illustrious Men, stand in a row and do not engage with one another, Raphael's authoritative solution consists of devoting a large area of the space to a broad group of enlivened figures set on three different planes with an extraordinarily rich variety of poses. Scant clues have been provided about who these individuals are. The artist identifies his cast of characters by their activities, books and tablets. Indeed, Raphael's original narrative is unprecedented in the way the artist has presented his ensemble of Illustrious Men.

The *School of Athens* also depends on the tradition of the Seven Liberal Arts which consisted of the *trivium* (grammar, rhetoric, dialectic) and the *quadrivium* (arithmetic, music, geometry and astronomy).[35] The *quadrivium*, in contrast to the *trivium*, was considered the more advanced part of the Seven Liberal Arts. This may be due in part to the theoretical basis of its subjects and their mathematical structure. During the early modern period, there was an emphasis on the necessity of mathematics in order to arrive at a correct understanding of the natural world.[36] In support of this orientation, many cited Boethius, the sixth-century Roman scholar, who was, in fact, the first to use the term *quadrivium*.[37] It is likely that he is among the Illustrious Men in the *School of Athens*. While Raphael has distinctly depicted the *quadrivium* disciplines in the foreground, the artist has not given the same clarity to the *trivium*. Instead, the *trivium* is informally distributed among the background figures.

The quadrivial disciplines are ornately depicted in frescos painted by Pinturicchio between 1492 and 1494 to decorate the living quarters of Pope Alexander VI Borgia.[38] In the *Sala delle Arti Liberali*, Pinturicchio painted a personification of Geometry who sits enthroned above a figure of Euclid seated on the floor with a compass in hand, inscribing a circle in a square. In late 1507, Julius II moved out of the apartments of his predecessor, and according to Ernst Gombrich, the pope evidently had nothing against the ideas painted by Pinturicchio, as the Stanza repeats the theme of the Borgia apartments.[39]

In a pioneering article, Julius von Schlosser presented a study on the representation of philosophy and the Seven Liberal Arts and the tradition's medieval origins.[40] Schlosser presents a compelling analysis of the association of the personifications of the individual arts and virtues with their exponents. The analogy between the parallel placement of Pythagoras and Euclid in the *School of Athens* and the medieval representation of these two mathematicians is striking and points to a long-established tradition. In the twelfth-century miniature from Herrad of Landsberg's *Hortus deliciarum*, Philosophy is enthroned and encircled by figures representing the Seven Liberal Arts which are nourished by streams that flow from her.[41] On the title page of Gregor Reisch's *Margarita Philosophica*, printed for the first time in 1503 in Freiburg, female personifications of the Seven Liberal Arts are arranged below Dame Philosophy. Two inscriptions, *Ph[ilosoph]ia Natur[alis]* and *Ph[ilosoph]ia Mora[lis]* appear next to images of Aristotle and Seneca, classical authors on natural philosophy and moral philosophy, at the bottom corners of the page, while at the top of the page the inscription *Ph[ilosoph]ia Divina*, referring to theology, is at the center of depictions of the four Doctors of the Church.

The symbolism of Philosophy nourishing the Seven Liberal Arts is even more vividly expressed in a 1508 edition of Reisch's *Margarita Philosophica*. Here Dame Philosophy is enthroned with a tree growing from her womb and the Seven Liberal Arts, depicted with their attributes, float on the leaves of the tree (Figure 9.2). She hands

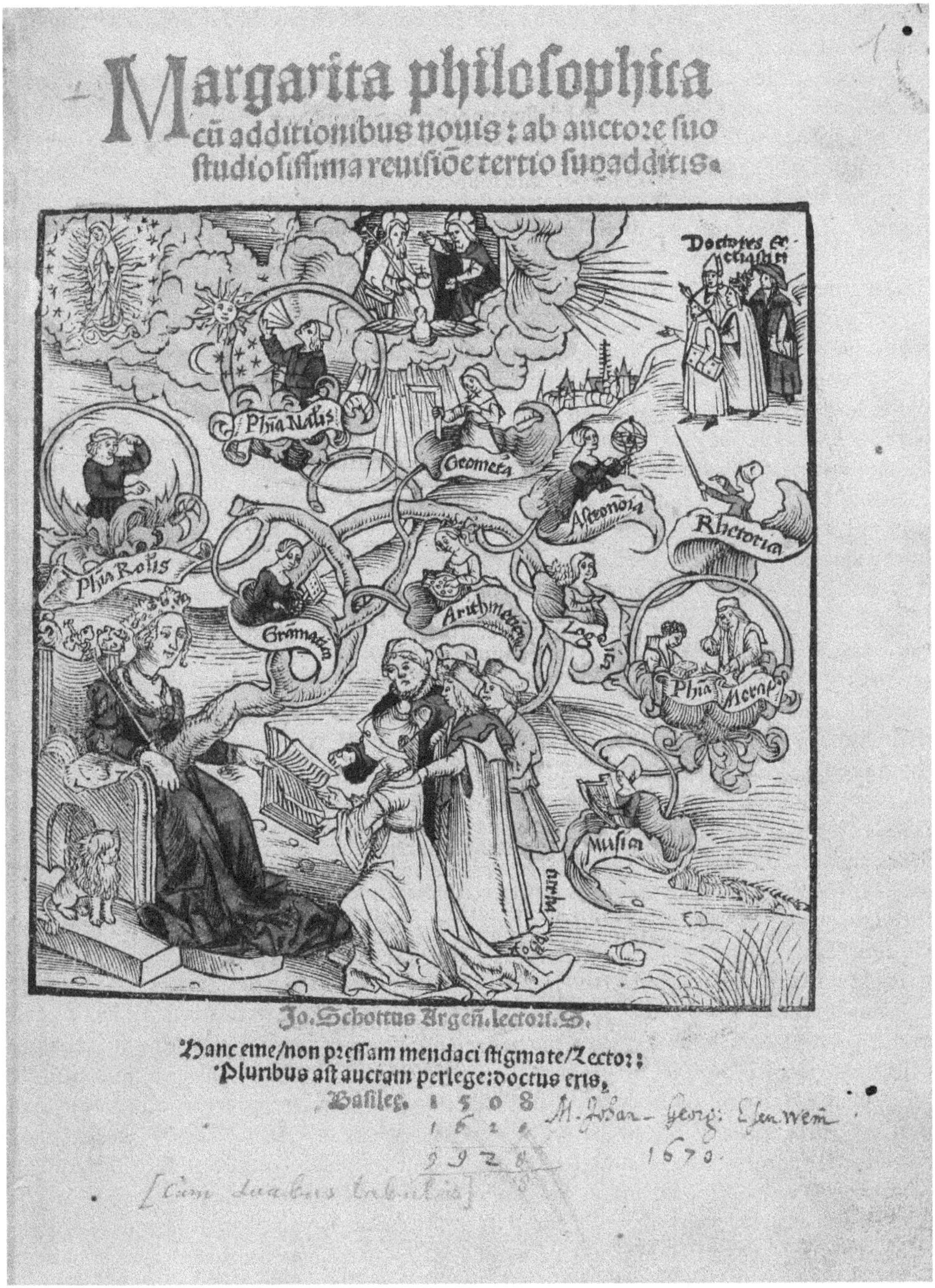

Figure 9.2 Philosophy and the Liberal Arts. Gregor Reisch. *Margarita Philosophica*. Title page. Johann Schott and Michael Furter: Basel, 1508. Universitätsbibliothek Freiburg i. Br./Historische Sammlungen.

Photo credit: Universitätsbibliothek Freiburg i. Br./Historische Sammlungen.

a book, possibly the *Margarita Philosophica*, with its pages fluttering, to a group of eager students in front of her.

In large circles on the branches sit representations of Rational Philosophy, Natural Philosophy and Moral Philosophy.[42] Images of God the Father, Christ and the Holy Spirit preside over the scene from above. To the left, the Virgin Mary appears, and to the right, the four Doctors of the Church look on. Thus Dame Philosophy nurtures the Seven Liberal Arts, and they are in turn enriched by Theology represented by the Trinity and the Doctors of the Church. The symbolism of these images is strikingly comparable to the *School of Athens* and the ceiling *tondo* where Philosophy is enthroned. It reminds us that Theology, represented by the *Disputa* on the opposite wall, guides Philosophy and its liberal arts, a connection that dates from the medieval period.

Philosophy reigning over the Seven Liberal Arts is often depicted on cathedral tympana. In the archivolt at Laon, for example, Philosophy has her head partly hidden by clouds. As Adolf Katzenellenbogen has argued, the partially hidden head symbolizes the highest part of philosophy, that is, theology "extending into the sphere of Divine Reason."[43] Given the close parallels between the medieval tradition of the juxtaposition of the Seven Liberal Arts with Philosophy and Raphael's fresco, with Philosophy in the ceiling above, it is surprising that scholars have seemingly abandoned the idea that Raphael has, in fact, reinterpreted this theme with striking monumentality and High Renaissance vigor.[44]

Although Vasari carefully described and named several of the figures in the *School of Athens* with an interesting narrative about the individuals who represent the mathematical sciences, he surprisingly gave no indication whether the figure drawing on the tablet on the right is Euclid or Archimedes. In 1695, Giovanni Pietro Bellori identified the figure as Archimedes. Bellori's corrections of some of Vasari's identifications may be the reason that later scholars believed the figure should be identified as Archimedes. Writing in 1839, Johann David Passavant noted that the *basamento* painting that simulates bronze reliefs executed by Perino del Vaga in the 1540s for Pope Paul III depicts the execution of Archimedes by a Roman soldier.[45] This image just below the right foreground of the fresco may be another reason scholars have thought that Archimedes was portrayed in the fresco. Writing in 1864, William Watkiss Lloyd was the first writer to propose Euclid.[46] Most scholars today identify the figure as Euclid.

It has often been claimed that the figure of Euclid is painted in the guise of Bramante, an idea introduced by Vasari and continued by Bellori.[47] Examining the biographies of Bramante and Raphael in Vasari's 1550 and 1568 editions of *Le Vite*, questioning why the artist would paint such an unflattering image of Bramante as a portrait, and looking at portraits of Bramante on medals and in other media, Hans W. Hubert has convincingly argued that the figure of Euclid in the fresco could not be a portrait of Bramante.[48] Furthermore, and as Hubert points out, Raphael's Euclid depends on the figure of Geometry painted by Pinturicchio in the Borgia apartments.[49]

That the figure is Euclid is indicated not only by the diagram he inscribes on a slate but also the correlation to the painting of Geometry by Pinturicchio in the Borgia apartments and the image of Euclid in the Urbino *studiolo*. The artist paints Euclid with a balding pate, not unlike the image in Pinturrichio's fresco. The physiognomy of Euclid in Pinturicchio's Borgia room and Euclid in the School of Athens are quite similar. Both have rather long, flowing hair around the balding pate. Furthermore, it is unlikely that Raphael would inscribe his name on the border of the garment of

Bramante, his mentor and compatriot, but rather it is more plausible that he would inscribe his name on the garment of Euclid, as the artist identified his work closely with geometry.[50]

Euclid, the Greek mathematician, was active c. 300 B.C. in Alexandria. Celebrated during the Renaissance, his most famous work, the *Elements*, became synonymous with geometry. Humanist libraries eagerly sought to acquire copies of his works. Euclid's *Elements*, which is devoted to plane and solid geometry, was among the most prized possession of collectors and those interested in the mathematical sciences. The *Elements* is made up of 13 books comprising definitions, postulates, common notions (axioms) and propositions.[51] Two additional books of uncertain authorship have been added to older editions of the *Elements*. The first six books are concerned with planar geometry—points, circles, triangles, right angles, straight lines—essentially the shapes in painting and architecture. Book I, proposition 47 presents the celebrated Pythagorean theorem, which asserts that in a right-angled triangle the square of the longest side (hypotenuse) is equal to the sum of the squares of the other two sides.

Book V, the most admired of Euclid's *Elements*, develops a theory of proportion first discovered by Pythagoras and so important for architects in constructing a right angle. It also had relevance for painters and musicians. Books VI through IX develop theories concerned with proportion but also deal with magnitudes and arithmetical problems. The remaining books are devoted to solid figures. Book XIII is devoted to the five regular solids constructed within a sphere. Since these five regular solids were discussed by Plato in his *Timaeus*, they are also known as the Platonic solids.

Euclid is also clearly linked to Aristotle; thus it is no surprise that the geometer is situated on the Aristotelian side of the fresco. For example, according to the Greek philosopher, the mere definition of an object does not establish its existence; this must be done by the method of proof.[52] Euclid clearly applies Aristotle's principle.[53] For his definition of a point, a straight line and a circle, Euclid applies Aristotle's idea and supposes their existence as the basis of his geometry and therefore proves the existence of every other figure that can be derived from these.

In the *School of Athens*, Euclid is surrounded by a group of enrapt students who show their interest and enthusiasm as well as their grasp of the Euclidean demonstration in different degrees (Figure 9.3). They express their individual experience of geometry.

The young man at the top of the group is shown to be the most enlightened student with his left hand seemingly resting on Euclid. His expression is one of captivated listener. And as Arnold Nesselrath has noted, he has become "one of the most famous figures of the *School of Athens*."[54] The earliest stages of learning are seen in the two kneeling figures, one pointing towards the slate while looking up at his confident companion. The two more advanced students are standing. The standing youth on the left points towards Euclid as he rests a reassuring hand on the back of his kneeling companion. The two standing students reveal through their gestures and postures a greater understanding of the geometric demonstration while they explain it to the other two.

It is interesting to reflect on how the postures of the young men in this group mirror processes in mathematics. They show a gradual progression of understanding. This brings to mind that in mathematics, the order in which things are done is important.[55] As they visualize a theory, they deduce logically, much in the way Euclidean thought proceeds. Euclid's students exhibit different levels of understanding, and, like

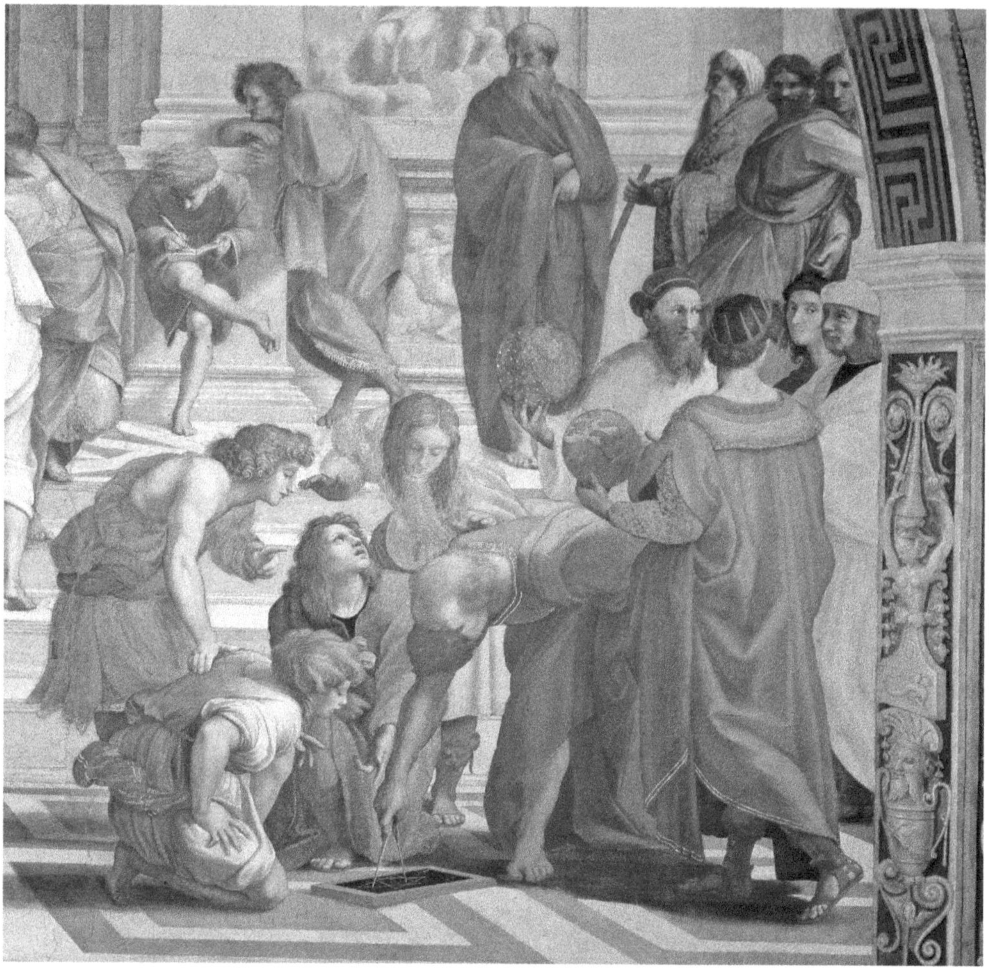

Figure 9.3 Raphael. *School of Athens*. Detail of Euclid and his pupils. Stanza della Segnatura, Vatican Palace.

Photo credit: Vatican Museums.

the viewer, who ranges from the layman to the specialist, understanding is changing and developing like the experience of Euclidean geometry in the *Elements*. The figures around Euclid illustrate the methodological approach used in Euclid's geometry. Conclusions follow a logical path that originates in definitions and propositions.

Contrary to what many scholars have suggested, the diagram on Euclid's slate is not about a particular figure or theory in geometry but rather about processes and general concepts in Euclidean geometry and illustrates how in the *Elements*, Euclid's methods come into being (Figure 9.4).

Furthermore, most scholars have failed to recognize that the diagram on the slate is not finished. The resting point of Euclid's compass is at a crossing of two triangles,

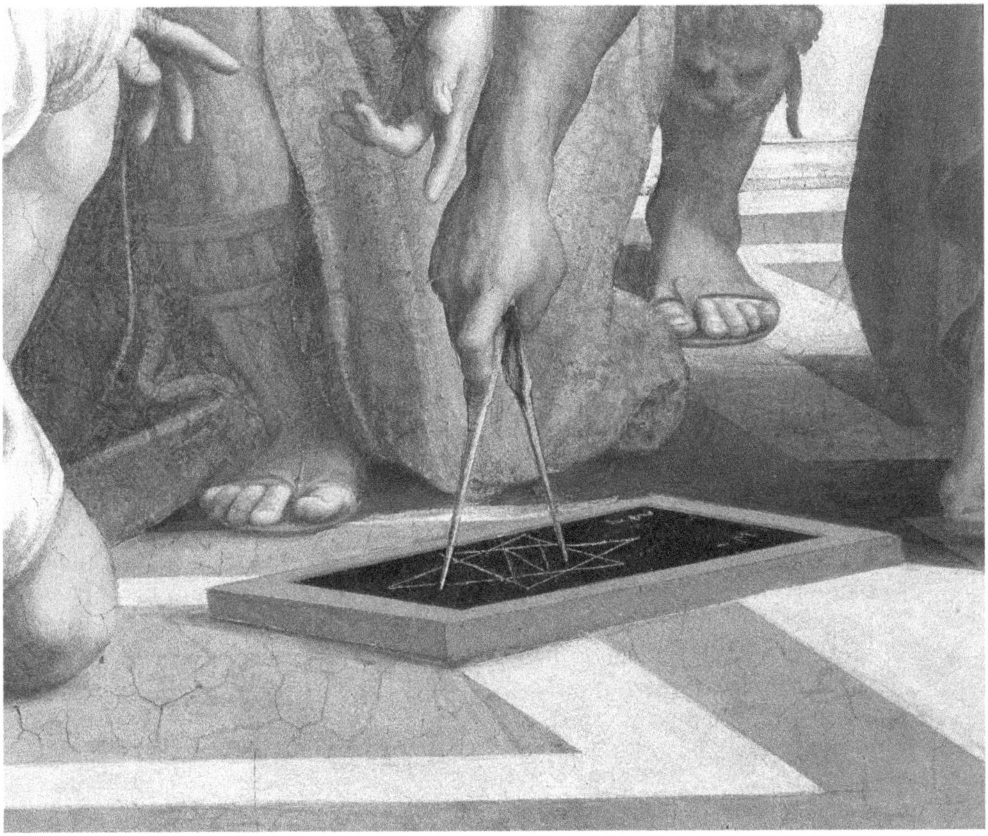

Figure 9.4 Raphael. *School of Athens*. Detail of Euclid's slate. Stanza della Segnatura, Vatican Palace.

Photo credit: Vatican Museums.

and he is ready to continue drawing and making another geometric figure. This further underlines the processes Euclid adopts in his step-by-step development of his propositions in the *Elements*. Moreover, the diagram on the slate points not only to the importance of triangles in Euclidean geometry but also to the importance of the Pythagorean theorem for architects, where constructing the right angle properly was essential. Painters used the Pythagorean theorem to enlarge drawings and calculate the scale in their works. The theorem was also important for astronomers and cartographers. Thus it is not surprising that Euclid with his slate and pupils is the focal point in a group of scientists. That Raphael included himself in this group indicates that the painter views his art as a mathematical activity governed by principles of geometry. He also pays tribute to the importance of mathematics and geometry for architecture. At the same time, while gazing out at the spectator, Raphael wants the viewer to see that the artist is not merely a craftsman but an inspired intellectual with knowledge of mathematical principles.

Like the figure of Heraclitus, which was added later and, according to Chastel, "*a tempera*," i.e. painted *a secco* after the fresco dried, the drawing on Euclid's slate also does not appear in the cartoon for the fresco and was also painted a *secco*.[56] It is undoubtedly a more sophisticated and complex diagram, in mathematical terms, when we compare it to the drawing on Euclid's tablet in the Borgia apartments and in the Urbino *studiolo* portrait of Euclid, where it seems that a drawing of a circle is noticeable.

In the *School of Athens*, Euclid is in a group of mathematicians that includes Ptolemy and Zoroaster and represents half of the *quadrivium*, i.e., geometry and astronomy. The presence of Zoroaster among the group of mathematicians may seem unusual, since he is primarily known as an astrologer. In the medieval and Renaissance period, however, astrology was taught together with mathematics and astronomy at the universities.[57] Moreover, Ptolemy's position in the fresco, facing Zoroaster, recalls the Greek cartographer/mathematician's attempts at reconciling astrology and astronomy in his *Tetrabiblos*.[58] In the fresco, Ptolemy's terrestrial globe is held quite closely to Zoroaster's celestial globe. This brings to mind that in the *Almagest*, an important text for astrology, Ptolemy wrote on predicting the position of heavenly bodies, while in the *Tetrabiblos*, which complements the *Almagest*, he wrote on how heavenly bodies affect terrestrial things.[59] With his back to the spectator, Ptolemy is positioned so that his golden robe overlaps Euclid's orange mantle, and by holding his terrestrial globe near Euclid indicates that his cartographic work depends on Euclidean geometry. Indeed, Ptolemy is best known for his *Cosmografia*, which opens with a mathematical discourse on cartography.[60] Longitude and latitude with their mathematical requirements are represented in a list of places. Ptolemy also drew on Euclid in his writings on optics. And both Euclid and Ptolemy applied mathematics to their theories of vision.[61]

By placing himself near Euclid and in the circle of the other mathematicians, Raphael clearly identifies himself with this group. His signature on the collar of Euclid's garment also visualizes the artist's admiration of Euclid in particular and Greek mathematics in general. Raphael's proximity to Euclid with his compass touches on the tangible aspects of the artist's work as a draftsman, painter and architect and his own use of the compass. In fact, traces of the use of the compass by the artist can be seen in the cartoon for the *School of Athens*.[62] For example, it is evident that Raphael used a compass to draw the terrestrial globe held by Ptolemy and the celestial globe held by Zoroaster. It is likely that he also used the compass together with a ruler to draw parallel lines, like those we see in the steps, the pavement and the architecture, clearly discernible in the cartoon. The compass is important in order to make right angles and other angles and to measure and construct distances in a perspectival system.

Similarly, ruled lines and arcs drawn with a compass are visible in a *modello* for the frescoes in the Piccolomini Library in Siena (1502–8).[63] Raphael's talent as a draftsman and inventive designer of large, complex scenes in fresco was already demonstrated in Siena before he came to Rome. Juxtaposing his organic flow of figures in the *School of Athens* next to his elegant mathematical solutions, Raphael deftly adopts the tools of geometry.

Vitruvius himself notes the importance of the compass and ruler when he writes on the training of an architect in his *De architectura*, written c. 30–20 B.C. In Book 1, Chapter 1, Vitruvius outlines the importance of geometry in the education of artists. He writes that geometry "hands down the technique of compass and rule."[64]

He continues: "Likewise, through knowledge of optics windows are properly designed so as to face particular regions of heaven . . . the difficult issues of symmetry are resolved by geometric principles and methods."[65] Among the subjects that Vitruvius recommends for the education of an architect was knowledge of letters, geometry, philosophy and draftsmanship, and these reflect the training for architects in Renaissance Italy. Draftsmanship was of particular importance to architects, and Raphael, even before his arrival in Rome, was a precocious draftsman as his early architectural drawings and the painted architectural experience attested in works such as the *Sposalizio* of 1504. And John Shearman suggested that as early as 1508–9, Raphael may have had his first building experience in the remodeling of the Stanza della Segnatura.[66] The artist carefully studied ancient texts and the monuments of Rome, and his interest in Vitruvian classicism is often noted in the letter he wrote to Pope Leo X.[67] While Raphael did not officially begin his career as an architect until 1514, the Vitruvian elements in the architecture of the *School of Athens* are unmistakable, and it would be just a few years later before he would have his own Italian copy of *De architectura*.

The importance of draftsmanship for the training of an architect is further underscored in the inscription on a Uffizi study by Maso Finiguerra; the artist writes, "vo esere uno buono disegnatore e vo (some letters missing near the margin) ventare uno buono archittetore" (I want to be a good draftsman to become a good architect)."[68] Furthermore, Alberti strove to emphasize a complete education for the architect and related this to the training of the artist, whose achievements should be viewed as both intellectual and practical.

The only fresco in the room with an architectural framework for the figures, the *School of Athens*, also reflects the artist's versatility.[69] Raphael painted an extraordinarily rich image of classical architecture in the fresco, visually expressing its mathematical foundations as well as a mastery of perspective and illusionism. Whereas the facing fresco of the *Disputa* is painted with a strict one-point perspective system, in the *School of Athens*, control of the fresco's perspective is also revealed by manipulating the rules of perspective to obtain particular visual effects. In the *Disputa,* the vanishing point is located at the Host in the center of the monstrance. In contrast, while situated between Plato and Aristotle, the vanishing point in the *School of Athens* is hidden. The two vanishing points, one visible and one concealed, which are apparently aligned with one another on a visual axis, may symbolize that truth is revealed in theology and that the highest part in philosophy, that is, theology, when uncovered reaches into the "sphere of Divine Reason."[70]

The figures of Plato and Aristotle are grander than their perspectival requirements in order to give them a larger presence in the fresco as they approach the spectator. And there are other notable areas where Raphael has manipulated perspectival principles. Euclid's slate, for instance, is askew and directed lower than its counterpart in front of Pythagoras. This enables the viewer to see the drawing more clearly. Heraclitus' stone block is tilted downwards and incongruous with the fresco's spatial requirements.

Let us turn our attention to the mathematics in the books that Plato and Aristotle carry. For Plato, mathematics could sharpen the mind and was important in order to arrive at knowledge of the Good.[71] Carrying the manuscript *Timeo* (*Timaeus*) under his arm, Plato points heavenward to indicate the source of all knowledge and his cosmology. Plato's interest in the world of nature is expressed in the *Timaeus*. While the

Republic contains a great deal on mathematics, the *Timaeus*, which can be considered somewhat of a sequel to it, consists of a more advanced mathematics and yet a type that would be understood by the ancient listeners.[72] In the *Timaeus*, Plato speaks about the stereometric features of the five bodies—tetrahedron, cube, icosahedron, octahedron, and dodecahedron. He associates these forms with the elements—fire, earth, water and air. For the fifth body, the dodecahedron, closest to a sphere, he states "and this one the god used for the whole universe embroidering figures on it."[73]

The faces of these five solid forms derive from triangles. For Plato, the elements can be seen as pure geometrical figures. In addition, Plato writes about molding gold into several shapes, including triangles.[74] What we have then is a mathematization of nature. In effect, Plato has undertaken the resolution of the Pythagorean principles. Pythagoras adopted mathematics in his study of harmony in music and the universe. To the left in the fresco, Raphael has depicted Pythagoras with a tablet illustrating two sets of perfect numbers. Four numbers, 1, 2, 3 and 4, are arranged so that they create an equilateral triangle. This recalls a tetrahedron, which is made up four triangular sides and is a direct link to Plato, as it symbolizes the number 4 and the four elements mentioned in *Timaeus*. In the lower part of the slate, we see that Pythagoras adopted mathematics to study music ratios.[75] In Book XIII of the *Elements*, Euclid discusses the so-called Platonic solids and shows that there are no others beyond the regular five geometrical solids. These geometric figures were significant for painters and architects for the depiction of three-dimensional objects. Piero della Francesca consulted Euclid's *Elements*, and there is evidence of this in the treatises he wrote.[76] He also came into contact with Greek mathematics in Rome where he had the opportunity to consult a Latin translation of the *Elements*.

It is likely that Raphael saw mathematical texts such as Euclid's *Elements* in Rome. The papal librarian and humanist Tommaso Inghirami may have assisted Raphael in learning about the mathematical codices in the Vatican Library.[77] It is also thought that Tommaso played an important role in the program of the *School of Athens*. His enthusiasm for Vitruvius, which he gained through a classical education from Pomponio Leto, he would also have shared with Raphael.[78] When we know that Raphael had some knowledge of Latin and pored over ancient texts, it is natural that he would be interested in Euclid's texts.

The diagram on Euclid's slate seems to indicate that Raphael had at some point seen a manuscript of the *Elements* whose margins are richly illustrated with geometric figures.[79] Without referring to a specific drawing, Raphael has captured the essence of Euclidean geometry. Indeed, he visualizes an essential part of the *Elements* by illustrating triangles, some of the most frequently encountered images in the *Elements*.

In the *School of Athens*, Aristotle holds in his left hand a tome of the *Etica* (*Ethics*). He holds out his right arm in a gesture that points to his philosophical system, which lies in the world of reality, i.e. is perceivable by the senses. Astronomy, cartography and optics, represented in the lower right group of figures, use mathematics to reveal the mysteries of nature. Even astrology adopted mathematical methods to carry out its inquiries. While Aristotle treats mathematics more thoroughly in the *Metaphysics* and to a lesser degree in the *Physics*, in the *Nicomachean Ethics*, he uses mathematical language for a discussion of ethical behavior; for example, one should seek the golden mean between two extremes to arrive at excellence.[80] Moreover, in Book 1, he discusses mathematics when he contrasts the differences between how a carpenter and a geometer look for right angles. It is interesting that he writes that the carpenter uses the right angle for practical purposes while the geometer looks for right angles for theoretical consideration, for he is a "spectator of the truth."[81] Aristotle gives further thought to mathematics in Book VI when he describes geometry as "the science of spatial magnitudes."[82] On account of its consistency, its accurate descriptions and use of logic, mathematics had great appeal. And whether explicit or implied, the fresco's mathematical themes energize discussion and stimulate the imagination.

The four large *tondi* in the ceiling represent the four areas of learning depicted in the room. In the *tondo* just above the *School of Athens*, a figure representing Philosophy sits enthroned holding two books inscribed *Mor[alis]* and *Na[tural]is* (Figure 9.5), referring to the divisions of philosophy established since antiquity.

In many ways, the third part of philosophy, rational philosophy, is invoked by the fresco below, that is to say, by the liberal arts. In *The City of God*, Saint Augustine writes about the tripartite division of philosophy, which he states comes from Plato, a division handed down to later authors such as Boethius and Isidore of Seville.[83] St. Bonaventure, the Franciscan theologian who followed Augustinian thought, also commented on philosophy's three divisions. In the *Margarita Philosophica*, rational philosophy is described in the first seven books, comprising the liberal arts. When we consider the *School of Athens*, we note that it evokes a tradition, long established, of

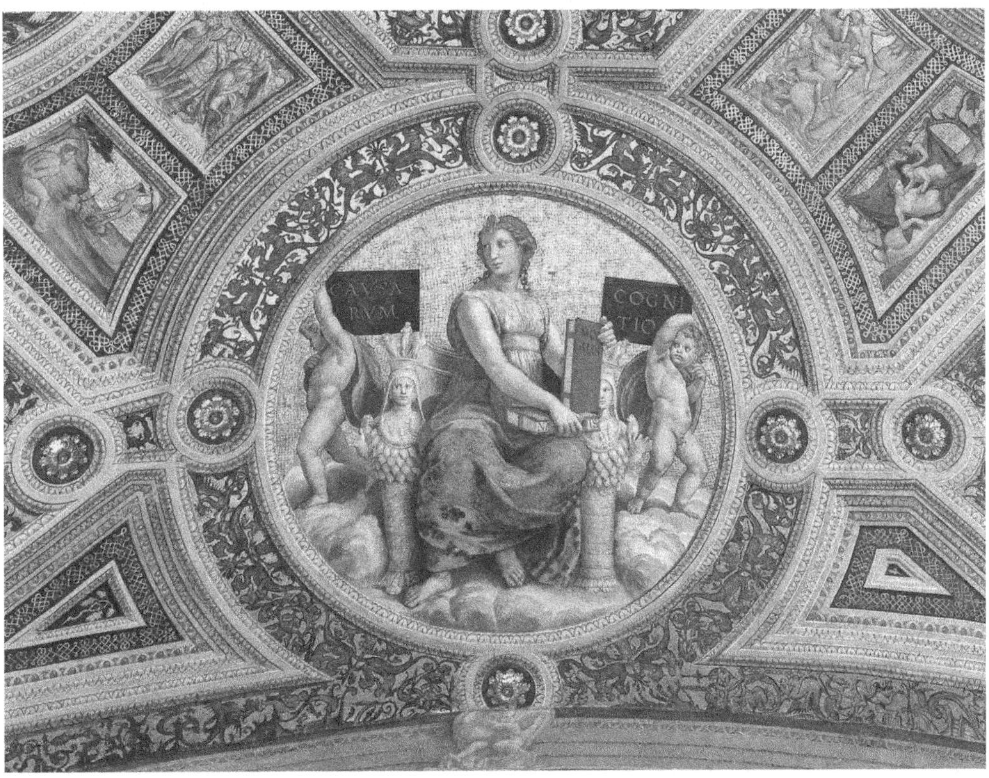

Figure 9.5 Raphael. *Philosophy*. Ceiling of the Stanza della Segnatura, Vatican Palace.

Photo credit: Vatican Museums.

the tripartite divisions of philosophy; rational philosophy is represented in the fresco by the liberal arts, and moral and natural philosophy are represented not only in the fresco, but also in the ceiling *tondo*.

Vasari informs us that the figure of Philosophy is dressed in the colors of the four elements—fire, air, earth and water—from her neck downwards.[84] As noted earlier, in Plato's mathematical discourse in the *Timaeus*, he associated these elements with four of the five regular bodies.

Just opposite and above the *Disputa*, the allegorical figure of Theology is accompanied by putti bearing inscriptions that read *Divinar[um] rer[um] notitia* ("Knowledge of the Divine"). Putti on each side of the figure of Philosophy carry an inscription, *causarum cognitio* ("Knowledge of causes"). Considered together, *causarum cognitio* can be revealed through knowledge of the Divine. This concordance between the two disciplines, reflected not only among the philosophy and theology books in the library but also in the *School of Athens* and the *Disputa*, is also found in the ceiling *tondi* texts, *causarum cognitio* and *Divinar[um] rer[um] notitia*.

The inscription *causarum cognitio*, held up by two animated putti and about which so much has been written, derives from Aristotle's *Metaphysics*, and several authors have paraphrased the Greek philosopher, notably in antiquity, Cicero and Virgil, and during the Renaissance, Egidio da Viterbo and Marsilio Ficino.[85] Aristotle writes:

> We have said in the *Ethics* what the difference is between art and science and the other kindred faculties; but the point of our present discussion is this, that all men suppose what is called wisdom to deal with the first causes and the principles of things. This is why, as has been said before, the man of experience is thought to be wiser than the possessors of any perception whatever, the artist wiser than the men of experience, the master-worker than the mechanic, and the theoretical kinds of knowledge to be more of the nature of wisdom than the productive. Clearly then wisdom is knowledge about certain causes and principles.[86]

Here he refers to causes which also have significance for mathematics. In the *Ethics*, in his explanation of knowledge, Aristotle writes that when we understand the essence of things we actually uncover its causes.[87] This is a process analogous to mathematical inquiry, such as Euclid adopted. In the *Metaphysics*, Aristotle includes mathematics among the theoretical sciences, which he regards as having an elevated status:

> Now all causes must be eternal, but especially these; for they are the causes of so much of the divine as appears to us. There must, then, be three theoretical philosophies, mathematics, natural science, and theology, since it is obvious that if the divine is present anywhere, it is present in things of this sort. And the highest science must deal with the highest genus, so that the theoretical sciences are superior to the other sciences, and this to the other theoretical sciences.[88]

Plato had written on the primacy of mathematics, particularly geometry, which he wrote it is "pursued for the sake of knowledge."[89] He referred to geometry as "knowledge of what always is. Then it draws the soul towards truth and produces philosophic thought by directing upwards what we now wrongly direct downwards."[90]

Egidio da Viterbo, the Augustinian prior and favorite preacher of Julius II, who several scholars have argued is likely to have been instrumental in formulating the program in the Stanza della Segnatura, wrote that "*Scientia humana est cognitio causarum*,"[91] While Marsilio Ficino, the Florentine philosopher, also added a theological meaning to the phrase when he wrote in his *Opera Omnia* that "*Theologi finis, supernarum causarum cognitio.*"[92] What we have then is that in the Stanza, *causarum cognitio* has taken on a theological significance, and this creates an interconnection between the two frescos, the *School of Athens* and the *Disputa*, and between philosophy and theology. And as Timothy Verdon expressed it, "What is new in the Stanza della Segnatura is Raphael's depiction of pagan thinkers 'in the Church.'"[93]

When we turn our attention to the technical and coloristic aspects of the fresco, we observe that optics, or the geometry of vision, was of great importance for Raphael, not only for spatial concerns of linear perspective but also to understand the properties of color.[94] Raphael adopted an extraordinary sense of color in the fresco given the limited number of pigments that can be used in *buon fresco* painting. Like Michelangelo in the Sistine Chapel, which was being painted while Raphael worked in the Stanza, the Umbrian artist used a mode of color that was suitable to the composition and the subject. As Marcia Hall described it, Raphael used a technique of coloring that combined the softened contours inherited from Leonardo's *sfumato* technique with the *bellezza di colore*, or beautiful color, associated with Florentine painting.[95] Raphael applied this *unione* of color to a remarkable degree in the Stanza. And, like Michelangelo in the Sistine ceiling, touches of *cangiantismo*, where modeling is achieved by shifting from one hue to another in a contrasting fashion, contributing to a greater sense of variety.[96] Since the frescoes in the room were restored, one can more fully appreciate the modes of coloring. And in order to expand the limited range of colors that are traditionally used for *buon fresco*, Raphael adopted malachite, for example, more usual in tempera painting, and glazes.

Raphael's mastery of the use of highlights on figures, such as the one ascending the steps draped in white over a green garment, adds a sculptural robustness to the figure. In the silverpoint drawing, the range of hatchings from deeply colored to nearly white enhance the powerful modeling in the figure (Figure 9.6).[97] This is carried through to the painted image. As Francis Ames-Lewis has noted, the particular features of silverpoint techniques provide a more volumetric form than the use of a pen, where the result is more planar and flattened.[98]

The asymmetrical poses of the two men are splendidly rhythmic. Notably, the highlighted features of the man draped in white as well as that of Euclid and his followers, and not least Pythagoras and the group around him, respond to the variations of the natural light flowing in from the windows in the room. The window on the right is closer to the fresco than the window on the left, and Raphael has taken into consideration the differences in the room's natural lighting. This is particularly noticeable in the figure ascending the stairs, with its emphatic play of light on the white mantle. Raphael painted the heavy folds in the drapery of his figures, particularly noticeable in the foreground group of mathematicians, with an elegant and vivid robustness. What emerges then is that the artist obviously had a remarkable understanding of the optical features of sunlight streaming through the windows, and he was able to transform this knowledge into color of an extraordinary radiance.

In the *School of Athens*, Raphael invites the viewer to reflect on the importance of mathematics as well as its elegance and rigor. The privileged arrangement of the

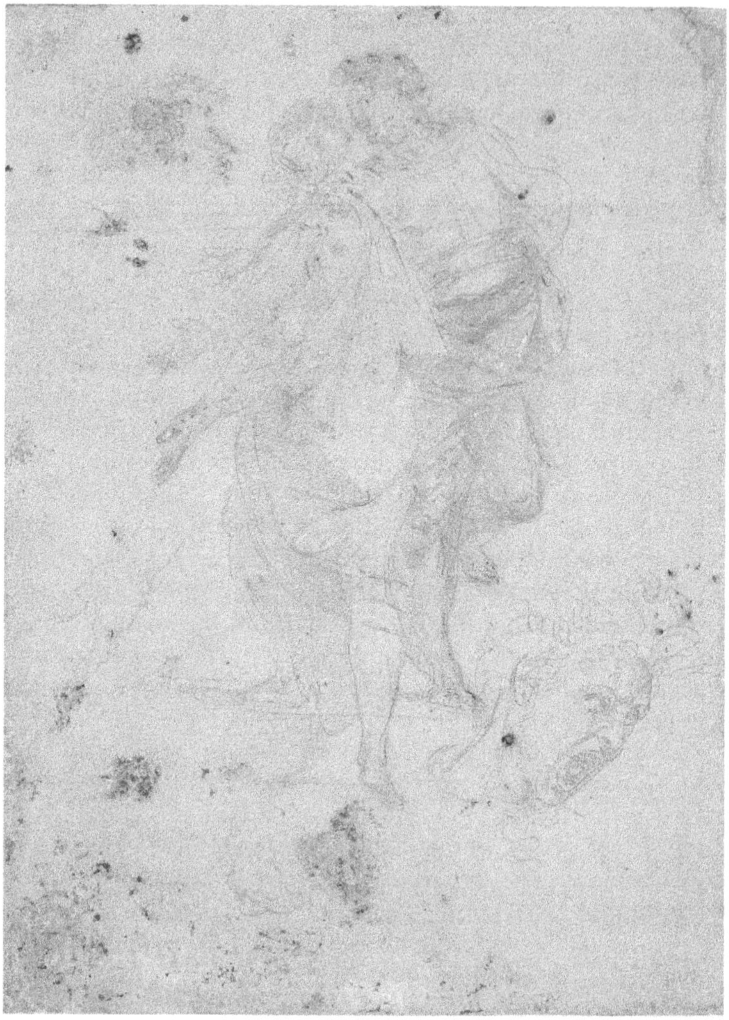

Figure 9.6 Raphael. *Two Men Conversing on a Flight of Steps*. (WA1846.191). A Study for the *School of Athens*. Silverpoint with white heightening on pink paper, 27.8 × 20 cm.

Photo credit: © Ashmolean Museum, University of Oxford.

quadrivium, arithmetic and music on the left and geometry and astronomy on the right, in the foreground emphasizes the role of mathematics to bring the spectator closer to the fresco's themes. Significantly, Raphael placed himself among the men of science, clearly viewing his own achievements from within the world of mathematics. Through the mathematical themes, activities and objects, the spectator is encouraged to investigate, contemplate and discover the power of mathematics to understand the world and to reflect on higher things.

Notes

I would like to extend my warmest thanks to Kolbjørn Skipnes, Markus Banagl and Hans W. Hubert for their helpful comments on my work and for discussing with me various aspects of this chapter. I would like to also thank the anonymous reviewer for helpful suggestions on an earlier draft. I am grateful for the kind help of the staff of the Biblioteca Apostolica Vaticana.

1 Giorgio Vasari, *Le vite de' più eccellenti pittori, scultori e architettori: nelle redazioni del 1550 e 1568*, ed. Rosanna Bettarini and Paola Barocchi, vol. 4 (Florence: Sansoni; Studio per Edizioni Scelte, 1976), 166. "Né si può esprimere la bellezza di quelli astrologi e geometri che disegnano con le seste in su le tavole moltissime figure e caratteri." Translations are mine unless otherwise indicated.
2 Vasari, *Le Vite*, 4: 166–167, "una figure che, chinata a terra, con un paio di seste in mano le gira sopra le tavole."
3 Ibid.: "Adornò ancora questa opera di una prospettiva e di molte figure, finite con tanto delicate e dolce maniera."
4 Giovanni Pietro Bellori, *Descrizzione delle imagini dipinte da Rafaelle d'Urbino nelle camere del Palazzo Apostolico Vaticano* (Rome: Stamparia di Gio; Giacomo Komarek Boëmo, 1695), 19.
5 The literature on the slate's drawing is extensive. According to Simonetta Valtieri, from the two triangles on the slate the entire building in the fresco can be constructed. See Simonetta Valtieri, "La Scuola d'Atene: 'Bramante' suggerisce un nuovo metodo per costruire in prospettiva un'architettura armonica," *Mitteilungen der Kunsthistorischen Institutes in Florenz* 16 (1972): 63–72. An elaborate scheme of a complexity that the artist is unlikely to have intended is presented in Konrad Oberhuber, *Polarität und Synthese in Raphaels 'Schule von Athen'* (Stuttgart: Urachhaus, 1983), 53–59; Richard Fichtner, *Der Verborgene Geometrie in Raffaels 'Schule von Athen'* (Munich: R. Oldenbourg, Deutsches Museum, 1984), 13–22; Enrico Gamba, *Raffaello e la Matematica* (Urbino: Centro Internazionale di Studi Urbino e la Prospettiva, 2009). For a discussion on incommensurability with regard to the fresco's mathematics, see Jürgen Schönbeck, *Euklid: Um 300 v. Chr.* (Basel, Boston and Berlin: Birkhäuser, 2003), 47–74. For a discussion of the slate in front of Pythagoras and the drawing on Euclid's slate see Conrad Doose, Jürgen Eberhardt, Hajo Lauenstein, Guido von Büren and Günter Bers, eds., *Das italienische Jülich: Grundzüge im Konzept Alessandro Pasqualinis für die Stadtanlage, die Zitadelle und das Residenzschloss* (Jülich: Förderverein Festung Zitadelle Jülich, 2009). Nicholas Temple sees a hexagon on Euclid's slate. See Nicholas Temple, *Disclosing Horizons: Architecture, Perspective and Redemptive Space* (London and New York: Routledge, 2007), 63–65. See also Temple, *Renovatio Urbis: Architecture, urbanism and ceremony in the Rome of Julius II* (London and New York: Routledge, 2011), 226–232 and Wolfgang Jung, *Über szenographisches Entwerfen: Raffael und die Villa Madama* (Braunschweig: Vieweg, 1997), 81–162.
6 John Shearman, *Raphael in Early Modern Sources, 1483–1602*, vol. 1 (New Haven and London: Yale University Press, 2003), 807–808.
7 Franz Wickhoff, "Die Bibliothek Julius' II," *Jahrbuch der königlich Preussischen Kunstsammlungen* 14 (1893): 49–64, especially 54–55. John Shearman provides an interesting narrative of the earlier accounts. See Shearman, "The Vatican Stanze: Functions and Decoration," in *Art and Politics in Renaissance Italy: British Academy Lectures*, ed. George Holmes (Oxford: Oxford University Press, 1993), 196–197. See also D. Redig de Campos, *Le stanze di Raffaello* (Rome: Del Turco, 1957), 5.
8 Shearman, "The Vatican Stanze," 196–200. See also Shearman's "Gli appartamenti di Giulio II e Leone X," in *Raffaello nell' appartamento di Giulio II e Leone X*, ed. Guido Cornini (Milan: Electa, 1993), 15–36. David Rijser is not convinced that the room was a library and questions whether the books in the fresco are there literally or virtually. See David Rijser, *Raphael's Poetics: Art and Poetry in High Renaissance Rome* (Amsterdam: Amsterdam University Press, 2012), 108–115. According to Ludwig von Pastor, the bookcases at the time were not placed against the wall but rather in free-standing cases as in the Laurentian Library in Florence ("die Bücher damals nicht in Wandschränken, sondern in

freistehenden Pulten (wie noch heute in der Laurentianischen Bibliothek zu Florenz) aufbewahrt wurden"). See Ludwig von Pastor, *Geschichte der Päpste im Zeitalter der Renaissance von der Wahl Innozenz' VIII. bis zum Tode Julius' II. 1484–1513*, vol. 3 (Freiburg and Rome: Verlag Herder, 1956), 1020.

9 Some of the manuscripts are astronomical in content. See Robert Devreesse, *Le fonds grec de la Bibliothèque Vaticane des origines à Paul V* (Vatican City: Bibliotheca Apostolica Vaticana, 1965), 9–42, especially 36. Eugène Müntz and Paul Fabre, *La bibliothèque du Vatican au XVe siècle d'après des documents inédits* (Paris: Ernest Thorin, 1887), 98–99. Nicholas V loaned several Greek mathematical texts, among them a "*Geomatria Euclidis*" (Vat. gr. 1040 ?) and a "*Historiographia Ptholomei* al[ias] *Mapa* [sic] *mundi*. Paul Lawrence Rose notes that two very interesting Greek mathematical texts seem to have not been included in the inventory but appear in a list of books loaned to Cardinal Bessarion. See Paul Lawrence Rose, *The Italian Renaissance of Mathematics: Studies on Humanists and Mathematicians from Petrarch to Galileo* (Geneva: Librairie Droz, 1975), 37–38. See also Devreesse, *Le fonds grec*, 41.

10 Devreesse, *Le fonds grec*, 59–61. Ptolemy referred to the *Almagest* as "The Mathematical Collection" because he thought that its focus on the motions of the heavenly bodies could be explained in mathematical terms. See Gerald J. Toomer, "Ptolemy (or Claudius Ptolemaeus)," in *Dictionary of Scientific Biography*, ed. Charles Coulston Gillispie, vol. 11 (New York: Charles Scribner's Sons, 1981), 186–197.

11 Devreesse, *Le fonds grec*, 61, 183. See also Rose, *Italian Renaissance of Mathematics*, 66, n. 131. Vat. gr. 184 contains *Ptolomei Almagestus cum Theone*. Theon of Alexandria is famously known for his commentaries on Ptolemy's *Almagest* and *Handy Tables*. The latter was used to calculate the position of heavenly bodies. See G. J. Toomer, "Theon of Alexandria," in *Dictionary of Scientific Biography*, ed. Charles Coulston Gillispie (New York: Charles Scribner's Sons, 1981), 13: 321–325.

12 Devreesse, *Le fonds grec*, 183–184. See also Eugène Müntz, *La bibliothèque du Vatican au XVIe siècle: Notes et Documents* (Paris: Ernest Leroux, 1886), 5–21. See also Johannes Mercati and Pius Franchi De' Cavalieri, *Codices Vaticani Graeci*, vol. 1 (Rome: Typis Polyglottis Vaticanis, 1923), 206–207, 210–212, 229–231. For more on the acquisition and translation of ancient mathematical codices in the Vatican Library, see N. M. Swerdlow, "The Recovery of the Exact Sciences of Antiquity: Mathematics, Astronomy, Geography," in *Rome Reborn: The Vatican Library and Renaissance Culture*, ed. Anthony Grafton (New Haven and London: Yale University Press, 1993), 125–167.

13 Devreesse, 59–61, 92–94, 104–105.

14 Rose, *Italian Renaissance of Mathematics*, 38. See also Müntz and Fabre, *La bibliothèque du Vatican au XVe siècle*, 217

15 Devreesse, *Le fonds grec*, 152–180, especially 158, 162.

16 Ibid., 162.

17 Rose, *Italian Renaissance of Mathematics*, 38.

18 "[F]or the pleasure of the eyes as well as practical use." See Paul Fabre, "La Vaticane de Sixte IV," *Mélanges d'archéologie et d'histoire* 15 (1895): 456.

19 Maria Bertòla, *I due primi registri di prestito della Biblioteca Apostolica Vaticana: Codici Vaticani latini 3964, 3966* (Vatican City: Biblioteca Apostolica Vaticana, 1942), 36–76. Ptolemy's *Almagest* was also loaned. For the borrowed Euclid's *Elements*, 43–44.

20 Bertòla, *I due primi registri di prestito*, 46. Another manuscript of interest for this study and included in Bertòla's register is a ". . . *librum grecum Aristotilis quo continetur Etica* . . ." (Vat. gr. 506). The codex also contained *historia animalium*. See Bertòla, *I due primi registri di prestito*, 45, n. 4. See also Devreesse, *Le fonds grec*, 55.

21 Bertòla, *I due primi' registri di prestito*, 55. Although Galen is primarily known as a physician, his writings adopt a mathematical-geometrical approach to medicine. He was educated, in part, by his father, who had a thorough training in mathematics.

22 Léon Dorez, "La bibliothèque privée du Pape Jules II," *Revue des Bibliothèques* 6 (1896): 97–126. His study follows the earlier work of Franz Wickhoff, see note 7.

23 Paul Taylor, "Julius II and the Stanza della Segnatura," *Journal of the Warburg and Courtauld Institutes* 72 (2009): 108.

24 Dorez, "La bibliothèque privée," 112.

25 The bookcases were not fixed to the wall.

26 Herman Grimm, *Das Leben Raphael's* (Berlin: Verlag von Wilhelm Hertz, 1886), 331–333. Léon Dorez suggests that the popular myth that Julius II was no scholar may have been started by humanists at the time who criticized the pope because he lacked knowledge of Greek. See Dorez, "La bibliothèque privée," 100.

27 See André Chastel, *I centri del Rinascimento: Arte italiana 1460–1500* (Milan: Biblioteca Universale Rizzoli, 1965), 41–51. For the interest in science in Urbino, see Enrico Gamba, "Pittura e storia della scienza," in *La ragione e il metodo: Immagini della scienza nell'arte italiana dal XVI al XIX secolo*, ed. Marco Bona Castellotti, Enrico Gamba and Fernando Mazzocca (Milan: Electa, 1999), 43–53; Enrico Gamba and Vico Montebelli, *Le scienze a Urbino nel tardo Rinascimento* (Urbino: Edizioni QuattroVenti, 1988), 11–34. See also the chapter by Renzo Baldasso and John Logan in this volume.

28 Dorez, "La bibliothèque privée," 119.

29 Visual representation of Illustrious Men began in the fourteenth century. For a discussion of earlier representations of *uomini illustri* see, Christiane L. Joost-Gaugier, "The Early Beginnings of the Notion of 'Uomini Famosi' and the 'De Viris Illustribus' in Greco-Roman Literary Tradition," *Artibus et Historiae* 3 (1982): 97–115; Idem., "Poggio and Visual Tradition: *Uomini Famosi* in Classical Literary Description," *Artibus et Historiae* 6 (1985): 57–74. Humanist interest in classical antiquity and the theme of Illustrious Men were energized by Petrarch through his *De viris illustribus*. The oldest known series, now lost, was painted by Giotto for King Robert in Naples at the Castelnuovo. Later depictions of the theme appear in frescoes painted c. 1370 for Francesco da Carrara in the Palazzo del Capitano in Padua. Petrarch may have been a consultant to the artists. See Theodor E. Mommsen, "Petrarch and the Decoration of the Sala Virorum Illustrium in Padua," *Art Bulletin* 34 (1952): 95–116. The rich cultural environment in Avignon also inspired Giovanni Colonna to write his *De viris illustribus*. See W. Braxton Ross Jr., "Giovanni Colonna, Historian at Avignon," *Speculum* 45 (1970): 533–563. See Joost-Gaugier, "'Uomini Famosi' and the 'De Viris Illustribus,'" 100. On the frescoes painted c. 1413 by Taddeo di Bartolo see Nicolai Rubinstein, "Political Ideas in Sienese Art: The Frescoes by Ambrogio Lorenzetti and Taddeo di Bartolo in the Palazzo Pubblico," *Journal of the Warburg and Courtauld Institutes* 21 (1958): 179–207, especially 194–197. Paintings of heroes by Melozzo da Forlì and Piero della Francesca were in the Vatican Palace before Raphael's arrival. See Paul Schubring, "Uomini famosi," in *Repertorium für Kunstwissenschaft*, ed. Henry Thode and Hugo von Tschudi, vol. 23 (Berlin and Stuttgart: W. Spemann, 1900): 424–425.

30 For an account of Petrarch's sources for his story on Archimedes, see Marshall Clagett, *Archimedes in the Middle Ages*, vol. 3, *The Fate of the Medieval Archimedes 1300 to 1565*, Part III (Philadelphia: American Philosophical Society, 1978), 1336–1341. See Guido Martellotti, ed., *Francesco Petrarca, De viris illustribus* (Florence: Sansoni, 1964), 111–137. Giovanni Colonna referred to Archimedes as "*maximus in geometria*" in his *De viribus illustribus*. See Clagett, *Archimedes in the Middle Ages*, 1335; Ross, "Giovanni Colonna," 545. Common philological interests enriched the friendship between Petrarch and Colonna. Both writers seem to have written their biographies of Archimedes at about the same time. Martellotti, *Francesco Petrarca*, 123. Petrarch goes on to lament Archimedes' death at Syracuse. His comments seem to be based on Livy, his favorite ancient author. See also Clagett, *Archimedes in the Middle Ages*, 1336–1337; Walter Roy Laird, "Archimedes among the Humanists," *Isis* 82 (1991): 628–638.

31 Joanna Woods-Marsden, *Renaissance Self-Portraiture: The Visual Construction of Identity and the Social Status of the Artist* (New Haven and London: Yale University Press, 1998), 122–124.

32 Hugo Chapman, Tom Henry and Carol Plazzotta, *Raphael: From Urbino to Rome* (London: National Gallery Company, 2004), 45.

33 Ingrid Alexander-Skipnes, " 'Bound with Wond'rous Beauty': Eastern Codices in the Library of Federico da Montefeltro," *Mediterranean Studies* 19 (2010): 78.

34 Pietro Scarpellini, ed., *Il Collegio del Cambio in Perugia* (Cinisello Balsamo: Silvana Editoriale, 1998), 67–106; Laura Teza, "Osservazioni sulla decorazione del Collegio del Cambio," in *Perugino: il divin pittore*, ed. Vittoria Garibaldi and Francesco Federico Mancini

(Cinisello Balsamo, Milano: Silvana Editoriale, 2004), 115–127. Pope Julius II may have seen the Collegio del Cambio on his visit to Perugia on 13 September 1506. See Christine Shaw, *Julius II: The Warrior Pope* (Oxford and Cambridge, MA: Blackwell, 1993), 154–158. After Perugia, he traveled to Urbino.

35 Ludwig H. Heydenreich, "La ripresa 'critica' di rappresentazioni medievali delle 'septem artes liberales' nel Rinascimento," in *Il mondo antico nel Rinascimento. Atti del V convegno internazionale di studi sul Rinascimento, Firenze-Palazzo Strozzi, 2–6 Settembre, 1956* (Florence: Istituto Nazionale di Studi sul Rinascimento, 1958), 265–273.

36 On the certainty of mathematics see Judith V. Grabiner, "The Centrality of Mathematics in the History of Western Thought," *Mathematics Magazine* 61 (1988): 220–230.

37 For more on the development of mathematics within the *quadrivium* see Olaf Pedersen, "Du Quadrivium à la Physique," in *Artes Liberales von der Antiken Bildung zur Wissenschaft des Mittelalters*, ed. J. Koch (Leiden and Cologne: Brill, 1959), 107–123; Paul Oskar Kristeller, *Renaissance Thought and the Arts: Collected Essays* (Princeton, NJ: Princeton University Press, 1964, reprinted 1980), 163–196. For an interesting study on the mathematical sciences and music, see Leo Schrade, "Music in the Philosophy of Boethius," *The Music Quarterly* 33 (1947): 188–200. Michael Masi provides a concise examination of Boethius' mathematics. See Michael Masi, "The Influence of Boethius' *De Arithmetica* on Late Medieval Mathematics," in *Boethius and the Liberal Arts: A Collection of Essays*, ed. Michael Masi (Bern: Peter Lang, 1981), 81–95. Boethius was undoubtedly inspired by Plato's writings on the four mathematical disciplines, which Boethius named the *quadrivium*. See Plato's *Republic*, 7: 525a–531c. Furthermore, in *The City of God*, Saint Augustine praises the "science of numbers." He wrote: "And therefore, we must not despise the science of numbers, which, in many passages of holy Scripture, is found to be of eminent service to the careful interpreter. Neither has it been without reason numbered among God's praises." Saint Augustine, *The City of God*, trans. Marcus Dods (New York: Modern Library, 1950, reprinted 2000), 11: 30.

38 Arnold Nesselrath, *Raphaël et Pinturicchio: Les grands décors des appartements du pape au Vatican* (Paris: Hazan and Musée du Louvre, 2012).

39 Ernst Gombrich, "Raphael's Stanza della Segnatura and the Nature of Its Symbolism," in *Gombrich on the Renaissance. Volume 2: Symbolic Images*, 3rd ed. (London: Phaidon Press, 1985, repr. 1993), 85–87. Furthermore, from the sixteenth century, the popes preferred to live on the second floor of the Vatican, since it was airy and filled with light. See Franz Ehrle and Enrico Stevenson, *Les fresques du Pinturicchio dans les salles Borgia au Vatican* (Rome: Danesi, 1898), 27. Moreover, Julius II's uncle, Pope Sixtus IV, had his tomb decorated with images of the Seven Liberal Arts, see L. D. Ettlinger, "Pollaiuolo's Tomb of Pope Sixtus IV," *Journal of the Warburg and Courtauld Institutes* 16 (1953): 239–274.

40 Julius von Schlosser, "Giusto's Fresken in Padua und die Vorläufer der Stanza della Segnatura," *Jahrbuch der Kunsthistorischen Sammlungen des Allerhöchsten Kaiserhauses* 17 (1896): 13–100. The idea that the liberal arts are illustrated in the *School of Athens* was first advanced by Anton Springer, "Raffaels Schule von Athen," *Die Graphischen Künste* 5 (1883): 53–106. See also *Arts libéraux et philosophie au Moyen Âge. Actes du quatrième Congrès international de philosophie médiévale*. Université de Montréal, Montréal, Canada, 27 août—2 septembre 1967 (Montreal: Institut d'études médiévales; Paris: Librairie philosophique J. Vrin, 1969).

41 L. D. Ettlinger, "Muses and Liberal Arts: Two Miniatures from Herrad of Landsberg's *Hortus Deliciarum*," in *Essays in the History of Art Presented to Rudolf Wittkower*, ed. Douglas Fraser, Howard Hibbard and Milton J. Lewine (Bath: Phaidon, 1967, reprinted 1969), 29–35; Mary D. Garrard, "The Liberal Arts and Michelangelo's First Project for the Tomb of Julius II (With a Coda on Raphael's 'School of Athens')," *Viator* 15 (1984): 366–376.

42 As in the 1503 edition, these divisions of philosophy appear on a border around the figure of Philosophy. In *The City of God*, Saint Augustine writes about the tripartite division of philosophy, which he states comes from Plato. See *The City of God*, 8: 4. Isidore of Seville also writes on the branches of philosophy. See *Sancti Isidori Hispalensis Episcoli, Opera Omnia, Patrologia Latina* (Paris: J.-P. Migne, 1841–1855), 82: 140–141.

43 Adolf Katzenellenbogen, "The Representation of the Seven Liberal Arts," in *Twelfth-Century Europe and the Foundations of Modern Society*, ed. Marshall Clagett, Gaines Post and

Robert Reynolds (Madison: University of Wisconsin Press, 1961), 43–44. The image of Philosophy's head hidden by clouds comes from Boethius. See Ludovicus Bieler, ed., *Anicii Manlii Severini Boethii Philosophiae Consolatio* (Turnhout: Brepols, 1984), I, prosa 1. See also Myra L. Uhlfelder, "The Role of the Liberal Arts in Boethius' *Consolatio*," in *Boethius and the Liberal Arts: A Collection of Essays*, ed. Michael Masi (Bern: Peter Lang, 1981), 17–34.

44 Glenn W. Most, however, sees a close connection between Raphael's fresco and medieval representations of philosophy. Glenn W. Most, "Reading Raphael: The School of Athens and Its Pre-Text," *Critical Inquiry* 23 (1996): 145–182. And for the role that Marsilio Ficino may have played in the fresco's program, see ibid., 162–171. First noted by Bellori, apse decorations in medieval churches also inspired the pictorial themes in the *Disputa*. See Bellori, *Descrizzione delle imagini dipinte da Rafaelle*, 8–14; Chapman, Henry, and Plazzotta, *Raphael: From Urbino to Rome*, 288.

45 Johan David Passavant, *Rafael von Urbino und Sein Vater Giovanni Santi*, vol. 1 (Leipzig: F. A. Brockhaus, 1839), 158–159, n. 1. See also Elena Parma Armani, *Perin del Vaga: L'anello mancante: Studi sul Manierismo* (Genoa: Sagep Editrice, 1986), 191, 195, 286–287.

46 W. W. Lloyd, "Raphael's School of Athens," *The Fine Arts Quarterly Review* 2 (1864): 62–63. Passavant stated, however, that it is difficult to decide whether Archimedes or Euclid is the geometer on the right. Anton Springer argued that the figure was Archimedes. Anton Springer, "Raffaels Schule von Athen," *Die Graphischen Künste* 5 (1883): 80. Both André Chastel and Ernst Gombrich based their suggestion that Euclid was the geometer depicted on the significance of the figure of Euclid as geometer in the *Sala delle Arti Liberali* in the Borgia apartment frescoes. In a 1993 essay, Matthias Winner believed the figure to be Euclid but subsequently identified the figure as Archimedes based on the *basamento* painting by Perino del Vaga. Matthias Winner, "Progetti ed esecuzione nella Stanza della Segnatura," in *Raffaello nell'appartamento di Giulio II e Leone X*, 268; Idem., "The Mathematical Sciences in Raphael's *School of Athens*," in *The Power of Images in Early Modern Science*, ed. Wolfgang Lefèvre, Jürgen Renn and Urs Schoepflin (Basel: Birkhäuser, 2003), 302–303.

47 Bellori, *Descrizzione delle imagini dipinte da Rafaelle*, 18.

48 I am grateful to Professor Hubert for discussing with me why the figure of Euclid is not a portrait of Bramante and for kindly giving me a copy of his essay before it was published. See Hans W. Hubert, "Perspektiven auf Bramantes Virtus in Wort und Bild," in *Die Virtus in Kunst und Kunsttheorie der italienischen Renaissance*, ed. Thomas Weigel, Britta Kusch-Arnhold and Candida Syndikus (Münster: Rhema, 2014), 115–126.

49 Ibid., 116–121.

50 Raphael wrote the letters RVSM (*RAPHAEL VRBINAS SVA MANV*) in gold on the neckline of Euclid's garment, embellishing the border with additional markings to create a decorative edge. See D. Redig de Campos, "La firma siglata di Raffaello nella Scuola d'Atene," *Arti figurative: rivista d'arte antica e moderna* 1 (1945): 151. In spite of the fact that Rona Goffen recognizes Euclid as a portrait of Bramante, she writes that it would be an unusual instance of an artist's signing his name to another artist's portrait, given the fact of Raphael's own presence in the fresco. The idea that the artist's signature is a tribute and acknowledgment of Raphael's debt to Bramante is untenable. See Rona Goffen, "Raphael's Designer Labels: From the Virgin Mary to *La Fornarina*," *Artibus et Historiae* 24 (2003): 124–130.

51 Thomas L. Heath, trans. and Dana Densmore, ed., *Euclid's Elements* (Sante Fe: Green Lion Press, 2002, reprinted 2007). For a colorful study of Euclid's mathematics and its impact, see David Berlinski, *The King of Infinite Space: Euclid and His Elements* (New York: Basic Books, 2013); John E. Murdoch, "Euclid," in *Dictionary of Scientific Biography*, ed. Charles Coulston Gillispie (New York: Charles Scribner's Sons, 1981), 4: 414–459. For a discussion of the importance of Euclid's *Elements*, particularly the first six books, in universities, summarized in an introduction, see Sabine Rommevaux, "La réception des Éléments d'Euclide au Moyen Âge et à la Renaissance," *Revue d'histoire des sciences* 56 (2004): 267–273.

52 Aristotle, *Posterior Analytics*, 1: 10, 76a31–77a4. See also Charles B. Schmitt, *Aristotle and the Renaissance* (Cambridge, MA and London: Harvard University Press, 1983), 1–9; 10–33.

53 Ivor Bulmer-Thomas, "Euclid," in *Dictionary of Scientific Biography*, ed. Charles Coulston Gillispie, vol. 3 (New York: Charles Scribner's Sons, 1980), 416.

54 Arnold Nesselrath, *Raphael's School of Athens* (Vatican City: Edizioni Musei Vaticani, 1996), 18.
55 I would like to thank Professor Markus Banagl for kindly discussing Euclid and mathematics with me.
56 André Chastel, *Art et Humanisme a Florence au Temps de Laurent le Magnifique: Études sur la Renaissance et l'Humanisme platonicien* (Paris: Presses Universitaires de France, 1959), 470. The figure of Heraclitus (Michelangelo) was added to a new area of *intonaco* when the painting was completed. See Redig de Campos, *Le stanze di Raffaello*, 19. There is no drawing on Pythagoras' tablet in the cartoon.
57 Paul F. Grendler, *The Universities of the Italian Renaissance* (Baltimore and London: Johns Hopkins University Press, 2001), 408–429.
58 Toomer, "Ptolemy," 197–198.
59 Ibid., 198.
60 The literature on Ptolemy's *Geography* is substantial. See Leo Bagrow, "The Origin of Ptolemy's *Geographia*," *Geografiska Annaler* 27 (1945): 318–387. See also Patrick Gautier Dalché, *La Géographie de Ptolémée en Occident (IVe-XVIe Siècle)* (Turnhout: Brepols, 2009).
61 Like Euclid, Ptolemy wrote on optics. In his *Optics*, he examines the nature of sight. See A. Mark Smith, *Ptolemy and the Foundations of Ancient Mathematical Optics: A Source Based Guided Study* (Philadelphia: American Philosophical Society, 1999).
62 Paul Joannides, *The Drawings of Raphael* (Oxford: Phaidon, 1983), 82–83. Eve Borsook examines Raphael's skill at working on a monumental cartoon. According to Borsook, in order to facilitate handling such a large composition, the cartoon may have been hung on an adjacent wall or placed on the floor of the Stanza. See Borsook, "Technical Innovation and the Development of Raphael's Style in Rome," *Canadian Art Review* 12 (1985): 127–136.
63 Chapman, Henry and Plazzotta, *Raphael: From Urbino to Rome*, 23–26.
64 Vitruvius, *Ten Books on Architecture*, ed. Ingrid D. Rowland and Thomas Noble Howe (Cambridge: Cambridge University Press, 1999, reprinted 2007), 22.
65 Ibid.
66 John Shearman, "Raphael as Architect," *Journal of the Royal Society of Arts* 116 (1968): 394, 396–397.
67 Vincenzo Golzio, *Raffaello nei documenti nelle testimonianze dei contemporanei e nella letteratura del suo secolo* (Vatican City: Pontificia Insigne Accademia Artistica dei Virtuosi al Pantheon, 1936), 78–92. On the role of Angelo Colocci and Baldassare Castiglione in writing the letter to Pope Leo X, see Ingrid D. Rowland, "Raphael, Angelo Colocci, and the Genesis of the Architectural Orders," *Art Bulletin* 76 (1994): 81–104.
68 Cited in Hans H. Hubert, "In der Werkstatt Filaretes: Bemerkungen zur Praxis des Architekturzeichnens in der Renaissance," *Mitteilungen des Kunsthistorischen Institutes in Florenz* 47 (2003–2004): 331. For a study of the changes in techniques used in Quattrocento architectural drawing, see ibid., 311–344.
69 Statues in the niches recall ancient baths and triumphal arches. The sculptural details are barely visible, with the exception of Apollo, god of music, on the left (close to the *Parnassus* fresco), above Pythagoras, and Minerva, goddess of wisdom, on the right, above Euclid.
70 Here I am basing this comparison on Adolf Katzenellenbogen's comments on the archivolt at Laon Cathedral. See Katzenellenbogen, "Representation of the Liberal Arts," 43–44. For a study on the symbolic meaning of the vanishing point see Charles H. Carman, *Leon Battista Alberti and Nicholas Cusanus: Towards an Epistemology of Vision for Italian Renaissance Art and Culture* (Burlington, VT: Ashgate, 2014).
71 David Fowler, *The Mathematics of Plato's Academy: A New Reconstruction* (Oxford: Oxford University Press, 1999), 8–13. For an interesting study which covers abstract reasoning and mathematics, see M. F. Burnyeat, "Plato on Why Mathematics Is Good for the Soul," in *Mathematics and Necessity: Essays in the History of Philosophy: Proceedings of the British Academy*, ed. Timothy Smiley, vol. 103 (Oxford: Oxford University Press, 2000), 1–81. On Plato's views on the importance of mathematical sciences in education see James A. Weisheipl, "The Concept of Scientific Knowledge in Greek Philosophy," in *Mèlanges à la mèmoire de Charles De Koninck* (Quebec: Les presses de L'Université Laval, 1968), 487–507. For a discussion of Plato's emphasis on mathematical studies and its

connection to the liberal arts, see Hardy Grant, "Mathematics and the Liberal Arts," *The College Mathematics Journal* 30 (1999): 96–105.

72 Burnyeat, "Plato on Why Mathematics Is Good," 65.

73 Plato, *Complete Works*, ed. John M. Cooper (Indianapolis and Cambridge: Hackett Publishing Company, 1997), *Timaeus*, 55c. See also Fowler, *Mathematics of Plato's Academy*.

74 Plato, *Timaeus*, 50ab.

75 For a comprehensive study of the importance of Pythagoras in the Renaissance see Joost-Gaugier, *Pythagoras and Renaissance Europe*. See also Thomas L. Heath, *A Manual of Greek Mathematics* (Mineola, NY: Dover, 1963, reprinted 2003), 38–45; Bellori, *Descrizzione delle imagini dipinte da Rafaelle*, 16; Harry B. Gutman, "The Medieval Content of Raphael's 'School of Athens,'" *Journal of the History of Ideas* 2 (1941): 420–429.

76 Piero's three treatises, *Trattato d'abaco* (*Abacus Treatise*), *Libellus de quinque corporibus* (*Book on the Five Regular Bodies*) and *Prospectiva pingendi* (*On painting perspective*) cite Euclid's *Elements*. See James R. Banker, *Piero della Francesca: Artist & Man* (Oxford: Oxford University Press, 2014), 79–95. See also Perry Brooks' chapter in this volume. On Piero in Rome, see Ingrid Alexander-Skipnes, "Greek Mathematics in Rome and the Aesthetics of Geometry in Piero della Francesca," in *Early Modern Rome, 1341–1667*, ed. Portia Prebys (Ferrara: Editore Edisai, 2001), 176–186.

77 It has been suggested that Paul of Middelburg was a consultant on mathematics to Raphael. See Christiane L. Joost-Gaugier, *Raphael's Stanza della Segnatura: Meaning and Invention* (Cambridge: Cambridge University Press, 2002), 84, 167–168; Christiane L. Joost-Gaugier, "Raphael as Intellectual: Paulus de Middelbourg, an Early Pythagorean Model," in *Lezioni di metodo: Studi in onore di Lionello Puppi*, ed. Loredana Olivato e Giuseppe Barbieri (Vicenza: Terra Ferma, 2002), 347–351. Paul of Middelburg is best known, however, as an astrologer at the court of Urbino. For a biography and more on his astrological work see Dirk Jan Struik, "Paulus van Middelburg (1445–1533)," *Mededelingen van her Nederlands Historisch Institut te Rome* 5 (1925): 79–118; Bernardino Baldi, *Le vite de' matematici*, ed. Elio Nenci (Milan: Francoangeli, 1998).

78 Joost-Gaugier, *Raphael's Stanza della Segnatura*; Roger Jones and Nicholas Penny, *Raphael* (New Haven and London: Yale University Press, 1983), 199–205.

79 These manuscripts of the *Elements*, Vat. gr. 190, Vat. gr. 191, Vat. gr. 192, Vat. gr. 193, are notably rich in drawings. A Latin translation of the *Elements*, Vat. lat. 2224 has many geometrical drawings, and the manuscript has fine, colorful illuminations. On folio 50v, a young man is drawing with his left hand and holding a compass in his right hand.

80 Although Aristotle does not use the word "golden" it has become common practice to use it. See the *Ethics* 2.6. 1106a14-b6. For a study of how Aristotle in the *Physics* contrasts mathematical activity with the study of nature and his general philosophy of mathematics in the *Metaphysics*, see Jonathan Lear, "Aristotle's Philosophy of Mathematics," *The Philosophical Review* 91 (1982): 161–192; Richard Pettigrew, "Aristotle on the Subject Matter of Geometry," *Phronesis* 54 (2009): 239–260, See also Alfonso Gómez-Lobo, "Aristotle's Hypotheses and the Euclidean Postulates," *The Review of Metaphysics* 30 (1977): 430–439.

81 Aristotle, *Ethics*, 1.1098a29–32.

82 Aristotle, *Ethics*, 6.10.1143a4. For an interesting discussion of Aristotle's views on mathematical objects, see John J. Cleary, *Aristotle and Mathematics: Aporetic Method in Cosmology and Metaphysics* (Leiden: E. J. Brill, 1995).

83 See *The City of God*, 8: 4. Isidore of Seville also writes on the branches of philosophy. See *Sancti Isidori Hispalensis*, PL, 140–141.

84 Vasari, *Le Vite*, 4: 168; see also Gombrich, "Raphael's Stanza della Segnatura," 95–96.

85 Aristotle, *Metaphysics*, 1: 981b25–982a2. See also Cicero, *Topica*, 18, 67; Virgil, *Georgics*, 2, 490; Marsilio Ficino, *Opera Omnia*, 2: 1949 in Marsilio Ficino, *Opera Omnia con una lettera introduttiva di Paul Oskar Kristeller e una premessa di Mario Sancipriano* (Turin: Bottega d'Erasmo, 1962); Egidio da Viterbo, Vat. lat. 6325, folio 113r. The various sources are brought together in Heinrich Pfeiffer, *Zur Ikonographie von Raffaels Disputa* (Rome: Università Gregoriana Editrice, 1975), 153–160. In the *Timaeus*, Plato writes, "Now everything that comes to be must of necessity come to be by the agency of some cause, for it is impossible for anything to come to be without a cause." See Plato, *Timaeus*, 28a.

86 Aristotle, *Metaphysics*, 981b25–982a2. By "techne" Aristotle means craft activity and by "science" he means knowledge.
87 Aristotle, *Ethics*, 1139b19–35.
88 Aristotle *Metaphysics*, 1026a17–22. See also Hippocrates G. Apostle, *Aristotle's Philosophy of Mathematics* (Chicago: The University of Chicago Press, 1952), 49–54.
89 Plato, *Republic*, 527b.
90 Ibid.
91 "Human knowledge is the knowledge of causes." (Vat. lat. 6325, folio 113v." See Pfeiffer, *Zur Ikonographie*, 154. For a study on the important role Egidio da Viterbo played at the Julian court, see John W. O'Malley, "Giles of Viterbo: A Reformer's Thought on Renaissance Rome," *Renaissance Quarterly* 20 (1967): 1–11; Idem., *Giles of Viterbo on Church and Reform: A Study in Renaissance Thought* (Leiden: Brill, 1968); Ingrid D. Rowland, "The Intellectual Background of the *School of Athens*: Tracking Divine Wisdom in the Rome of Julius II," in Raphael's *School of Athens*," in *The Cambridge Campanion to Raphael*, ed. Marcia B. Hall (Cambridge: Cambridge University Press, 2005), 131–170.
92 "The aim of theologians, the knowledge of heavenly causes," see Marsilio Ficino, *Opera Omnia*, con una lettera introduttiva di Paul Oskar Kristeller e una premessa di Mario Sancipriano, vol. 2 (Turin: Bottega d'Erasmo, 1962), 1949; Pfeiffer, *Zur Ikonographie*, 154.
93 Timothy Verdon, "Pagans in the Church: The School of Athens in Religious Context," in *Raphael's School of Athens*, ed. Marcia Hall (Cambridge: Cambridge University Press, 1997), 122, 114–130.
94 See Erwin Panofsky, *Perspective as Symbolic Form* (New York: Urzone, 1991), originally published as "Die Perspektive als symbolische Form," *Vorträge der Bibliothek Warburg*, 24 (Leipzig and Berlin, 1927), 258–230. Richard Tobin challenges Panofsky's interpretation of Euclid's *Optics*. See Richard Tobin, "Ancient Perspective and Euclid's *Optics*," *Journal of the Warburg and Courtauld Institutes* 53 (1990): 14–42. On the fresco's perspective see Christian K. Kleinbub, *Vision and the Visionary in Raphael* (University Park: Pennsylvania State University Press, 2011), 48–58. Corrado Maltese points to the unusual chromatic features of the fresco, a technique he describes as "*scienza luministico-cromatica.*" See Corrado Maltese, "Raffaello e la cultura scientifica e tecnologica del suo tempo," in *Studi su Raffaello: Atti del congress internazionale di studi*, ed. Micaela Sambucco Hamoud and Maria Letizia Strocchi (Urbino: QuattroVenti, 1987), 445–446.
95 Marcia B. Hall, *Color and Meaning: Practice and Theory in Renaissance Painting* (Cambridge and New York: Cambridge University Press, 1992), 18–23.
96 Marcia B. Hall, "Introduction," in *The Cambridge Companion to Raphael*, 5–6.
97 For a detailed study of Raphael's silverpoint technique, see Francis Ames-Lewis, *The Draftsman Raphael* (New Haven: Yale University Press, 1986), 73–94; Joannides, *The Drawings of Raphael*, 80–81. For more on Raphael's graphic style see Joachim Jacoby and Martin Sonnabend, eds., *Raffael: Zeichnungen* (Munich: Hirmer, 2012).
98 Ames-Lewis, *The Draftsman Raphael*, 92.

Bibliography

Agostini, Amedeo. "Il *De viribus quantitatis* di Luca Pacioli." *Periodico di Matematiche* 4 (1924): 165–192.

Alberti, Leon Battista. *De pictura*. Basel, 1540.

———. *On the Art of Building in Ten Books (De re aedificatoria)*. Translated by Joseph Rykwert, Neil Leach and Robert Tavenor. Cambridge, MA: MIT Press, 1988.

———. *On Painting and On Sculpture: The Latin Texts of De pictura and De statua*. Edited and translated by Cecil Grayson. London: Phaidon, 1972.

Alexander-Skipnes, Ingrid. "'Bound with Wond'rous Beauty': Eastern Codices in the Library of Federico da Montefeltro." *Mediterranean Studies* 19 (2010): 67–85.

———. "Greek Mathematics in Rome and the Aesthetics of Geometry in Piero della Francesca." In *Early Modern Rome, 1341–1667*. Edited by Portia Prebys. Ferrara: Editore Edisai, 2011, 176–186.

Alpers, Svetlana. *The Art of Describing: Dutch Art in the Seventeenth Century*. Chicago: University of Chicago Press, 1983.

Ames-Lewis, Francis. *The Draftsman Raphael*. New Haven and London: Yale University Press, 1986.

Andersen, Kirsti. *The Geometry of an Art: The History of the Mathematical Theory of Perspective from Alberti to Monge*. New York: Springer, 2007.

———. "The Mathematical Treatment of Anamorphoses from Piero della Francesca to Niceron." In *History of Mathematics: States of the Art: Flores Quadrivii—Studies in Honor of Christoph J. Scriba*. Edited by Joseph W. Dauben, Menso Folkerts, Eberhard Knobloch and Hans Wussing. San Diego: Academic Press, 1996, 3–28.

Angelini, Alessandro. "Jacometto Veneziano e gli umanisti. Proposta per il 'Ritratto di Luca Pacioli e di Guidobaldo da Montefeltro' del Museo di Capodimonte." *Prospettiva* 147–148 (2012): 126–149.

Anonimo Fiorentino. *Trattato di Geometria Pratica: dal Codice L.IV.18 (sec.XV) della Biblioteca Comunale di Siena*. Edited by Annalisa Simi. Siena: Università degli Studi di Siena, 1993.

Apostle, Hippocrates G. *Aristotle's Philosophy of Mathematics*. Chicago: The University of Chicago Press, 1952.

Arasse, Daniel. " 'Oltre le scienze di sopra': Piero della Francesca et la vision de l'histoire." In *Piero della Francesca and His Legacy*. Edited by Marilyn Aronberg Lavin. Washington, DC: National Gallery of Art, 1995, 105–13.

Aristotle. *The Complete Works of Aristotle: The Revised Oxford Translation*. Edited by Jonathan Barnes. 2 vols. Bollingen Series. Princeton, NJ: Princeton University Press, 1984.

———. *Metaphysics*. Translated by W. D. Ross. In *The Works of Aristotle*. Chicago: Encyclopaedia Britannica, 1952.

Arts libéraux et philosophie au Moyen Âge. Actes du quatrième Congrès international de philosophie médiévale. Université de Montréal, Montréal, Canada, 27 août—2 septembre 1967. Montreal: Institut d'études médiévales; Paris: Librairie philosophique J. Vrin, 1969.

Augustine, Saint. *The City of God*. Translated by Marcus Dods. New York: Modern Library, 1950, repr. 2000.

Baader, Hannah. "Das fünfte Element oder Malerei als achte Kunst das Porträt des Mathematikers Fra Luca Pacioli." In *Der stumme Diskurs der Bilder: Reflexionsformen des Ästhetischen in der Kunst der Frühen Neuzeit*. Edited by Valeska von Rosen, Klaus Krüger and Rudolph Preimesberger. Berlin: Deutscher Kunstverlag, 2003, 177–203.

Bagrow, Leo. "The Origin of Ptolemy's *Geographia*." *Geografiska Annaler* 27 (1945): 318–387.

Baldasso, Renzo. "Luca Pacioli and Disciple: A New, Mathematical Look." *The Art Bulletin* 102 (2010): 83–102.

Baldi, Bernardino. *Le vite de' matematici*. Edited by Elio Nenci. Milan: Franco Angeli, 1998.

———. *Vita e fatti di Federigo di Montefeltro, Duca di Urbino*. 3 vols. Rome: Salvioni, 1824.

———. *Cronica dei matematici*. Urbino: Monticelli, 1707.

Bambach, Carmen Cappel. "On '*La Testa Proportionalmente* Degradata': Luca Signorelli, Leonardo da Vinci and Piero della Francesca's *De Prospectiva* Pingendi." In *Florentine Drawing at the Time of Lorenzo the Magnificent*. Edited by Elizabeth Cropper. Bologna: Nuova Alfa Editoriale, 1994, 17–41.

Banker, James R. "A Manuscript of the Works of Archimedes in the Hand of Piero della Francesca." *Burlington Magazine* 147 (2005): 165–169.

———. *Piero della Francesca: Artist & Man*. Oxford: Oxford University Press, 2014.

Barany, Michael J. "'That Small and Unsensible Shape': Visual Representations of the Euclidean Point in Sixteenth-Century Print." *Spontaneous Generations: A Journal for the History and Philosophy of Science* 6 (2012): 148–159.

Bardeschi, Marco Dezzi. "Nuove ricerche sul S. Sepolcro nella Cappella Rucellai a Firenze." *Marmo* 2 (1963): 134–61.

Barnes, Carl F., Jr., ed. *The Portfolio of Villard de Honnecourt (Paris, Bibliothèque nationale de France, Ms Fr 19093): A New Critical Edition and Color Facsimile*. Farnham: Ashgate, 2009.

Baron, Roger, ed. *Hugonis de Sancto Victore Opera Propaedeutica: Practica geometriae, De grammatica, Epitome Dindimi in philosophiam*. Notre Dame, IN: University of Notre Dame, 1966.

Battisti, Eugenio. *Piero della Francesca*. 2 vols. Milan: Istituto Editoriale Italiano, 1971.

Baxandall, Michael. *The Limewood Sculptors of Renaissance Germany*. New Haven and London: Yale University Press, 1980.

———. *Painting and Experience in Fifteenth Century Italy*. Oxford: Oxford University Press, 1972.

———. *Patterns of Intention*. New Haven and London: Yale University Press, 1985.

———. *Words into Pictures*. New Haven: Yale University Press, 2003.

Bede. "De locis sanctis, II, 2." In *Corpus Christianorum Series Latina CLXXV: Itineraria et alia geographica*. Turnhout: Brepols, 1965.

Beham, Sebald. *Das Kunst und Lere Büchlin Sebalden Behems*. Malen vnd Reissen zu lernen, Nach rechter Proportion, Masz und auszteylung des Circkels. Angehnden Malern vnd Kunstbaren Werckleuten dienlich. Frankfurt, 1552.

Bellori, Giovanni Pietro. *Descrizzione delle imagini dipinte da Rafaelle d'Urbino nelle camere del Palazzo Apostolico Vaticano*. Rome: Stamparia di Gio; Giacomo Komarek Boëmo, 1695.

Benelli, Francesco. "Diversification of Knowledge: Military Architecture as a Political Tool in the Renaissance: The Case of Francesco di Giorgio." *Res* 57/58 (2010): 140–155.

Berlinski, David. *The King of Infinite Space: Euclid and His Elements*. New York: Basic Books, 2013.

Bertòla, Maria. *I due primi registri di prestito della Biblioteca Apostolica Vatican: Codici Vaticani latini, 3964, 3966*. Vatican City: Biblioteca Apostolica Vaticana, 1942.

Betts, Richard. "On the Chronology of Francesco di Giorgio's Treatises: New Evidence from an Unpublished Manuscript." *Journal of the Society of Architectural Historians* 36 (1977): 3–14.

———. "Structural Innovation and Structural Design in Renaissance Architecture." *Journal of the Society of Architectural Historians* 52 (1993): 5–25.

Bieler, Ludovicus, ed. *Anicii Manlii Severini Boethii Philosophiae Consolatio*. Turnhout: Brepols, 1984.

Billingsley, Henry and John Dee. *The Elements of Geometrie of the Most Ancient Philosopher Euclide of Megara*. London: John Daye, 1570.

Bisticci da, Vespasiano. *Commentario de la vita del Signore Federico d'Urbino*, in *Le* Vite. Edited by Aulo Greco. 2 vols. Florence: Istituto nazionale di studi sul Rinascimento, 1970–76.

Boncompagni, Baldassare, ed. *Scritti di Leonardo Pisano, matematico del secolo decimoterzo*. 2 vols. Rome: Tipografia delle Scienze Matematiche e Fisiche, 1867.

Borsi, Franco. *Leon Battista Alberti*. New York: Harper&Row, 1977.

Borsook, Eve. "Technical Innovation and the Development of Raphael's Style in Rome." *Canadian Art Review* 12 (1985): 127–136.

Bos, Henk J. M. *Redefining Geometrical Exactness: Descartes' Transformation of the Early Modern Conception of Construction*. New York: Springer Verlag, 2001, 119–134.

Boskovits, Miklos. "'Quello ch'e dipintori oggi dicono prospettiva': Contribution to Fifteenth Century Italian Art Theory." *Acta Historiae Artium Academiae Scientiarum Hungaricae 9* (1963): 139–162.

Bosse, Abraham. *Traité des pratiques géometrales et perspectives: enseignées dans l'Academie royale de la peinture et sculpture*. Paris: Chez l'auteur, 1665.

Brachert, Thomas. "A Musical Canon of Proportion in Leonardo da Vinci's Last Supper." *Art Bulletin* 53 (1971): 461–466.

Brooks, Perry. "Observations on Piero della Francesca's Frescoes in Ferrara." In *Watching Art: Writings in Honor of James Beck: Studi di storia dell'arte in onore di James Beck*. Edited by Lynn Catterson and Mark Zucker. Todi: Ediart, 2006, 95–102.

———. *Piero della Francesca: The Arezzo Frescoes*. Rizzoli Art Series. New York: Rizzoli, 1992.

———. *Ut Pictura Mathesis: Studies in the Art of Piero della Francesca*. Ann Arbor: UMI, 1990.

Bull, George, trans. *Giorgio Vasari: Lives of the Artists*. 3 vols. London: Folio Society, 1993.

Bulmer-Thomas, Ivor and John Murdoch. "Euclid." *Dictionary of Scientific Biography*. Edited by Charles Coulston Gillispie. 18 vols. New York: Charles Scribner's Sons, 1981, 4: 414–59.

Burnyeat, M. F. "Plato on Why Mathematics Is Good for the Soul." In *Mathematics and Necessity: Essays in the History of Philosophy: Proceedings of the British Academy*. Edited by Timothy Smiley. Vol. 103. Oxford: Oxford University Press, 2000, 1–81.

Burtt, Edwin Arthur. *The Metaphysical Foundations of Modern Physical Science*. London: Routledge and Kegan Paul Ltd, 1949.

Büttner, Jochen, Peter Damerow, Jürgen Renn and Mattias Schemmel. "The Challenging Images of Artillery: Practical Knowledge at the Roots of the Scientific Revolution." In *The Power of Images in Early Modern Science*. Edited by Wolfgang Lefèvre, Jürgen Renn and Urs Schoepflin. Basel: Birkhäuser Verlag, 2003, 3–27.

Calvesi, Maurizio. *Piero della Francesca*. New York: Rizzoli, 1998.

Camerota, Filippo. *Nel segno di Masaccio: L'invenzione della prospettiva*. Florence: Giunti, 2001.

———. "Teaching Euclid in a Practical Context: Linear Perspective and Practical Geometry." *Science & Education* 15 (2006): 323–334.

Capannelli, Spartaco, ed. *Il Palazzo Ducale di Gubbio e Francesco di Giorgio Martini*. Gubbio: TMM, 2008.

Carman, Charles H. *Leon Battista Alberti and Nicholas Cusanus: Towards an Epistemology of Vision for Italian Renaissance Art and Culture*. Burlington, VT: Ashgate, 2014.

Cassiodorus. "Psalterium Expositio." *Patrologia Latina*. Edited by J.-P. Migne. 221 vols. Paris: J.-P. Migne, 1841–1855, 70: 3.

Cassirer, Ernst. *The Individual and the Cosmos in Renaissance Philosophy*. Translated by Mario Domandi. Philadelphia: University of Pennsylvania Press, 1963.

Castelli, Patrizia. "La proporzione 'divina'. Alcuni aspetti della cultura scientifica alla corte di Federico da Montefeltro." In *Presenze Folosofiche in Umbria II. Dal Medievo all'Età Contemporanea*. Edited by Antonio Pieretti. Milan: Mimesis, 2012, 127–146.

———. "Matematici e astrologi tedeschi alla 'corte' dei Montefeltro" In *Die Kunst und das Studium der Natur vom 14. zum 16. Jahrhundert*. Edited by Wolfram Prinz and Andreas Beyer. Weinheim: Acta Humaniora–VCH, 1987, 237–251.

Cennini, Cennino. *Il libro dell' arte*. Edited by Franco Brunello. Vicenza: N. Pozza, 1971.

———. *The Craftsman's Handbook: The Italian 'Il libro dell'arte'*. Translated by Daniel V. Thompson, Jr. New York: Dover Publications, 1960, repr. of Yale University Press, 1933.

Chapman, Hugo, Tom Henry and Carol Plazzotta, eds. *Raphael: From Urbino to Rome*. London: National Gallery Company, 2004.

Chastel, André. *Art et Humanisme a Florence au Temps de Laurent le Magnifique: Études sur la Renaissance et l'Humanisme platonicien*. Paris: Presses Universitaires de France, 1959.

———. *I centri del Rinascimento: Arte italiana 1460–1500*. Milan: Biblioteca Universale Rizzoli, 1965.

———. *Renaissance Méridionale: Italie 1460–1500*. Paris: Gallimard, 1965.

Chojecka, Ewa. "Johann Kepler und die Kunst: Zum Verhältnis von Kunst und Naturwissenschaften in der Spätrenaissance." *Zeitschrift für Kunstgeschichte* 30 (1967): 55–72.

Ciapponi, Lucia A. "A Fragmentary Treatise on Epigraphic Alphabets by Fra Giocondo da Verona." *Renaissance Quarterly* 32 (1979): 18–40.

Ciocci, Argante. "Il Doppio Ritratto del Poliedrico Luca Pacioli." *De computis. Revista Española de Historia de la Contabilidad* 15 (2011): 107–130.

———. *Luca Pacioli e la matematizzazione del sapere nel Rinascimento*. Bari: Cacucci, 2003.

———. *Luca Pacioli tra Piero della Francesca e Leonardo*. Sansepolcro: Aboca Museum, 2009.

Clagett, Marshall. *Archimedes in the Middle Ages*. Vol 3, *The Fate of the Medieval Archimedes 1300 to 1565, Part III*. Philadelphia: American Philosophical Society, 1978.

———. "The Medieval Latin Translations from the Arabic of the *Elements* of Euclid, with Special Emphasis on the Versions of Adelard of Bath." *Isis* 44 (1953): 16–42.

Clark, Kenneth. *Piero della Francesca*. London: Phaidon, 1951, 2nd ed. 1969.

Cleary, John J. *Aristotle and Mathematics: Aporetic Method in Cosmology and Metaphysics*. Leiden: E. J. Brill, 1995.

Clements, Robert. *Michelangelo's Theory of Art*. New York: New York University Press, 1961.

Cohen, Matthew A. "Introduction: Two Kinds of Proportion." *Architectural Histories* (Special Collection, Objects of Belief: Proportional Systems in the History of Architecture) 2(1): 21 (2014): 1–25. DOI: http://dx.doi.org/10.5334/ah.bv. Uploaded April 17, 2016.

———. *Beyond Beauty: Reexamining Architectural Proportion through the Basilicas of San Lorenzo and Santo Spirito in Florence*. Venice: Marsilio, 2013.

Coleschi, Lorenzo. *Storia della città di Sansepolcro*. Città di Castello: S. Lapi, 1886.

Conticelli, Valentina. *"Guardaroba di cose rare et preziose": lo studiolo di Francesco I de' Medici: arte, storia e significati*. La Spezia: Agorà, 2007.

Cornford, F. M. *From Religion to Philosophy*. London: Edward Arnold, 1912.

———. *Plato's Cosmology*. 1937, repr. Indianapolis: Bobbs-Merrill, n.d.

Crous, Ernst. *Dürer und die Schrift: Dürers Abhandlung aus seiner 'Unterweisung der Messung'*. Berlin: Aldus Druck, 1933.

Curchin, Leonard and Roger Herz-Fischler. "De quand date le premier rapprochement entre la suite de Fibonacci et la division en extrême et moyenne raison?" *Centaurus* 28 (1985): 129–138.

Cusa de, Nicolai. *De coniecturis*. Hamburg: In Aedibus Felicis Meiner, 1972.

———. *De Docta Ignorantia*. Leipzig: Felicis Meiner, 1932.

Cusanus, Nicolaus. *Nicholas of Cusa: On Learned Ignorance, a Translation and an Appraisal of De Docta Ignorantia*. By Jasper Hopkins. Minneapolis: Arthur J. Banning Press, 1981.

———. *Opera*. 3 vols. Paris: Badius Ascensius, 1514.

Dacos, Nicole. *The Loggia of Raphael: A Vatican Art Treasure*. Translated by Josephine Bacon. New York: Musei Vaticani and Abbeville Press, 2008.

Dalché, Patrick Gautier. *La Géographie de Ptolémée en Occident (IVe-XVIe Siècle)*. Turnhout: Brepols, 2009.

Dal Poggetto, Maria Grazia Ciardi Dupré. "Il Ritratto di Luca Pacioli e di Guidubaldo da Montefeltro." In *Piero e Urbino, Piero e le Corti rinascimentali*. Edited by Paolo Dal Poggetto. Venice: Marsilio, 1992, 197–200.

Daston, Lorraine. "Historical Epistemology." In *Questions of Evidence: Proof, Practice, and Persuasion across the Disciplines*. Edited by James K. Chandler, Arnold Ira Davidson and Harry D. Harootunian. Chicago: University of Chicago Press, 1994, 282–289.

Davies, Elton M. and Dean Snyder. "Piero della Francesca's Madonna of Urbino." *Gazette des Beaux-Arts* 75 (1970): 193–212.

Da Novaria, Francisco Torniello. *Opera del modo de fare le littere maiuscole antique*. Milan: Gotardo da Ponte, 1517.

Da Vinci, Leonardo. *Leonardo on Painting*. Translated by Martin Kemp and Margaret Walker. New Haven and London: Yale University Press, 1989.

———. *The Madrid Codices*. Translated by Ladislao Reti 5 vols. New York: McGraw-Hill, 1974.

———. *Trattato della pittura/Traité de la peinture*. Paris: Langlois, 1651.

———. *Treatise on Painting: Codex Urbinas Latinus 1270 by Leonardo da Vinci*. Translated by Philip McMahon. Princeton: Princeton University Press, 1956.

———. *A Treatise of Painting by Leonardo da Vinci: Translated from the Original Italian*. London: Printed for J. Senex . . . and W. Taylor . . ., 1721.

Davis, Margaret Daly. "Carpaccio and the Perspective of Regular Bodies," in *La prospettiva rinascimentale: codificazioni e trasgressioni*. Edited by Marisa Dalai Emiliani, vol.1. Florence: Centro Di, 1980, 183–200.

———. *Piero della Francesca's Mathematical Treatises: The 'Trattato d'abaco' and 'Libellus de quinque corporibus regularibus'*. Ravenna: Longo Editore, 1977.

———. "Piero's Treatises: The Mathematics of Form." In *The Cambridge Companion to Piero della Francesca*. Edited by Jeryldene M. Wood. Cambridge: Cambridge University Press, 2002, 134–151.

Dee, John. "John Dee his Mathematicall Praeface," *The elements of geometrie of the most ancient philosopher Euclide of Megara*. Translated by Henry Billingsley. London: John Daye, 1570.

Dehio, Georg. "Zur Geschichte der Buchstabenreform in der Renaissance." *Repertorium für Kunstwissenschaft* 4 (1881): 269–279.

Della Torre, Arnaldo. *Storia dell'Accademia Platonica di Firenze*. Florence: G. Carnesecchi and Sons Typography, 1902.

Del Vita, A. "Il volto di Piero della Francesca." *Rassegna d'arte* 7 (1920): 109–112.

De Tolnay, Charles. "Conceptions religieuses dans la peinture de Piero della Francesca." *Arte antica e moderna* 6: 23 (1963): 205–241.

Devreesse, Robert. *Le fonds grec de la Bibliothèque Vaticane des origines à Paul V*. Vatican City: Bibliotheca Apostolica Vaticana, 1965.

Dezzi Bardeschi, Marco. "Nuove ricerche sul S. Sepolcro nella Cappella Rucellai a Firenze." *Marmo* 2 (1963): 134–161.

Dixon, Laurinda. *The Dark Side of Genius: The Melancholic Persona in Art, ca. 1500–1700.* University Park: Pennsylvania State University, 2013.

Doose, Conrad, Jürgen Eberhardt, Hajo Lauenstein et al., eds. *Das italienische Jülich: Grundzüge im Konzept Alessandro Pasqualinis für die Stadtanlage, die Zitadelle und das Residenzschloss.* Jülich: Förderverein Festung Zitadelle Jülich, 2009.

Dorez, Léon. "La bibliothèque privée du Pape Jules II." *Revue des Bibliothèques* 6 (1896): 97–126.

Drake, Stillman, trans. *Discoveries and Opinions of Galileo.* Garden City, NY: Doubleday, 1957.

Duhem, Pierre. "Leonard de Vinci et les deux infinis." *Études sur Léonard de Vinci* (Paris: A. Hermann, 1906–1913), 2: 3–53.

Dürer, Albrecht. *The Painter's Manual: A Manual of Measurement of Lines, Areas and Solids by Means of Compass and Ruler.* Translated by Walter L. Strauss. New York: Abaris Books, 1977.

———. *Underweysung der Messung mit dem Zirckel un[d] Richtscheyt in Linien, Ebnen und gantzen Corporen.* Gedrückt zu Nuremberg, [s.n.] 1525.

Edgerton, Samuel Y. *The Mirror, the Window, and the Telescope: How Renaissance Linear Perspective Changed Our Vision of the Universe.* Ithaca and London: Cornell University Press, 2009.

———. *The Renaissance Rediscovery of Linear Perspective.* New York: Basic Books, Inc., 1974.

Ehrle, Franz and Enrico Stevenson. *Les fresques du Pinturicchio dans les salles Borgia au Vatican.* Rome: Danesi, 1898.

Eisler, János. "Remarks On Some Aspects of Francesco di Giorgio's Trattato." *Acta Historiae Artium (Academiae Scientiarum Hungaricae)* 18 (1972): 193–231.

Ettlinger, L. D. "Muses and Liberal Arts: Two Miniatures from Herrad of Landsberg's *Hortus Deliciarum.*" In *Essays in the History of Art Presented to Rudolf Wittkower.* Edited by Douglas Fraser, Howard Hibbard and Milton J. Lewine. Bath: Phaidon, 1967, repr. 1969, 29–35.

———. "Pollaiuolo's Tomb of Pope Sixtus IV." *Journal of the Warburg and Courtauld Institutes* 16 (1953): 239–274.

Euclid. *Elementorum Geometricorum. Lib.XV. Cum Expositione Theonis in Priores XIII à Bartholomaeo Veneto Latinitate donata, Campani in omnes, & Hybsiclis Alexandrini in duos postremos.* Basel: J. Herwagen, 1537.

———. *Euclid's Elements.* Edited by Dana Densmore and translated by Thomas L. Heath. Santa Fe: Green Lion Press, 2007.

Fabre, Paul. "La Vaticane de Sixte IV." *Mélanges d'archéologie et d'histoire* 15 (1895): 455–483.

Fanti, Sigismondo. *Theorica et Practica de modo scribendi.* Venice: Joannes Rubeus Vercellensis, 1514.

Farago, Claire, ed. *Re-Reading Leonardo: The Treatise on Painting across Europe, 1550–1900.* Burlington: Ashgate, 2009.

Fehrenbach, Frank. "Leonardo da Vinci: Leonardo's Point." The Warburg Institute. http://www.youtube.com/watch?v=XwwkvjCpauQ. Uploaded February 16, 2012.

Felfe, Robert. *Naturform und bildnerische Prozesse: Elemente einer Wissensgeschichte in der Kunst des 16. und 17. Jahrhunderts.* Berlin and Boston: Walter de Gruyter, 2015.

Ferrari, Simone. *Jacopo de' Barbari: Un protagonista del Rinascimento tra Venezia e Dürer.* Milan: Mondadori, 2006.

———. "Dürer e Jacopo de' Barbari: persistenza di un rapporto." In *Dürer, l'Italia e l'Europa.* Edited by Sybille Ebert-Schifferer and Kristina Hermann Fiore. Milano: Silvana Editoriale, 2011, 39–46.

Fibonacci, Leonardo Pisano. *The Book of Squares.* Translated by L. E. Sigler. New York: Academic Press, 1987.

Fichtner, Richard. *Der Verborgene Geometrie in Raffaels 'Schule von Athen'*. Munich: R. Oldenbourg; Deutsches Museum, 1984.

Ficino, Marsilio. *Commentary on Plato's Symposium on Love (De amore)*. Translated by Sears Jayne. Dallas: Spring Publications, 1985.

———. *Opera Omnia*. Basel: Heinrich Petri, 1576.

———. *Opera Omnia con una lettera introduttiva di Paul Oskar Kristeller e una premessa di Mario Sancipriano*. Turin: Bottega d'Erasmo, 1962.

Field, J. V. *The Invention of Infinity: Mathematics and Art in the Renaissance*. Oxford: Oxford University Press, 1997, repr. 2005.

———. "A Mathematician's Art." In *Piero della Francesca and His Legacy*. Edited by Marilyn Aronberg Lavin. Hanover and London: University Press of New England, 1995, 177–197.

———. "Mathematics and the Craft of Painting: Piero della Francesca and Perspective." In *Renaissance and Revolution: Humanists, Scholars, Craftsmen and Natural Philosophers in Early Modern Europe*. Edited by J. V. Field and Frank A.J.L. James. Cambridge: Cambridge University Press, 1993, repr. 1997, 73–95.

———. *Piero della Francesca: A Mathematician's Art*. New Haven and London: Yale University Press, 2005.

Filippi, Elena. *Umanesimo e misura viva: Dürer tra Cusano e Alberti*. Venice: Arsenale, 2011.

Finocchiaro, Maurice A., ed. *The Essential Galileo*. Cambridge and Indianapolis: Hackett Publishing Company, 2008.

Fiore, Francesco Paolo. *Città e Macchine del 400: nei disegni di Francesco di Giorgio Martini*. Florence: Leo S. Olschki, 1978.

Folkerts, Menso. *The Development of Mathematics in Medieval Europe: The Arabs, Euclid, Regiomontanus*. Aldershot and Burlington: Ashgate, 2006.

———. "Piero della Francesca and Euclid." In *Piero della Francesca: tra arte e scienza. Atti del convegno internazionale di studi, Arezzo, 8–11 ottobre 1992*. Edited by Marisa Dalai Emiliani and Valter Curzi. Venice: Marsilio, 1996, 293–312.

———. "Zur Frühgeschichte der magischen Quadrate in Westeuropa." *Sudhoffs Archiv Kiel* 65 (1981): 313–338.

Fowler, David. *The Mathematics of Plato's Academy: A New Reconstruction*. Oxford: Oxford University Press, 1999.

Fowler, David and Eleanor Robson. "Square Root Approximations in Old Babylonian Mathematics: YBC 7289 in Context." *Historia Mathematica* 25 (1998): 366–378.

Franceschi Marini, Evelyn. *Piero della Francesca*. Città di Castello: Lapi, 1912.

Frommel, Christoph. "Francesco del Borgo: Architekt Pius' II. und Pauls II." *Römisches Jahrbuch für Kunstgeschichte* 20 (1983): 108–154.

Gal, Ofer and Raz Chen-Morris. *Baroque Science*. Chicago and London: University of Chicago Press, 2013.

Galilei, Galileo. "The Assayer." In *Discoveries and Opinions of Galileo*. Translated by Stillman Drake. New York: Anchor Books, 1957, 231–280.

Gallo, F. Alberto. *Music in the Castle: Troubadours, Books, and Orators in Italian Courts of the Thirteenth, Fourteenth, and Fifteenth Centuries*. Translated by Anna Herklotz and Kathryn Krug. Chicago: University of Chicago Press, 1995.

Gamba, Enrico. "La matematica dei tecnici tra oralità e scrittura, tra esattezza e approssimazione." In *L'arte della matematica nella prospettiva*. Edited by Rocco Sinisgalli. Florence: C. B. Cartei & Bianchi, 2009, 229–246.

———. "L'Umanesimo matematico a Urbino." In *La città ideale: l'utopia del Rinascimento a Urbino tra Piero della Francesca e Raffaello*. Edited by Alessandro Marchi and Maria Rosaria Valazzi. Milan: Electa, 2012, 233–247.

———. "Pittura e storia della scienza." In *La ragione e il metodo: Immagini della scienza nell'arte italiana dal XVI al XIX secolo*. Edited by Marco Bona Castellotti, Enrico Gamba and Fernando Mazzocca. Milan: Electa, 1999, 43–53.

———. "Proviamo a rileggere il 'Doppio ritratto' di Luca Pacioli." In *Le tre facce del poliedrico Luca Pacioli*. Edited by Francesca Maria Cesaroni, Massimo Ciambotti, Enrico Gamba and Vico Montebelli. Urbino: Quaderni del Centro Internazionale di Studi Urbino e la Prospettiva, 2010, 81–97.

———. *Raffaello e la Matematica*. Urbino: Centro Internazionale di Studi Urbino e la Prospettiva, 2009.

Gamba, Enrico and Vico Montebelli. *Le scienze a Urbino nel tardo Rinascimento*. Urbino: Edizioni Quattro Venti, 1988.

Garrard, Mary D. "The Liberal Arts and Michelangelo's First Project for the Tomb of Julius II (With a Coda on Raphael's School of Athens)." *Viator* 15 (1984): 335–404.

Gautier Dalché, Patrick. *La Géographie de Ptolémée en Occident (IVe-XVIe Siècle)*. Turnhout: Brepols, 2009.

Geoffroy Tory de Bourges. *Champ Fleury*. Paris: Geoffroy Tory de Bourges, 1529.

George, William and Emily Waters, trans. *The Vespasiano Memoirs: Lives of Illustrious Men of the XVth Century*. 1926, repr. Toronto: University of Toronto Press, 1997.

Ghiberti, Lorenzo. *Commentari*. Edited by Ottanio Morisani. Naples: R. Ricciardi, 1947.

Ghyka, Matila. *Esthétique des proportions dans la nature et dans les arts*. Paris: Gallimard, 1927.

———. *The Geometry of Art and Life*. New York: Sheed & Ward, 1946.

———. *Le nombre d'or: rites et rythmes pythagoriciens dans le développement de la civilisation occidentale*. Introduced by Paul Valéry. 2 vols. Paris: Gallimard, 1931.

Gilbert, Creighton. *Change in Piero della Francesca*. Locust Valley, NY: J. J. Augustin, 1968.

Gillispie, Charles Coulston, ed. *Dictionary of Scientific Biography*. Vols. 1–16. New York: Charles Scribner's Sons, 1970–1980, repr. 8 vols (1981). Supplements, vols. 17 and 18. Edited by Frederic L. Holmes. (1990).

Ginzburg, Carlo. *The Enigma of Piero*. Translated by Martin Ryle and Late Soper. London: Verso, 2000.

Goffen, Rona. "Raphael's Designer Labels: From the Virgin Mary to *La Fornarina*." *Artibus et Historiae* 24 (2003): 123–142.

Golzio, Vincenzo. *Raffaello nei documenti nelle testimonianze e I contemporanei e nella letteratura del suo secolo*. Vatican City: Pontificia insigne academia Artistica dei Virtuosi al Pantheon, 1936.

Gombrich, Ernst. "Raphael's Stanza della Segnatura and the Nature of its Symbolism." In *Gombrich on the Renaissance. Volume 2: Symbolic Images,* 3rd ed. London: Phaidon, 1985, repr. 1993, 85–101.

Gómez-Lobo, Alfonso. "Aristotle's Hypotheses and the Euclidean Postulates." *The Review of Metaphysics* 30 (1977): 430–439.

Grabiner, Judith V. "The Centrality of Mathematics in the History of Western Thought." *Mathematics Magazine* 61 (1988): 220–30.

Grant, Hardy. "Mathematics and the Liberal Arts." *The College Mathematics Journal* 30 (1999): 96–105.

———. "Mathematics and the Liberal Arts—II." *The College Mathematics Journal* 30 (1999): 197–203.

Grassi, Liliana and Anna Maria Finoli, eds. Antonio Averlino, *Trattato di Architettura*. Milan: Il Polifilo, 1972.

Greenstein, Jack M. "On Alberti's 'Sign': Vision and Composition in Quattrocento Painting." *The Art Bulletin* 79: 4 (1997): 669–698.

Grendler, Paul F. *The Universities of the Italian Renaissance*. Baltimore and London: Johns Hopkins University Press, 2001.

Grimm, Herman. *Das Leben Raphael's*. Berlin: Verlag von Wilhelm Hertz, 1886.

Gutman, Harry B. "The Medieval Content of Raphael's 'School of Athens'." *Journal of the History of Ideas* 2 (1941): 420–429.

Hahn, Nan L., ed. *Medieval Mensuration: 'Quadrans Vetus' and 'Geometrie Due Sunt Partes Principales. . .'*. Philadelphia: The American Philosophical Society, 1982.

Hall, Marcia B. *Color and Meaning: Practice and Theory in Renaissance Painting*. Cambridge and New York: Cambridge University Press, 1992.

———, ed. *The Cambridge Companion to Raphael*. Cambridge: Cambridge University Press, 2005.

———, ed. *Raphael's School of Athens*. Cambridge: Cambridge University Press, 1997.

Hankins, James. *Plato in the Italian Renaissance*. Leiden, New York and Cologne: E. J. Brill, 1994.

Harries, Karsten. *Infinity and Perspective*. Cambridge, MA: MIT Press, 2001.

Hart, George. http://www.georgehart.com/virtual-polyhedra/pacioli.html. Uploaded April 16, 2016.

Heath, Thomas L. *A Manual of Greek Mathematics*. Mineola, NY: Dover, 1963, repr. 2003.

———. *The Thirteen Books of Euclid's Elements*. 3 vols. New York: Dover, 1956, repr. of 2nd edition (1926).

Heiderhoff, Horst. "Dürers Beitrag zur technischen Reproduzierbarkeit der Schrift." *Imprimatur* 8 (1976): 299–304.

Heliodorus of Larissa, transc. *Danti's proem to Euclid, Optics*. Italian translation by Egnazio Danti. Florence: Bernardo Giunta, 1573.

Hellmann, Günter. "Proportionsverfahren des Francesco di Giorgio Martini." In *Miscellanea Bibliothecae Hertzianae zu Ehren von Leo Bruhns, Franz Graf Wolff Metternich, Ludwig Schudt*. Edited by Hanno Hahn. Munich: A. Scholl, 1961, 157–166.

Hendrix, John. *Platonic Architectonics: Platonic Philosophies and the Visual Arts*. New York: Peter Lang, 2004.

Henricus Cornelius Agrippa ab Nettesheym. *De Occulta philosophia sive De Magia Libri Tres*, 1533. Facsimile reprint with commentary by Karl Anton Nowotny, Graz: Akademische Druck- und Verlagsanstalt, 1967.

Hernad, Béatrice and Karl Dachs, ed. *Die Graphiksammlung des Humanisten Hartmann Schedel*. Munich: Prestel-Verlag, 1990.

Hersey, George L. *Pythagorean Palaces: Magic and Architecture in the Italian Renaissance*. Ithaca, NY: Cornell University Press, 1976.

Herz-Fischler, Roger. *A Mathematical History of Division in Extreme and Mean Ratio*. Waterloo, Ontario: Wilfrid Laurier University Press, 1987.

Hesychius Presbyter Hierosolymitanus. "Martyrium Sancti Longini Centurionis." *Patrologia Graeca*. Edited by J.-P. Migne. 161 vols. Paris: J.-P. Migne, 1857–1866, 93: 1545–1560.

Heydenreich, Ludwig H. "La ripresa 'critica' di rappresentazioni Medievali delle 'septem artes liberales' nel Rinascimento." In *Il mondo antico nel Rinascimento. Atti del V convegno internazionale di studi sul Rinascimento*, Firenze-Palazzo Strozzi, 2–6 Settembre, 1956 . Florence: Istituto Nazionale di Studi sul Rinascimento, 1958, 265–273.

Hofmann, Joseph. "Nicholas of Cusa." In *Dictionary of Scientific Biography*. Edited by Charles Coulston Gillispie. 18 vols. New York: Scribner, 1970–80, 3: 512–16.

Hogarth, William. *The Analysis of Beauty*. Edited by Ronald Paulson. New Haven and London: Yale University Press, 1997.

Homann, Frederick A. S. J., trans. *Practical Geometry: [Practica Geometriae]-Attributed to Hugh of St. Victor*. Milwaukee: Marquette University Press, 1991.

Hopkins, Jasper. *Nicholas of Cusa on Wisdom and Knowledge*. Minneapolis: Arthur J. Banning Press, 1996.

Horster, Marita. "Mantuae Sanguis Preciosus." *Wallraf-Richartz Jahrbuch* 25 (1963): 151–180.

Hubert, Damisch. *The Origin of Perspective*. Translated by John Goodman. London and Cambridge: MIT Press, 1994.

Hubert, Hans W. "In der Werkstatt Filaretes: Bemerkungen zur Praxis des Architekturzeichnens in der Renaissance." *Mitteilungen des Kunsthistorischen Institutes in Florenz* 47 (2003): 311–344.

———. "Perspektiven auf Bramantes Virtus in Wort und Bild." In *Die Virtus in Kunst und Kunsttheorie der italienischen Renaissance*. Edited by Thomas Weigel, Britta Kusch-Arnhold and Candida Syndikus. Münster: Rhema, 2014, 111–138.

Hudson, Hugh. *Paolo Uccello: Artist of the Florentine Renaissance Republic*. Saarbrücken: VDM Verlag, 2008.

Huylebrouck, Dirk. "Lost in Triangulation: Leonardo da Vinci's Mathematical Slip-Up," *Scientific American* March 29 (2011).

Iamblichus. *On the Pythagorean Way of Life (De vita pythagorica)*. Translated by John Dillon and Jackson Hershbell. Atlanta: Scholars Press, 1991.

Isidore of Seville. *Sancti Isidori Hispalensis Episcopi, Opera Omnia, Patrologia Latina*. Edited by J.-P. Migne. 221 vols. Paris: J.-P. Migne, 1841–1855, 82: 140–141.

Iversen, Margaret. "The Discourse of Perspective in the Twentieth Century: Panofsky, Damisch, Lacan." *Oxford Art Journal* 28: 2 (2005): 193–202.

Ivins, William M., Jr. *Art & Geometry: A Study in Space Intuitions*. Cambridge, MA: Harvard University Press, 1946, repr. 1964.

———. *On the Rationalization of Sight: With an Examination of Three Renaissance Texts on Perspective*. New York: Metropolitan Museum of Art, 1938.

Jacoby, Joachim and Martin Sonnabend, eds. *Raffael: Zeichnungen*. Munich: Hirmer, 2012.

Joannides, Paul. *The Drawings of Raphael*. Oxford: Phaidon, 1983.

Jones, Roger and Nicholas Penny. *Raphael*. New Haven and London: Yale University Press, 1983.

Joost-Gaugier, Christiane L. "The Early Beginnings of the Notion of 'Uomini Famosi' and the 'De Viris Illustribus' in Greco-Roman Literary Tradition." *Artibus et Historiae* 3 (1982): 97–115.

———. "Poggio and Visual Tradition: Uomini Famosi in Classical Literary Description." *Artibus et Historiae* 6 (1985): 57–74.

———. *Pythagoras and Renaissance Europe: Finding Heaven*. Cambridge: Cambridge University Press, 2009.

———. "Raphael as Intellectual: Paulus de Middelbourg, an Early Pythagorean Model." In *Lezioni di metodo: Studi in onore di Lionello Puppi*. Edited by Loredana Olivato e Giuseppe Barbieri. Vicenza: Terra Ferma, 2002, 347–51.

———. *Raphael's Stanza della Segnatura: Meaning and Invention*. Cambridge: Cambridge University Press, 2002.

Jung, Wolfgang. *Über szenographisches Entwerfen: Raffael und die Villa Madama*. Braunschweig: Vieweg, 1997.

Katz, Victor J. and Karen Hunger Parshall. *Taming the Unknown: A History of Algebra from Antiquity to the Early Twentieth Century*. Princeton: Princeton University Press, 2014.

Katzenellenbogen, Adolf. "The Representation of the Seven Liberal Arts." In *Twelfth-Century Europe and the Foundations of Modern Society*. Edited by Marshall Clagett, Gaines Post and Robert Reynolds. Madison: University of Wisconsin Press, 1961, 39–55.

Kemp, Martin. "Leonardo e lo spazio dello scultore." *Lettura Vinciana* 27 (1988): 5–23.

——— ed. *Leonardo on Painting*. Translated by Martin Kemp and Margaret Walker. New Haven and London: Yale University Press, 1989.

———. *The Science of Art: Optical Themes in Western Art from Brunelleschi to Seurat*. New Haven and London: Yale University Press, 1990.

Kinne, Mara-Lisa, Roland Krischel, and Stefan Neuner. "Annäherung an ein Rätsel: Das *Bildnis des Luca Pacioli* von Jacopo de' Barbari." *Wallraf-Richartz-Jahrbuch* 76 (2015): 79–115.

Kittler, Friedrich. "Perspective and the Book." *Grey Room* 5 (2001): 38–53.

Klein, Jacob. *Greek Mathematical Thought and the Origin of Algebra*. Translated by Eva Braun. Cambridge, MA and London, England: MIT Press, 1968.
Kleinbub, Christian K. *Vision and the Visionary in Raphael*. University Park: Pennsylvania State University Press, 2011.
Klibansky, Raymond, Erwin Panofsky and Fritz Saxl. *Saturn and Melancholy: Studies in the History of Natural Philosophy, Religion and Art*. London: Thomas Nelson & Sons, 1964.
Kolb, Carolyn. "The Francesco di Giorgio Material in the Zichy Codex." *Journal of the Society of Architectural Historians* 47 (1988): 132–159.
Koyré, Alexandre. *Metaphysics and Measurement Essays in Scientific Revolution*. London: Chapman & Hall, 1968.
Krautheimer, Richard. *Lorenzo Ghiberti*. Princeton: Princeton University Press, 1956.
Kristeller, Paul Oskar. *Renaissance Thought and the Arts: Collected Essays*. Princeton, NJ: Princeton University Press, 1964, repr. 1980.
Krüger, Peter. "Der Kompositionsbegriff und die Rhetorik: Alberti und Dürer." In *Dürers 'Apokalypse': Zur poetischen Struktur einer Bilderzählung der Renaissance*. Wiesbaden: Harrassowitz Verlag, 1996, 89–98.
Lacan, Jacques. *The Four Fundamental Concepts of Psycho-Analysis*. Translated by Alan Sheridan. New York and London: W. W. Norton & Company, 1981.
Laird, Walter Roy. "Archimedes among the Humanists." *Isis* 82 (1991): 628–638.
Landrus, Matthew. *Leonardo da Vinci's Giant Crossbow*. Heidelberg: Springer, 2010.
———. "The Proportions of Leonardo da Vinci's Last Supper." *Raccolta Vinciana* 32 (2007): 43–100.
Lautensack, Heinrich. *Deß Circkels und Richtschents auch der Perspectiva und Proportion der Menschen*. Frankfurt, 1564.
Lavin, Marilyn Aronberg. *Piero della Francesca: The Flagellation*. New York: Viking Press, 1972.
Lear, Jonathan. "Aristotle's Philosophy of Mathematics." *The Philosophical Review* 91 (1982): 161–192.
Lefaivre, Liane. *Leon Battista Alberti's Hypnerotomachia Poliphili: Re-Cognizing the Architectural Body in the Early Italian Renaissance*. Cambridge: MIT Press, 1997.
Levenson, Jay A., ed. *Circa 1492: Art in the Age of Exploration*. Washington, D.C.: National Gallery of Art, 1991.
Lindberg, David C. *The Beginnings of Western Science*. 2nd ed. Chicago: The University of Chicago Press, 2007.
Livio, Mario. *The Golden Ratio: The Story of Phi, the World's Most Astonishing Number*. New York: Broadway Books, 2002.
Lloyd, W. W. "Raphael's School of Athens." *The Fine Arts Quarterly Review* 2 (1864): 42–68.
Lockwood, D. P. "De Rinucio Arentino graecarum litterarum interprete." *Harvard Studies in Classical Philology* 24 (1913): 51–119.
Lomazzo, Giovanni Paolo. *Trattato dell'arte della pittura, scultura ed architettura*. Milan, 1584, repr. Rome, 1844.
Long, Pamela O. *Openness, Secrecy, Authorship: Technical Arts and the Culture of Knowledge from Antiquity to the Renaissance*. Baltimore: Johns Hopkins University Press, 2001.
Lowic, Lawrence. "Francesco di Giorgio on the Design of Churches: The Use and Significance of Mathematics in the *Trattato*." *Architectura (Zeitschrift für Geschichte der Baukunst)* 12 (1982): 151–163.
Lucas, Edouard. *Théorie des nombres*. 2 vols. Paris: Gauthier-Villars, 1891.
Lucchesi, G. "Longino." In *Bibliotheca Sanctorum*. Vol. 7. Rome: Istituto Giovanni XXIII nella Pontificia Università lateranense, 1966, 90–95.
Malet, Antoni. "Euclid's Swan Song: Euclid's *Elements* in Early Modern Europe." In *Greek Science in the Long Run: Essays on the Greek Scientific Tradition (4th c. BCE–17th c. CE)*. Edited by Paula Olmos. Newcastle: Cambridge Scholars Publishing, 2012, 205–235.

———. "Renaissance Notions of Number and Magnitude." *Historia Mathematica* 33 (2006): 63–81.

Maltese, Corrado. "Raffaello e la cultura scientifica e tecnologica del suo tempo." In *Studi su Raffaello: Atti del congress internazionale di studi*. Edited by Micaela Sambucco Hamoud and Maria Letizia Strocchi. Urbino: QuattroVenti, 1987, 441–453.

——— ed. *Trattati di architectura ingegneria e arte militare* 2 vols. Milan: Il Polifilo, 1967.

Mancini, G., ed. "L'opera 'De corporibus regularibus' di Pietro Franceschi detto Della Francesca usurpata da Fra Luca Pacioli." In *Atti della Reale Accademia dei Lincei, anno CCCVI. Serie quinta. Memorie della Classe di Scienze morali, storiche e filologiche*. Vol. 14. Rome, 1909–16, 446–580.

Maor, Eli. *The Pythagorean Theorem: A 4000-Year History*. Princeton: Princeton University Press, 2007.

Marani, Pietro, ed. *La Crocifissione di Bramantino: Storia e Restauro*. Florence: Centro Di, 1992.

Marini, Evelyn Franceschi. *Piero della Francesca*. Città di Castello: Lapi, 1912.

Mardersteig, Giovanni. *The Alphabet of Francesco Torniello da Novara*. Verona: Officina Bodoni, 1971.

———. "Leon Battista Alberti e la rinascita del carattere lapidario romano nel Quattrocento." *Italia Medioevale e Umanistica* 2 (1959): 285–307.

Marinoni, Augosto. "L'Essere del Nulla." *Letture Vinciane* 1–12 (1960–1972): 9–28.

Marr, Alexander. *Between Raphael and Galileo: Mutio Oddi and the Mathematical Culture of Late Renaissance Italy*. Chicago and London: University of Chicago Press, 2011.

Martellotti, Guido, ed. *Francesco Petrarca, De viris illustribus*. Florence: Sansoni, 1964.

Martini, Angelo. *Manuale di Metrologia*. Turin: E. Loescher, 1883.

Martini, Francesco di Giorgio. *Trattati di architectura ingegneria e arte militare*. Edited by Corrado Maltese. 2 vols. Milan: Il Polifilo, 1967.

Masi, Michael. "The Influence of Boethius *De Arithmetica* on Late Medieval Mathematics." In *Boethius and the Liberal Arts: A Collection of Essays*. Edited by Michael Masi. Bern: Peter Lang, 1981, 81–95.

Massey, Lyle. "Configuring Spatial Ambiguity: Picturing the Distance Point from Alberti to Anamorphosis." In *The Treatise on Perspective: Published and Unpublished*. Edited by Lyle Massey. New Haven and London: Yale University Press, 2007, 161–176.

Matz, Friedrich. *Die antiken Sarkophagreliefs*. Vol. 4. Berlin: G. Grote, 1975.

McGinn, Bernard. *The Harvest of Mysticism in Medieval Germany (1300–1500)*. New York: Crossroad Publishing Company, 2005.

Medri, Giovanni. "Le opere calligrafiche a stampa. IV. Damiano Moile." *All'insegna del libro* 1 (1928): 172–179.

Mentasti, Rosa Barovier, ed. *Trasparenze e riflessi: il vetro italiano nella pittura*. Verona: Banca Popolare di Verona e Novara, 2006.

Mercati, Johannes and Pius Franchi De' Cavalieri. *Codices Vaticani Graeci*. Rome: Typis Polyglottis Vaticanis, 1923.

Merrill, Elizabeth. "*Trattato* as Textbook: Francesco di Giorgio's Vision for the Renaissance Architect." *Architectural Histories* 1(1): 20 (2013): 1–19. DOI: http://dx.doi.org/10.5334/ah.at.

Meuthen, Erich. *Nicholas of Cusa: A Sketch for a Biography*. Translated by David Crowner and Gerald Christianson. Washington, DC: Catholic University of America Press, 2010.

Milanesi, Gaetano. *Operette istoriche edite ed inedite di Antonio Manetti*. Florence: Successori Le Monnier, 1887.

Millon, Henry. "The Architectural Theory of Francesco di Giorgio." *Art Bulletin* 40 (1958): 257–261.

Mommsen, Theodor E. "Petrarch and the Decoration of the Sala Virorum Illustrium in Padua." *Art Bulletin* 34 (1952): 95–116.

Morison, Stanley. *A Newly Discovered Treatise on Classical Letter Design, Printed at Parma by Damianus Moyllus circa 1480*. Reproduced in facsimile, with an introduction by Stanley Morison. Paris: Pegasus Press, 1927.

Most, Glenn W. "Reading Raphael: The School of Athens and Its Pre-Text." *Critical Inquiry* 23 (1996): 145–182.

Müntz, Eugène. *La bibliothèque du Vatican au XVIe siècle: Notes et Documents*. Paris: Ernest Leroux, 1886.

Müntz, Eugène and Paul Fabre. *La bibliothèque du Vatican au XVe siècle d'après des documents inédits*. Paris: Ernest Thorin, 1887.

Murdoch, J. E. "The Medieval Euclid: Salient Aspects of the Translation of the Elements by Adelard of Bath and Campanus of Novara." *Revue de Synthèse* 49–52 (1968): 67–94.

Murdoch, John. "Euclid." In *Dictionary of Scientific Biography*. Edited by Charles Coulston Gillispie. Vol. 4. New York: Charles Scribner's Sons, 1981, 414–459.

Mussini, Massimo. *Francesco di Giorgio e Vitruvio: Le Traduzioni Del 'De Architectura' nei Codici Zichy, Spencer 129 e Magliabechiano II.I.141*. 2 vols. Florence: Leo S. Olschki, 2003.

———. "La trattatistica di Francesco di Giorgio: un problema critico aperto." In *Francesco di Giorgio Architetto*. Edited by Francesco Paolo Fiore and Manfredo Tafuri. Milan: Electa, 1993, 358–379.

Neal, Katherine. *From Discrete to Continuous: The Broadening of Number Concepts in Early Modern England*. Dordrecht, Boston and London: Kluwer Academic Publishers, 2002.

Nesselrath, Arnold. *Raphaël et Pinturicchio: Le grands décors des appartements du pape au Vatican*. Paris: Hazan and Musée du Louvre, 2012.

———. *Raphael's School of Athens*. Vatican City: Edizioni Musei Vaticani, 1997.

Oberhuber, Konrad. *Polarität und Synthese in Raphaels 'Schule von Athen'*. Stuttgart: Urachhaus, 1983.

O'Malley, John W. "Giles of Viterbo: A Reformer's Thought on Renaissance Rome." *Renaissance Quarterly* 20 (1967): 1–1.

———. *Giles of Viterbo on Church and Reform: A Study in Renaissance Thought*. Leiden: Brill, 1968.

Osborne, Harold. "Perspective." In *The Oxford Companion to Western Art. Oxford Art Online*. Oxford University Press http://www.oxfordartonline.com/subscriber/article/opr/t118/e2024.

Pacioli, Luca. *De divina proportione*. Edited by Franco Riva. *Fontes Ambrosiani*. Vol. 31. Milan: Mediobanca, 1956.

———. *De divina proportione*. Introduced by Augusto Marinoni. *Fontes ambrosiani* 72. 2 vols. Milan: Silvana, 1982.

———. *De divina proportione*. Milan: Casa Scriptorium Editrice, 1988; ristampa Anastastica dell'edizione del 1509.

———. *De viribus quantitatis*. Biblioteca Universitaria, Bologna Ms 250. In facsimile reproduction by Aboca. Sansepolcro: Aboca, 2009.

———. *Divina proportione, Opera a tutti glingegni perspicaci e curiosi necessaria ove ciascum studioso di philosophia prospectiva pictura sculptura architectura musica e altre mathematice suavissima: sottile: e admirabile doctrina consequira: e delectarassi: con varie questione de secretissima scientia*. Venice: Paganius Paganinus, 1509.

——— ed. Euclides, *Opera*. Venice: Paganino Paganini, 1509.

———. *Summa de arithmetica geometria. Proportioni: et proportionalita: nouamente impressa . . .* 2nd ed. of 1494 original. Tusculano: Paganino de Paganini, 1523.

Palisca, Claude V. *Humanism in Italian Renaissance Musical Thought*. New Haven and London: Yale University Press, 1985.

Panofsky, Erwin. *Life and Art of Albrecht Dürer*. Princeton: Princeton University Press, 1955.

———. *Perspective as Symbolic Form*. New York: Urzone, 1991, originally published as "Die Perspektive als symbolische Form." *Vorträge der Bibliothek Warburg*. 24, Leipzig and Berlin, 1927.

Paolucci, Antonio, Luciana Prati, and Stefano Tumidei, eds. *Marco Palmezzano: il Rinascimento nelle Romagne*. Milan: Silvana Editoriale, 2005.

Pappus. *The Commentary of Pappus on Book X of Euclid's Elements*. Translated by William Thomson. Vol. 8. Harvard Semitic Series. 1930, repr. New York: Johnson Reprint Corp., 1968.

Parma Armani, Elena. *Perin del Vaga: L'anello mancante: Studi sul Manierismo*. Genoa: Sagep Editrice, 1986.

Passavant, Johan David. *Rafael von Urbino und Sein Vater Giovanni Santi*. 4 vols. Leipzig: F. A. Brockhaus, 1839–1858.

Pastor, Ludwig von. *Geschichte der Päpste im Zeitalter der Renaissance von Wahl Innozenz' VIII bis zum Tode Julius' II. 1484–1513*. Vol. 3. Freiburg i. Br: Herder, 1956.

Patetta, Luciano. *Bramante e la sua cerchia: a Milano e in Lombardia 1480–1500*. Milan: Skira, 2001.

Pedersen, Olaf. "Du Quadrivium à la Physique." In *Artes Liberales von der Antiken Bildung zur Wissenschaft des Mittelalters*. Edited by J. Koch. Leiden and Cologne: Brill, 1959, 107–123.

Peebles, R. *The Legend of Longinus in Ecclesiastical Tradition and in English Literature, and Its Connection with the Grail*. Baltimore: J. H. Furst Co., 1911.

Peiffer, Jeanne. "Le style mathématique de Dürer et sa conception de la géométrie." In *History of Mathematics: States of the Art: Flores Quadrivii – Studies in honor of Christoph J. Scriba*. Edited by Joseph Warren Dauben, Menso Folkers, Eberhard Knobloch and Hans Wussing. San Diego, CA: Academic Press,1996, 49–61.

Peroni, Adriano ed. *Il Duomo di Pisa/The Cathedral of Pisa*. Modena: Panini, 1995.

Peterson, Mark A. *Galileo's Muse: Renaissance Mathematics and the Arts*. Cambridge, MA and London: Harvard University Press, 2011.

Pettigrew, Richard. "Aristotle on the Subject Matter of Geometry." *Phronesis* 54 (2009): 239–260.

Pfeiffer, Heinrich. *Zur Ikonographie von Raffaels Disputa*. Rome: Università Gregoriana Editrice, 1975.

Pichi, G. F. *La vita e le opere di Piero della Francesca*. Sansepolcro: Becamorti & Boncompagni, 1892.

Pieretti, Antonio. "Platonismo e Aristotelismo tra Umanesimo e Rinascimento." In *Presenze filosofiche in Umbria I. Dal Medioevo all'età contemporanea*. Edited by Antonio Pieretti. Milan: Mimesis, 2010, 111–134.

Piero della Francesca. *De Prospectiva Pingengi*. Edited by Giusta Nicco Fasola. 2 vols. Florence: Sansoni, 1942. Reprint introduced by Eugenio Battisti. Florence: Le lettere, 1984.

Piles, Roger de. *Les premiers elements de la peinture pratique*. Paris: Chez Nicolas Langlois, 1684.

Plato. *Complete Works*. Edited by John M. Cooper. Indianapolis and Cambridge: Hackett, 1997.

———. *Timaeus and Critias*. Translated by Desmond Lee. London: Penguin Books, 1965.

Pliny the Elder. *Natural History*. Translated by H. Rackham. Cambridge, MA and London: Harvard University Press, 1964.

Pollali, Angeliki. "Human Analogy in Trattati I: The *Ragione* of Modern Architecture." In *Reconstruction Francesco di Giorgio Architect*. Edited by Berthold Hub and Angeliki Pollali, 59–84. Frankfurt am Main: Peter Lang, 2011.

Prinz, W. *Vasaris Sammlungen von Künstlerbildnissen*. Florence: Kunsthistorisches Institut, 1966.

Pseudo-Dionysius. *The Complete Works*. Translated by Colin Luibheid. New York: Paulist Press, 1987.

Rackusin, Byrna. "The Architectural Theory of Luca Pacioli: *De Divina Proportione*, Chapter 54." *Bibliothèque d'Humanisme et Renaissance* 7 (1977): 479–502.

Radford, Luis. "On the Epistemological Limits of Language: Mathematical Knowledge and Social Practice during the Renaissance." *Educational Studies in Mathematics* 52: 2 (2003): 123–150.

Ragghianti, Carlo. "Il Maestro di S. Apollinare, 2." *Critica d'arte* 20, fasc. 135 (May–June 1974): 35–50.

Raggio, Olga. *The Gubbio Studiolo and its Conservation*. Vol. 1, *Federico da Montefeltro's Palace at Gubbio and its Studiolo*. Vol. 2, Antoine M. Wilmering, *Italian Renaissance Intarsia and the Conservation of the Gubbio Studiolo*. New York: Metropolitan Museum of Art, 1999.

Redig de Campos, D. "La firma siglata di Raffaello nella Scuola d'Atene." *Arti figurative: Rivista d'arte antica e moderna* 1 (1945): 151.

———. *Le stanze di Raffaello*. Rome: Del Turco, 1957.

Regiomontanus. *On Triangles. De triangulis omnimodis*. Translated by Barnabas Hughes. Madison: University of Wisconsin Press, 1967.

Rekveld, Joost. "The Ghost of Luca Pacioli." http://www.joostrekveld.net/?p=615.

Reti, Ladislao, trans. *Leonardo da Vinci: The Madrid Codices*. 5 vols. New York: McGraw-Hill, 1974.

Riahi, Pari. *Ars et Ingenium: The Embodiment of Imagination in Francesco di Giorgio Martini's Drawings*. London: Routledge, 2015.

Richter, Jean Paul. "History of the 'Trattato della Pittura'." In *Leonardo's Writings and Theory of Art*. Edited by Claire Farago. New York and London: Garland Publishing, 1999, 315–322.

Rijser, David. *Raphael's Poetics: Art and Poetry in High Renaissance Rome*. Amsterdam: Amsterdam University Press, 2012.

Rohault de Fleury, C. *Mémoire sur les instruments de la Passion*. Paris: L. Lesort, 1870.

Rommevaux, Sabine. "La réception des *Éléments* d'Euclide au Moyen Âge et à la Renaissance. Introduction." Revue d'histoire des sciences 56.2 (2003): 267–273.

Rommevaux, Sabine, Philippe Vendrix and Vasco Zara, eds. *Proportions: Science-Musique-Peinture & Architecture. Actes du LIe Colloque International d'Études Humanistes, 30 juin–4 juillet 2008*. Turnhout: Brepols, 2011.

Rosand, David. "Raphael's School of Athens and the Artist of the Modern Manner." In *The World of Savonarola: Italian Élites and Perceptions of Crisis*. Edited by Stella Fletcher and Christine Shaw. Aldershot: Ashgate, 2000, 212–232.

Rose, Paul Lawrence. *The Italian Renaissance of Mathematics: Studies on Humanists and Mathematicians from Petrarch to Galileo*. Geneva: Librairie Droz, 1975.

Ross, W. Braxton, Jr. "Giovanni Colonna, Historian at Avignon." *Speculum* 45 (1970): 533–563.

Rothstein, Bret. "Making Trouble: Strange Wooden Objects and the Pursuit of Difficulty ca. 1596." *The Journal for Early Modern Cultural Studies* 13 (2013): 96–129.

Rowland, Ingrid D. "The Intellectual Background of the *School of Athens*: Tracking Divine Wisdom in the Rome of Julius II." In *Raphael's School of Athens*. Edited by Marcia B. Hall. Cambridge: Cambridge University Press, 2005, 131–170.

———. "Raphael, Angelo Colocci, and the Genesis of the Architectural Orders." *Art Bulletin* 76 (1994): 81–104.

Rubinstein, Nicolai. "Political Ideas in Sienese Art: The Frescoes by Ambrogio Lorenzetti and Taddeo di Bartolo in the Palazzo Pubblico." *Journal of Warburg and Courtauld Institutes* 21 (1958): 179–207.

Rupprich, Hans. "Die kunsttheoretischen Schriften L.B. Albertis und ihre Nachwirkung bei Dürer." *Schweizer Beiträge zur allgemeinen Geschichte* 18/19 (1960/61): 219–239.

———. *Dürer: Schriftlicher Nachlass*. 3 vols. Berlin: Deutscher Verein für Kunstwissenschaft, 1956–69.

Rykwert, Joseph, ed. *Leon Battista Alberti*. London: Architectural Design, 1979.
Saalman, Howard, ed. and trans. by Catherine Enggass. *The Life of Brunelleschi, by Antonio di Tuccio Manetti*. University Park, Pennsylvania State University Press, 1970.
Sander, Jochen, ed. *Albrecht Dürer: His Art in Context*. Munich: Prestel, 2013.
Sangiorgi, Fert. *Documenti Urbinati: inventari del Palazzo Ducale (1582–1631)*. Urbino: Accademia Raffaello, 1976.
Scaglia, Gustina. *Francesco di Giorgio: Checklist and History of Manuscripts and Drawings in Autographs and Copies from ca. 1470 to 1687 and Renewed Copies (1764–1839)*. Bethlehem, PA: Lehigh University Press; London and Toronto: Associated University Presses, 1992.
Scarpellini, Pietro, ed. *Il Collegio del Cambio in Perugia*. Cinisello Balsamo: Silvana Editoriale, 1998.
Schäffner, Wolfgang. "The Point: The Smallest Venue of Knowledge in the 17th Century (1585–1665)." In *Collection, Laboratory, Theater: Scenes of Knowledge in the 17th Century*. Edited by Helmar Schram, Ludger Schwarte and Jan Lazardig. Berlin: Walter de Gruyter, 2005, 57–74.
Schiller, Gertrud. *Ikonographie der christlichen Kunst*. 5 vols. Gütersloh: Gütersloher Verlagshaus G. Mohn, 1968–1991.
Schlosser, Julius von. "Giusto's Fresken in Padua und die Vorläufer der Stanza della Segnatura." *Jahrbuch der Kunsthistorischen Sammlungen des Allerhöchsten Kaiserhauses* 17 (1896): 13–100.
Schmitt, Charles B. *Aristotle and the Renaissance*. Cambridge, MA and London: Harvard University Press, 1983.
Schoch, Rainer, Matthias Mende and Anna Scherbaum, ed. *Albrecht Dürer. Das druckgraphische Werk*. Vol. 3, *Buchillustrationen*. Munich, Berlin, London, and New York: Prestel Verlag, 2004.
Schönbeck, Jürgen. *Euklid: Um 300 v. Chr*. Basel, Boston and Berlin: Birkhäuser, 2003.
Schrade, Leo. "Music in the Philosophy of Boethius." *The Music Quarterly* 33 (1947): 188–200.
Schubring, Paul. "Uomini famosi." In *Repertorium für Kunstwissenschaft*. Edited by Henry Thode and Hugo von Tschudi 23. Berlin and Stuttgart: W. Spemann, 1900, 424–425.
Scribner, Bob. "Ways of Seeing in the Age of Dürer." In *Dürer and his Culture*. Edited by Dagmar Eichberger and Charles Zika. Cambridge: Cambridge University Press, 1998, 93–117.
Séquin, Carlo and Raymond Shiau. "Physically Correct Rendering of Pacioli's Rhombicuboctahedron." http://www.cs.berkeley.edu/~sequin/BIO/talksSorted.html. Uploaded April 17, 2016.
Serjeantson, R. W. "Proof and Persuasion." In *The Cambridge History of Science*. Vol. 3, *Early Modern Science*. Edited by Katherine Park and Lorraine Daston. Cambridge: Cambridge University Press, 2006, 132–175.
Serras, Herman. "Mathematics on the Ritratto di Frà Luca Pacioli." http://cage.ugent.be/~hs/pacioli/pacioli.html. Uploaded April 17, 2016.
Serres, Michel. *Le système de Leibniz et ses modèles mathématiques*. Paris: Presses Universitaires de France, 1968.
Shaw, Christine. *Julius II: The Warrior Pope*. Oxford and Cambridge, MA: Blackwell, 1993.
Shearman, John. "Gli appartamenti di Giulio II e Leone X." In *Raffaello nell'appartamento di Giulio II e Leone X*. Edited by Guido Cornini. Milan: Electa, 1993, 15–37.
———. "Raphael as Architect." *Journal of the Royal Society of Arts* 116 (1968): 388–409.
———. *Raphael in Early Modern Sources, 1483–1602*. 2 vols. New Haven and London: Yale University Press, 2003.
———. "The Vatican Stanze: Functions and Decoration." In *Art and Politics in Renaissance Italy: British Academy Lectures*. Edited by George Holmes. Oxford: Oxford University Press, 1993, 185–240.
Shelby, Lon R. "The Geometrical Knowledge of Medieval Master Masons." *Speculum* 47: 3 (1972): 395–421.

———. "Geometry." In *The Seven Liberal Arts in the Middle Ages*. Edited by David L. Wagner. Bloomington: Indiana University Press, 1983, 196–217.

———, ed. *Gothic Design Techniques: The Fifteenth-Century Design Booklets of Mathes Roriczer and Hanns Schmuttermayer*. Carbondale: Southern Illinois University Press, 1977.

Shell, Janice and Liana Castelfranchi, eds. *Giovanni Antonio Amadeo. Scultura e Architettura del suo tempo*. Milan: Cisalpino, 1993.

Sinisgalli, Rocco. *Il Nuovo De pictura di Leon Battista Alberti / The New De pictura of Leon Battista Alberti*. Edizioni Kappa: Roma, 2006.

———. *Leon Battista Alberti: On Painting*. Edited and translated by Rocco Sinisgalli. Cambridge: University of Cambridge Press, 2011.

———. "La sezione aurea nell'*Uomo vitruviano* di Leonardo." In *I disegni di Leonardo da Vinci e della sua cerchia nel Gabinetto dei Disegni e Stampe delle Gallerie dell'Accademia di Venezia*. Edited by Carlo Pedretti. Florence: Giunti, 2003, 179–181.

———. "Scuola di Atene: Note sulla lavagnetta di Euclide-Bramante," published on-line by http://urbinoelaprospettiva.uniurb.it/

———. *Perspective in the Visual Culture of Classical Antiquity*. Cambridge: Cambridge University Press, 2012.

Smith, A. Mark. *Ptolemy and the Foundations of Ancient Mathematical Optics: A Source Based Guided Study*. Philadelphia: American Philosophical Society, 1999.

Smith, Jeffrey Chipps. *Dürer*. London and New York: Phaidon, 2012.

Snow, C. P. *The Two Cultures: And a Second Look*. Cambridge: Cambridge University Press, 1964.

Sparrow, John. *Visible Words: A Study of Inscriptions in and as Books and Works of Art*. 2nd ed. Cambridge and New York: Cambridge University Press, 1969, repr. 2011.

Sparti, Donatella Livia. "Cassiano Dal Pozzo, Poussin and the Making and Publication of Leonardo's 'Trattato'." *Journal of the Warburg and Courtauld Institutes* 66 (2003): 143–188.

Spencer, John. Leon Battista Alberti. *On Painting*. Translated with Introduction and Notes. New Haven: Yale University Press, 1956; revised edition, 1966.

Springer, Anton. "Raffaels Schule von Athen." *Die Graphischen Künste* 5 (1883): 53–106.

Steinitz, Kate Traumann. *Leonardo da Vinci's 'Trattato della pittura': A Bibliography of the Printed Editions 1651–1956*. Copenhagen: Munksgaard, 1958.

Stevin, Simon. *The Principal Works of Simon Stevin*. Vol. IIB. Edited by D. J. Striuk. Amsterdam: C. V. Sets & Zeitlinger, 1958.

Strauss, Walter. *The Painter's Manual: A Manual of Measurement of Lines, Areas, and Solids by Means of Compass and Ruler Assembled by Albrecht Dürer for the Use of All Lovers of Art with Appropriate Illustrations Arranged to be Printed in the Year MDXXV*. New York: Abaris Books, 1977.

Struik, Dirk Jan. "Paulus van Middelburg (1445–1533)." *Mededelingen van her Nederlands Historisch Institut te Rome* 5 (1925): 79–118.

Swerdlow, N. M. "The Recovery of the Exact Sciences of Antiquity: Mathematics, Astronomy, Geography." In *Rome Reborn: The Vatican Library and Renaissance Culture*. Edited by Anthony Grafton. New Haven and London: Yale University Press, 1993, 125–167.

Symeon Metaphrastes. "*Martyrium Sancti et Gloriosi Martyris Longini Centurionis*." *Patrologia Graeca*. Edited by J.-P. Migne. 161 vols. Paris: J.-P. Migne, 1857–1866, *P.G.* 116: 31–44.

Tartaglia, Niccolò. *Euclide megarense philosopho: solo introduttore delle scientie mathmatice*. Venice, 1543.

Taylor, Paul. "Julius II and the Stanza della Segnatura." *Journal of the Warburg and Courtauld Institutes* 72 (2009): 103–141.

Taylor, R. Emmett. *No Royal Road: Luca Pacioli and His Times*. Chapel Hill: University of North Carolina Press, 1942.

Temple, Nicholas. *Disclosing Horizons: Architecture, Perspective and Redemptive Space*. London and New York: Routledge, 2007.

———. *Renovatio Urbis: Architecture, urbanism and ceremony in the Rome of Julius II*. London and New York: Routledge, 2011.

Teza, Laura. "Osservazioni sulla decorazione del Collegio del Cambio." In *Perugino: Il divin pittore*. Edited by Vittoria Garibaldi and Francesco Federico Mancini. Cinisello Balsamo, Milano: Silvana Editoriale, 2004, 115–127.

Tobin, Richard. "Ancient Perspective and Euclid's *Optics*." *Journal of the Warburg and Courtauld Institutes* 53 (1990): 14–42.

Toomer, Gerald J. "Ptolemy (or Claudius Ptolemaeus)." *Dictionary of Scientific Biography*. Edited by Charles Coulston Gillispie. 18 vols. New York: Charles Scribner's Sons, 1981, 11: 186–206.

———. "Theon of Alexandria." *Dictionary of Scientific Biography*. Edited by Charles Coulston Gillispie. 18 vols. New York: Charles Scribner's Sons, 1981, 13: 321–25.

Torrini, Annalisa Perissa. "Leonardo da Vinci: "L'armonica proporzionalità, la quale è composta de divine proporzioni." In *Ut Pictura Poësis: Per una storia delle arti visive/For a history of the visual arts*. Edited by Rocco Sinisgalli. Poggio a Caiano: CB Edizioni, 2012, 187–200.

Uhlfelder, Myra L. "The Role of the Liberal Arts in Boethius *Consolatio*." In *Boethius and the Liberal Arts: A Collection of Essays*, ed. Michael Masi (Bern: Peter Lang, 1981), 17–34.

Uzielli, Gustavo. *Le misure lineari medioevali e l'effigie di Cristo*. Florence: Seeber, 1899.

Valtieri, Simonetta. "La Scuola d'Atene: Bramante suggerisce un nuovo metodo per costruire in prospettiva un'architettura armonica." *Mitteilungen der Kunsthistorischen Institutes in Florenz* 16 (1972): 63–72.

Van der Schoot, Anthony. *De ontstelling van Pythagoras*. Baarn, The Netherlands: Kok Agora, 1999.

Van der Waerden, B. L. *Science Awakening*. Translated by Arnold Dresden. New York: Science Editions, John P. Wiley & Sons, 1963.

Van Egmond, Warren. "The Commercial Revolution and Beginnings of Western Mathematics in Renaissance Florence, 1300–1500." PhD diss., Indiana University, 1976.

———. "The Contributions of the Italian Renaissance to European Mathematics." *Symposia Mathematica* 27 (1986): 51–67.

———. *Practical Mathematics in the Italian Renaissance: A Catalog of Italian Abbacus Manuscripts and Printed Books to 1600*. Florence: Istituto e Museo di Storia della Scienza, 1980.

Vasari, Giorgio. *Le vite de' più eccellenti pittori, scultori e architettori: nelle redazioni del 1550 e 1568*. Edited by Rosanna Betttarini and Paola Barocchi. 6 vols. Florence: Sansoni; Studio per Edizioni Scelte, 1966–87.

———. *Le Vite de' più eccellenti pittori, scultori, ed archittettori*. Edited by Gaetano Milanesi. 9 vols. Florence: Sansoni, 1878–85.

Venerella, J. trans. and anno. *The Manuscripts of Leonardo da Vinci in the Institute de France: Manuscript A*. Milan: Castello Sforzesco, 1999.

Verdon, Timothy. "Pagans in the Church: The School of Athens in Religious Context." In *Raphael's School of Athens*. Edited by Marcia Hall. Cambridge: Cambridge University Press,114–130.

Vermeule, Corneliius. *European Art and the Classical Past*. Cambridge, MA: Harvard University Press, 1964.

Vescovini, Graziella Federici. *Studi sulla prospettiva medieval*. Turin: G. Giappichelli, 1965.

Victor, Stephen K., ed. *Practical Geometry in the Middle Ages: 'Artis Cuiuslibet Consummatio' and 'The Pratike de Geometrie'*. Philadelphia: The American Philosophical Society, 1979.

Vita, A. del. "Il volto di Piero della Francesca." *Rassegna d'arte* 7 (1920): 109–112.

Vitruvius. *On Architecture: Books 1-V*. Translated by Frank Granger. Cambridge, MA: Harvard University Press, 1931.

———. *Ten Books on Architecture*. Edited by Ingrid D. Rowland and Noble Howe. Cambridge: Cambridge University Press, 1999, repr. 2007.

Volpe, Gianni. "Scienza, arte e architettura a Urbino. Proposta per un itinerario." In *L'arte della matematica nella prospettiva*. In *L'arte della matematica nella prospettiva*. Edited by Rocco Sinisgalli. Foligno: Cartei & Bianchi, 2009, 101–113.

Wardrop, James. *The Script of Humanism: Some Aspects of Humanistic Script, 1460–1560*. Oxford: Clarendon Press, 1963.

Weisheipl, James A. "The Concept of Scientific Knowledge in Greek Philosophy." In *Mélanges à la mèmoire de Charles De Koninck*. Quebec: Les presses de L'Université Laval, 1968, 487–507.

Weller, Allen S. "Francesco di Giorgio, Siena 1439–1501." PhD diss., University of Chicago. Chicago: University of Chicago Press, 1943.

Welliver, Warman. "The Symbolic Architecture of Domenico Veneziano and Piero della Francesca." *Art Quarterly* 36 (1973): 1–30.

Wickhoff, Franz. "Die Bibliothek Julius' II." *Jahrbuch der königlich Preussischen Kunstsammlungen* 14 (1893): 49–64.

Winner, Matthias. "The Mathematical Sciences in Raphael's *School of Athens*." In *The Power of Images in Early Modern Science*. Edited by Wolfgang Lefèvre, Jürgen Renn and Urs Schoepflin. Basel: Birkhäuser, 2003, 289–308.

———. "Progetti ed esecuzione nella Stanza della Segnatura." In *Raffaello nell'appartamento di Giulio II e Leone X*. Edited by Guido Cornini. Milan: Electa, 1993, 246–291.

Wittkower, Rudolph. *Architectural Principles in the Age of Humanism*. London: The Warburg Institute, 1949.

———. "The Changing Concept of Proportion." *Daedalus* 89 (1960): 199–215.

———. "International Congress on Proportion in the Arts." *Burlington Magazine* 94 (1952): 52–55.

———. "Systems of Proportion." *Architect's Yearbook* 5 (1953): 9–18.

Wittkower, Rudolph and B. A. R. Carter. "The Perspective of Piero della Francesca's 'Flagellation'." *Journal of the Warburg and Courtauld Institutes* 16 (1953): 292–302.

Wölfflin, Heinrich. *The Art of Albrecht Dürer*. Translated by Alastair and Heide Grieve and edited by Kurt Gerstenberg. New York: Phaidon, 1971.

Wood, Jeryldene M., ed. *The Cambridge Companion to Piero della Francesca*. Cambridge: Cambridge University Press, 2002.

Woods-Marsden, Joanna. *Renaissance Self-Portraiture: The Visual Construction of Identity and the Social Status of the Artist*. New Haven and London: Yale University Press, 1998.

Contributors

Ingrid Alexander-Skipnes is an Associate Professor Emerita of Art History at the University of Stavanger, Norway. She currently lectures in the Kunstgeschichtliches Institut of the Albert-Ludwigs-Universität Freiburg. She received her Ph.D. from the Université catholique de Louvain, Belgium. She is the editor of the volume, *Cultural Exchange between the Low Countries and Italy, 1400–1600* (Brepols, 2007). Her research focuses on artistic and cultural contacts between Italy and northern Europe. Among her publications is a study on Ambrogio Bergognone in the volume, *Culture figurative a confronto tra Fiandre e Italia dal XV al XVII secolo*. She has also published in the area of Renaissance mathematics including, "Greek Mathematics in Rome and the Aesthetics of Geometry in Piero della Francesca," in the volume, *Early Modern Rome, 1341–1667*.

Renzo Baldasso is an Assistant Professor at Arizona State University, where he teaches Renaissance and Baroque art and history of the book. He studied mathematics and physics for his bachelor, and history of science and history of art in graduate school. After receiving his Ph.D. (Columbia University, 2007), he held several fellowships including at the Library of Congress, The Metropolitan Museum of Art, and The Newberry Library. His research interests are diverse and interdisciplinary, including art theory, naturalism, early prints and printing and the relationship between art and science. His larger projects address the emergence of the visuality of the printed page in the fifteenth century, and the history of diagrams and visual thinking in the early stages of the Scientific Revolution. He has published several articles in edited collections and journals, including in *The Art Bulletin*, *La Bibliofilìa*, *Centaurus*, and *Gutenberg Jahrbuch*.

Perry Brooks, an art historian based in Manhattan and Columbia County, New York, teaches Italian Renaissance art at Stony Brook University (State University of New York). He earned the Ph.D. *with distinction* from Columbia University. His doctoral dissertation, *Ut pictura mathesis: Studies in the Art of Piero della Francesca*, used mathematics as a key for the extraction of pictorial meaning in the work of the fifteenth-century painter/mathematician. He held Presidential and Whiting Fellowships at Columbia and a Chester Dale Fellowship at the Center for Advanced Studies in the Visual Arts (National Gallery, Washington, DC). He is the author of *Piero della Francesca: The Arezzo Frescoes* (New York, 1992) and of articles on Renaissance, Baroque, and Twentieth-Century art.

Caroline O. Fowler received her Ph.D. from the Department of Art and Archaeology at Princeton University. She is currently Postdoctoral Associate in the Physical History of Art at Yale University. Her scholarship focuses on the interconnecting plains of work on paper, history of the book, intellectual history, and philosophy. Her book *Drawing and the Senses: An Early Modern History* (Harvey Miller Series in Baroque Art, forthcoming 2016) examines intersections among drawing pedagogy, natural philosophy and early modern epistemologies of the senses. She has received fellowships from the Center for Advanced Study in the Visual Arts, the Getty Research Institute and the Historians of Netherlandish Art.

John Hendrix is a Professor at Roger Williams University in Bristol, Rhode Island. He has also been a professor at the Rhode Island School of Design, the University of Lincoln in the UK, and John Cabot University in Rome. He has published many books and articles on visual culture, aesthetics, and art and architectural history, including *The Contradiction between Form and Function in Architecture*, *The Cultural Role of Architecture* (co-edited with Jane Lomholt and Paul Emmons), *Architecture as Cosmology*, *Renaissance Theories of Vision* (co-edited with Charles H. Carman), *Architecture and Psychoanalysis*, *Aesthetics and the Philosophy of Spirit*, *Platonic Architectonics*, and *Neoplatonic Aesthetics* and *Neoplatonism and the Arts* (both co-edited with Liana De Girolami Cheney).

Matthew Landrus is a Research Fellow in the History of Art at Wolfson College, University of Oxford. He is also a Lecturer at the Rhode Island School of Design Department of the History of Art and Visual Culture. Having published widely on Leonardo da Vinci, his books include *Leonardo da Vinci's Giant Crossbow, Le armi e le macchine da guerra. Il De re militari di Leonardo*, and *The Treasures of Leonardo da Vinci.* His primary research addresses engagements between the practical arts and natural philosophy during the fourteen through eighteenth centuries. As a specialist on the working methods and intellectual interests of artist/engineers, he studies cross-disciplinary solutions to investigative and inventive developments in the histories of ideas, science and technology. Much of this work covers the histories of artisan notebooks, the art academy, and early European engagements with the East.

John Logan is an independent researcher who studies the connections between art history and the history of mathematics. He is a graduate of La Salle University and has studied art history at Vanderbilt University and the Centro Lingua Italiana Calvino in Florence. He has researched extensively the career of Fra Luca Pacioli. He retired from a twenty-eight year career in investment banking in 1998 and has since taught machining and mathematics at the Tennessee Technology School at Nashville. He taught business and market related subjects for fifteen years at the Stonier Graduate School of Banking (University of Pennsylvania). He has written on Trophime Bigot (Candlelight Master?) and is preparing a journal of his travels for publication. His research on the portrait of Fra Luca Pacioli has been aided by photographs which he was graciously permitted to take of the portrait at the Museo e Gallerie Nazionale di Capodimonte.

Angeliki Pollali is an Assistant Professor of art history at Deree-The American College of Greece, where she also studied history and art history. She received her MA in art history from Columbia University and her Ph.D. in art history and theory from the University of Essex, UK. She has published on Renaissance architectural theory and has co-edited with Berthold Hub the anthology *Reconstructing Francesco di Giorgio Architect* (2011). Her research interests include architectural history and theory of the Italian Renaissance; gender studies; historiography and methodology of art history.

Rangsook Yoon is an art historian and curator specialized in the print culture of early modern Europe. She wrote her dissertation on Albrecht Dürer's early career as printmaker and self-publisher at the Institute of Fine Arts, New York University. Her article, "Distribution and Sales of Dürer's *Apocalypse*," appeared in *Gutenberg Jahrbuch* in 2010, and "Dürer's First Journey to Venice: Revisiting and Reframing the Old Question," in *New Studies on Old Masters* in 2011. At the Cornell Fine Arts Museum in Winter Park, FL, she curated exhibitions, such as *Allure of Ancient Rome* (2014) and *Fashionable Portraits in Europe* (2015). She is currently preparing several publications on Dürer's *Apocalypse*.

Index

For Product Safety Concerns and Information please contact our EU
representative GPSR@taylorandfrancis.com
Taylor & Francis Verlag GmbH, Kaufingerstraße 24, 80331 München, Germany

www.ingramcontent.com/pod-product-compliance
Ingram Content Group UK Ltd.
Pitfield, Milton Keynes, MK11 3LW, UK
UKHW050054200926
475702UK00019B/2835

* 9 7 8 0 3 6 7 3 3 4 1 6 1 *